Lecture Notes in Computer Science 15368

Founding Editors

Gerhard Goos
Juris Hartmanis

Editorial Board Members

Elisa Bertino, *Purdue University, West Lafayette, IN, USA*
Wen Gao, *Peking University, Beijing, China*
Bernhard Steffen, *TU Dortmund University, Dortmund, Germany*
Moti Yung, *Columbia University, New York, NY, USA*

The series Lecture Notes in Computer Science (LNCS), including its subseries Lecture Notes in Artificial Intelligence (LNAI) and Lecture Notes in Bioinformatics (LNBI), has established itself as a medium for the publication of new developments in computer science and information technology research, teaching, and education.

LNCS enjoys close cooperation with the computer science R & D community, the series counts many renowned academics among its volume editors and paper authors, and collaborates with prestigious societies. Its mission is to serve this international community by providing an invaluable service, mainly focused on the publication of conference and workshop proceedings and postproceedings. LNCS commenced publication in 1973.

Ruber Hernández-García · Ricardo J. Barrientos ·
Sergio A. Velastin

Editors

Progress in Pattern Recognition, Image Analysis, Computer Vision, and Applications

27th Iberoamerican Congress, CIARP 2024
Talca, Chile, November 26–29, 2024
Proceedings, Part I

Editors
Ruber Hernández-García 📵
Universidad Católica del Maule
Talca, Chile

Ricardo J. Barrientos 📵
Universidad Católica del Maule
Talca, Chile

Sergio A. Velastin 📵
Queen Mary University of London
London, UK

ISSN 0302-9743　　　　　　ISSN 1611-3349　(electronic)
Lecture Notes in Computer Science
ISBN 978-3-031-76606-0　　ISBN 978-3-031-76607-7　(eBook)
https://doi.org/10.1007/978-3-031-76607-7

© The Editor(s) (if applicable) and The Author(s), under exclusive license
to Springer Nature Switzerland AG 2025

This work is subject to copyright. All rights are solely and exclusively licensed by the Publisher, whether the whole or part of the material is concerned, specifically the rights of translation, reprinting, reuse of illustrations, recitation, broadcasting, reproduction on microfilms or in any other physical way, and transmission or information storage and retrieval, electronic adaptation, computer software, or by similar or dissimilar methodology now known or hereafter developed.
The use of general descriptive names, registered names, trademarks, service marks, etc. in this publication does not imply, even in the absence of a specific statement, that such names are exempt from the relevant protective laws and regulations and therefore free for general use.
The publisher, the authors and the editors are safe to assume that the advice and information in this book are believed to be true and accurate at the date of publication. Neither the publisher nor the authors or the editors give a warranty, expressed or implied, with respect to the material contained herein or for any errors or omissions that may have been made. The publisher remains neutral with regard to jurisdictional claims in published maps and institutional affiliations.

This Springer imprint is published by the registered company Springer Nature Switzerland AG
The registered company address is: Gewerbestrasse 11, 6330 Cham, Switzerland

If disposing of this product, please recycle the paper.

Preface

The 27th Iberoamerican Congress on Pattern Recognition (CIARP) was the 2024 edition of the annual international conference CIARP, focusing on all aspects of pattern recognition, computer vision, artificial intelligence, data mining, and related areas, with contributions covering a broad spectrum of theory and applications to foster international collaboration and knowledge. Over the years, CIARP has become a key research event and one of the most important in Pattern Recognition for the Iberoamerican community. We are pleased to acknowledge the endorsement of CIARP 2024 by the International Association for Pattern Recognition (IAPR) and the support of the Chilean Association for Pattern Recognition (ACHIRP).

As in previous editions, CIARP 2024 brought together worldwide researchers and experts to showcase and disseminate ongoing research on mathematical methods and computing techniques for Artificial Intelligence and Pattern Recognition, particularly in Bioinformatics, Biometrics, Cognitive and Humanoid Vision, Computer Vision, Image Analysis, and Intelligent Data Analysis, as well as their application in several diverse areas such as industry, health, robotics, data mining, opinion mining and sentiment analysis, telecommunications, document analysis, and natural language processing. Moreover, CIARP 2024 was a platform for the global scientific community to share their research experiences, disseminate novel insights, and foster collaborations among research groups specializing in artificial intelligence, pattern recognition, and related fields.

CIARP has always prided itself on its international character, and this edition received contributions from 16 countries. Among the Iberoamerican contributors were Chile, Brazil, Ecuador, Mexico, Portugal, Colombia, Uruguay, Peru, Costa Rica, and Spain. Other notable submissions came from France, the UK, the Netherlands, Hungary, Egypt, and South Africa.

After a rigorous double-blind reviewing process, where three highly qualified reviewers spent significant time and effort reviewing each of the 61 submissions, 38 papers were accepted for inclusion in these proceedings, reflecting an acceptance rate of 62%. All accepted papers achieved scientific quality scores exceeding the overall mean rating. The selection of reviewers was guided by their expertise, ensuring representation from diverse countries and institutions worldwide. We want to extend our heartfelt gratitude to all the members of the Program Committee for their work, which undoubtedly enhanced the quality of the selected papers.

The conference was held at Universidad Católica del Maule, Talca - Chile, from November 26–29, 2024, and comprised four days of engaging single-track paper sessions, tutorials, and keynotes. The keynotes were delivered by distinguished lecturers Josep LLADOS, Julian FIERREZ, Angel D. SAPPA, and Domingo MERY. Moreover, the 1st IAPR LATAM School on Advanced Biometrics Techniques (Hybrid Event) took place during the conference with the participation of leading international researchers, organized by the ANID FOVI230126 Project and held in conjunction with CIARP 2024, aiming to provide up-to-date skills to participating students, professionals, academics, and researchers from the Latin American region in technical, regulatory, and ethical aspects of advanced biometric systems.

CIARP 2024 also awarded the Aurora Pons-Porrata Medal, honoring an Iberoamerican woman with a prestigious career in Pattern Recognition and related fields, the IAPR Best Paper and the IAPR Best Student Paper, whose authors were invited to submit an extended paper for publication in the Pattern Recognition Letters journal.

CIARP 2024 was jointly organized by the Faculty of Engineering Sciences of Universidad Católica del Maule and the Chilean Association for Pattern Recognition (ACHIRP). We express our sincere gratitude for their valuable contributions to its success. Furthermore, we wish to acknowledge the dedication of all members of the Organizing and Local Committees in organizing an outstanding conference and proceedings.

We especially thank the EquinOCS and LNCS teams at Springer for their invaluable support and guidance throughout the preparation of this volume.

Finally, our deepest gratitude goes to all authors who submitted their work to CIARP 2024, including those whose papers could not be accepted. We trust these proceedings will serve as a valuable reference for the global pattern recognition research community.

November 2024

Ruber Hernández-García
Ricardo J. Barrientos
Sergio A. Velastin

Organization

Conference Chairs

Ruber Hernández-García   Universidad Católica del Maule, Chile
Ricardo J. Barrientos   Universidad Católica del Maule, Chile

Program Chair

Sergio A. Velastin   Queen Mary University of London, UK

CIARP Steering Committee

Álvaro Pardo   APRU, Uruguay
César A. Astudillo   ACHIRP, Chile
César Beltrán-Castañón   APeRP, Peru
Joan A. Sánchez   AERFAI, Spain
João Paulo Papa   SIGPR-BR, Brazil
José F. Martínez-Trinidad   MACVNR, Mexico
José Ruiz-Shulcloper   ACRP, Cuba
Hélder Oliveira   APRP, Portugal
Manuel G. Forero Vargas   ACORP, Colombia
Marta Mejail   SARP, Argentina

Local Committee

César A. Astudillo   Universidad de Talca, Chile
Cristian A. Martínez   Universidad Nacional de Salta, Argentina
Felipe Tirado   Universidad Católica del Maule, Chile
Felipe J. Valencia   Universidad Católica del Maule, Chile
Ingrid M. López   Universidad Católica del Maule, Chile
Juan Bekios-Calfa   Universidad Católica del Norte, Chile
Marco Mora   Universidad Católica del Maule, Chile
Viviana Torres   Universidad Católica del Maule, Chile
Wladimir E. Soto-Silva   Universidad Católica del Maule, Chile
Xaviera A. López-Cortés   Universidad Católica del Maule, Chile

Technical Support

Luis Morán   Universidad Católica del Maule, Chile

Program Committee

Abel Díaz Berenguer   Vrije Universiteit Brussel, Belgium
Adrián Pérez-Suay   University of Valencia, Spain
Agustín Moreno Cañadas   Universidad Nacional de Colombia, Colombia
Alessandro Bof   Universidade Federal do Pampa, Brazil
Alexei Machado   Pontifícia Universidade Católica de Minas Gerais, Brazil
Alfonso Estudillo-Romero   Université de Rennes I, France
Amel Tuama   Northern Technical University, Iraq
Ana María Bernardos   Universidad Politécnica de Madrid, Spain
Ana María Mendonça   Universidade do Porto, Portugal
Ana Sequeira   Institute for Systems and Computer Engineering, Technology and Science, Portugal
Annette Morales-González   CENATAV, Cuba
Antoine Manzanera   ENSTA-ParisTech, France
Antoni Grau   Universitat Politècnica de Catalunya, Spain
Antonio José Sanchez-Salmeron   Universitat Politècnica de València, Spain
Attallah Bilal   Mohamed Boudiaf University of M'Sila, Algeria
Aurelio Lopez-Lopez   National Institute of Astrophysics, Optics and Electronics, Mexico
Barbara Benato   Universidade Estadual de Campinas, Brazil
Billy Peralta   Universidad Andres Bello, Chile
Carlo Sansone   University of Naples Federico II, Italy
Carlos Valle   Universidad de Playa Ancha, Chile
Catarina Silva   Universidade de Coimbra, Portugal
César A. Astudillo   Universidad de Talca, Chile
Clovis Tauber   UMR U1253 iBrain, Université de Tours, Inserm, France
Cristian Martínez   Universidad Nacional de Salta, Argentina
Domingo Mery   Pontificia Universidad Católica de Chile, Chile
Edgar Roman Rangel   Instituto Tecnológico Autónomo de México, Mexico
Fabricio Lopes   Universidade Tecnológica Federal do Paraná, Brazil

Felipe Belém	ESIEE, France
Felipe Tirado	Universidad Católica del Maule, Chile
Francesc J. Ferri	Universitat de València, Spain
Gaurav Jaswal	IIT Mandi, India
Giorgio Fumera	Università degli Studi di Cagliari, Italy
Guillermo Sanchez-Diaz	Universidad Autónoma de San Luis Potosí, Mexico
Gustavo Fernandez Dominguez	Austrian Institute of Technology, Austria
Hasan Aljabbouli	New York University, USA
Heber Ivan Mejia Cabrera	Universidad Señor de Sipán, Peru
Helio Pedrini	Universidade Estadual de Campinas, Brazil
Jacques Facon	Universidade Federal do Espírito Santo, Brazil
Jesús Ariel Carrasco-Ochoa	National Institute of Astrophysics, Optics and Electronics, Mexico
João Paulo Papa	Universidade Estadual Paulista, Brazil
Joel Arrais	Universidade de Coimbra, Portugal
José Eladio Medina Pagola	Universidad de las Ciencias Informáticas, Cuba
José Francisco Martínez-Trinidad	National Institute of Astrophysics, Optics and Electronics, Mexico
José García Rodríguez	Universidad de Alicante, Spain
Jose M. Molina	Universidad Carlos III de Madrid, Spain
José Ruiz-Shulcloper	Universidad de las Ciencias Informáticas, Cuba
Juan Carlos Briñez de Leon	Universidad Nacional de Colombia, Colombia
Juan Zamora	Pontificia Universidad Católica de Valparaíso, Chile
Julio Madera	Universidad de Camagüey, Cuba
Kalman Palagyi	University of Szeged, Hungary
Larbi Boubchir	University of Paris 8, France
Laurent Heutte	Université de Rouen, France
Lazaro Bustio	Universidad Iberoamericana, Mexico
Leopoldo Altamirano	National Institute of Astrophysics, Optics and Electronics, Mexico
Luis Enrique Sucar	National Institute of Astrophysics, Optics and Electronics, Mexico
Luis Gomez Deniz	Universidad de Las Palmas de Gran Canaria, Spain
M. Angelica Pinninghoff	Universidad de Concepción, Chile
Manuel S. Lazo Cortés	Tecnológico Nacional de México, Mexico
Marcelo Mendoza	Universidad Técnica Federico Santa María, Chile
Marcos Antonio Levano	Universidad Católica de Temuco, Chile
Marie Beurton-Aimar	University of Bordeaux, France
Mario Bruno	Universidad de Playa Ancha, Chile
Marjory da Costa Abreu	Sheffield Hallam University, UK

Martha R. Ortiz-Posadas	Universidad Autónoma Metropolitana-Iztapalapa, Mexico
Martin Kampel	Technische Universität Wien, Austria
Matilde Santos Peñas	Universidad Complutense de Madrid, Spain
Michal Haindl	Institute of Information Theory and Automation, Czech Republic
Miguel Moctezuma-Flores	Universidad Nacional Autónoma de México, Mexico
Mohit Dua	National Institute of Technology Kurukshetra, India
Nguyen Anh Minh Mai	Valeo, France
Nicolas Torres	Universidad Técnica Federico Santa María, Chile
Pedro Bugatti	Universidade Federal de São Carlos, Brazil
Pedro Couto	Universidade de Trás-os-Montes e Alto Douro, Portugal
Pedro Real	Universidad de Sevilla, Spain
Pilar Gómez-Gil	National Institute of Astrophysics, Optics and Electronics, Mexico
Priscila Saito	Universidade Federal de São Carlos, Brazil
Qiao Wang	Southeast University, China
Ricardo Contreras	Universidad de Concepción, Chile
Ripudaman Singh Arora	Blue River Technology, John Deere, USA
Rodrigo Salas	Universidad de Valparaíso, Chile
Rosana Matuk	Universidad Nacional de Luján, Argentina
Samuel Silva	Universidade de Aveiro, Portugal
Sebastian Moreno	Universidad Adolfo Ibañez, Chile
Shridhar Devamane	KLE Institute of Technology, India
Sonia Gouveia	Universidade de Aveiro, Portugal
Teresa Gonçalves	Universidade de Évora, Portugal
Vinay Kumar Venkataramana	IVIS LABS Pvt Ltd., India
Vitaly Kober	Centro de Investigación Científica y de Educación Superior de Ensenada, Mexico
Vladimir Milián Núñez	Universidad de las Ciencias Informáticas, Cuba
Walter Gómez	Universidad de La Frontera, Chile
Wilson Rivera	University of Puerto Rico, Puerto Rico
Xaviera A. López-Cortés	Universidad Católica del Maule, Chile
Yaima Filiberto Cabrera	AMV Soluciones, Spain
Yanio Hernandez Heredia	Universidad de las Ciencias Informáticas, Cuba
Yunia Reyes González	Universidad de las Ciencias Informáticas, Cuba

Contents – Part I

Towards a Lightweight CNN for Semantic Food Segmentation 1
    Bastián Muñoz, Beatriz Remeseiro, and Eduardo Aguilar

Ensemble Approach to Adaptable Behavior Cloning for a Fighting Game AI ... 16
    José García, Carlos Castro, and Carlos Valle

TeleoWatch: Pose-Transformer-Based Advanced Action Recognition 31
    Hanno Jacobs and Thambo Nyathi

Fruit Deformity Classification Through Single-Input and Multi-input
Architectures Based on CNN Models Using Real and Synthetic Images 46
    Tommy D. Beltran, Raul J. Villao, Luis E. Chuquimarca,
    Boris X. Vintimilla, and Sergio A. Velastin

CNN Sensitivity Analysis for Land Cover Map Models Using Sparse
and Heterogeneous Satellite Data 63
    Sebastián Moreno, Javier Lopatin, Diego Corvalán,
    and Alejandra Bravo-Diaz

Video Game Joystick by Recognizing Breathing Patterns 78
    Diego Robles, Andrea Lira, Carla Taramasco, and Jorge Mauro

Recovering Latent Hierarchical Relationships in Image Datasets Through
Hyperbolic Embeddings 92
    Ian Roberts, Mauricio Araya, Ricardo Ñanculef, and Mario Mallea

SwinDehazing: Haze Removal Using U-Net and Swin Transformer 104
    Percy Maldonado-Quispe and Helio Pedrini

A Proposal for Explainable Fruit Quality Recognition Using Multimodal
Models 118
    Felipe Nuñez, Billy Peralta, Orietta Nicolis, Luis Caro, and Marco Mora

Negative Sampling for Triplet-Based Loss: Improving Representation
in Self-supervised Representation Learning 133
    Manuel Alejandro Goyo and Mauricio Hidalgo

Seed-Based Superpixel Re-Segmentation for Improving Object Delineation 148
   *Lucca S. P. Lacerda, Felipe C. Belém,
Zenilton Kleber Gonçalves do Patrocínio Júnior, Alexandre X. Falcão,
and Silvio J. F. Guimarães*

Towards Interactive Video Segmentation by Dynamic and Iterative
Spanning Forest 162
   *Danielle Vieira, Isabela Borlido Barcelos, Zenilton K. G. Patrocínio Jr,
Alexandre Falcão, and Silvio Jamil F. Guimarães*

Data-Driven Evolutionary Algorithms for Optimizing Pumping Stations
in Water Distribution Networks: Classifier-Guided Search Space Reduction 178
   *Thalía Faúndez-Lizama, Nicolás Gajardo-Sepúlveda,
Jimmy H. Gutiérrez-Bahamondes, Daniel Mora-Melia,
and César A. Astudillo*

Depth Map Completion Using a Specific Graph Metric and Balanced
Infinity Laplacian for Autonomous Vehicles 187
   Vanel Lazcano

Beta Distribution Approach for Outlier Exposure in Multi-class Text
Classification 198
   Camilo Maldonado, Carlos Valle, and Héctor Allende

Impact of Quantization on Large Language Models for Portuguese
Classification Tasks 213
   *Danilo Samuel Jodas, Gabriel Lino Garcia, Pedro Henrique Paiola,
João Renato Ribeiro Manesco, and João Paulo Papa*

GemBode and PhiBode: Adapting Small Language Models to Brazilian
Portuguese 228
   *Gabriel Lino Garcia, Pedro Henrique Paiola, Eduardo Garcia,
João Renato Ribeiro Manesco, and João Paulo Papa*

Hate Speech Detection in Portuguese Using BERTimbau 244
   *João Otávio Rodrigues Ferreira Frediani, Gabriel Lino Garcia,
Pedro Henrique Paiola, Leandro Aparecido Passos, João Paulo Papa,
and Aparecido Nilceu Marana*

An Effective Approach to Text Detection and Recognition in Degraded
Historical Documents 256
   Percy Maldonado-Quispe and Helio Pedrini

Author Index 271

Contents – Part II

Unmasking Phishing Attempts: A Study on Detection in Spanish Emails 1
    Vitali Herrera-Semenets, Lázaro Bustio-Martínez,
    Yamel Pérez-Guadarramas, Jorge Ángel González-Ordiano,
    and Jan van den Berg

Comparative Analysis of Spatial and Spectral Methods in GNN for Power
Flow in Electrical Power Systems 16
    Paulo A. Espinoza and Gonzalo A. Ruz

An Effective Artificial Intelligence Pipeline for Automatic Manatee Count
Using Their Tonal Vocalizations 30
    Fabricio Quirós-Corella, Priscilla Cubero-Pardo, Athena Rycyk,
    Beth Brady, César Castro-Azofeifa, Sebastián Mora-Ramírez,
    and Juan Pablo Ureña-Madrigal

Exploring Neural Joint Activity in Spiking Neural Networks for Fraud
Detection 45
    Dylan Perdigão, Francisco Antunes, Catarina Silva,
    and Bernardete Ribeiro

Rethinking the Quality of Synthetic Palm Vein Images from Spectral
Analysis 60
    Colton Clarke, Edwin H. Salazar-Jurado, and Ruber Hernández-García

An Uncertainty-Driven ScaledYOLOv4 for Open-Pit Mining Helmet
Detection 74
    Roger Calle and Eduardo Aguilar

A Generative Algorithm to Compute NanoFingerprints 90
    Francesc Serratosa

Impact of Agricultural Production on Climate Change in South America:
Comparative Analysis Between 1990 and 2020 104
    Carlos Miguel Aizaga Ruiz and Rafael Melgarejo-Heredia

VAVnets: Retinal Vasculature Segmentation in Few-Shot Scenarios 120
    Idris Dulau, Benoit Recur, Catherine Helmer, Cecile Delcourt,
    and Marie Beurton-Aimar

Remote-Sensing Based Precipitation Detection Using Conditional GAN
and Recurrent Neural Networks 135
   Pablo Negri, Alejo Silvarrey, Sergio Gonzalez, Juan Ruiz,
   and Luciano Vidal

Data-Driven Genetic Algorithm for the Optimization of Water Distribution
Networks: A New Surrogate Model for Estimating Investment
and Operational Costs in Pumping Stations 151
   Nicolás Gajardo-Sepúlveda, Thalía Faúndez-Lizama,
   Jimmy H. Gutiérrez-Bahamondes, Daniel Mora-Melia,
   and César A. Astudillo

Gene Regulatory Network for the Tryptophanase Operon Under
the Threshold Boolean Network Model 161
   Felipe Encina-Chacana and Gonzalo A. Ruz

Multilabel Classification of Intracranial Hemorrhages Using Deep
Learning and Preprocessing Techniques on Non-contrast CT Images 175
   Rodrigo Salas, Juan Sebastian Castro, Marvin Querales,
   Carolina Saavedra, Claudia Prieto, and Steren Chabert

Segmentation of Brain Tumor Parts from Multi-spectral MRI Records
Using Deep Learning and U-Net Architecture 191
   Szabolcs Csaholczi, Ágnes Györfi, Levente Kovács, and László Szilágyi

Exploiting the *Segment Anything Model* (SAM) for Lung Segmentation
in Chest X-ray Images 205
   Gabriel Bellon de Carvalho and Jurandy Almeida

Predicting Next Phases of Multi-Stage Network Attacks: A Comparative
Study of Statistical and Deep-Learning Models 219
   Antonia Severín, Claudio Canales, Romina Torres, César Roudergue,
   and Rodrigo Salas

Improving Suicide Ideation Screening with Machine Learning
and Questionnaire Optimization Through Feature Analysis 233
   Ignacio Martínez, César Astudillo, and Daniel Núñez

Aquila Optimizer for Hyperparameter Metaheuristic Optimization in ELM 244
   Philip Vasquez-Iglesias, David Zabala-Blanco, Amelia E. Pizarro,
   Juan Fuentes-Concha, and Paulo Gonzalez

Mixture of LSTM Experts for Sales Prediction with Diverse Features 259
  Matías Soto, Felipe Cortés, Tímar Contreras, and Billy Peralta

Author Index 275

Towards a Lightweight CNN for Semantic Food Segmentation

Bastián Muñoz[1], Beatriz Remeseiro[3], and Eduardo Aguilar[1,2(✉)]

[1] Department of Computing and Systems Engineering, Catholic University of the North, Antofagasta, Chile
`bastian.munoz01@alumnos.ucn.cl, eduardo.aguilar@ucn.cl`
[2] Departament de Matematiques i Informatica, Universitat de Barcelona, Barcelona, Spain
[3] Department of Computer Science, Universidad de Oviedo, Gijon, Spain
`bremeseiro@uniovi.es`

Abstract. Semantic food segmentation is an important task for the development of nutritional systems that effectively manage daily diets. Recent advances in semantic segmentation have brought great performance improvements. However, these methods require high computational resources that limit their use in a mobile application without relying on external servers to perform the segmentation. Lightweight Convolutional Neural Networks (CNNs) have emerged as an efficient alternative for deploying deep network-based models on mobile devices, a solution not yet applied to semantic food segmentation. In this paper, we propose a lightweight variant of DeepLabv3+, replacing the standard backbone with the lightweight CNN EfficientNet-B1 and the Atrous Spatial Pyramid Pooling (ASPP) with the Cascade Waterfall ASPP (CWASPP). Validation of the proposed lightweight DeepLabv3+ method, in terms of mIoU, parameters and FLOPs, was performed on two public food datasets: UNIMIB2016 and UECFoodPixComplete. The experimental results show a better tradeoff between segmentation performance and computational resources than the state-of-art methods.

Keywords: Deep Network · Lightweights CNN · DeepLabv3+ · Semantic Food Segmentation

1 Introduction

Deep Neural Networks (DNNs) are powerful techniques that have driven significant advances in the field of Artificial Intelligence (AI). They are capable of learning complex representations of data through multiple layers of processing. Their ability to model abstract and hierarchical features has radically transformed the way we tackle problems in pattern recognition, image analysis, natural language processing, etc. Over time, DNNs have offered better performance, but at the same time have increased computational complexity in their design, resulting in a higher number of parameters, floating point operations per second (FLOPS), and training time. Unlike, new trends in AI, such as green AI, have focused on the development of lightweight DNN methods with the goal of reducing the impact on the environment as measured by the carbon footprint associated with the training and execution of these methods [30]. Lightweight architectures, characterized by fewer parameters, FLOPS, and compact size compared to traditional DNNs, enable faster training and execution, leading to a reduced carbon footprint [36]. This environmentally conscious design not only supports sustainable practices but also enhances their applicability in mobile and embedded devices.

On the other hand, semantic image segmentation aims to assign a label to each pixel of the image, effectively performing pixel-wise classification. This means that each region of pixels that belong to the same category or class is grouped under a common label [28]. Several well-known models, such as U-Net [31] and the DeepLab family [7,8], have been proposed to address semantic segmentation problems. Recently, there have also been significant advances in the design of lightweight architectures for semantic segmentation [42,49]. Most of them proposed an encoder-decoder framework that incorporates innovative techniques to enhance performance while maintaining a low number of parameters, thus achieving an optimal balance between model efficiency and accuracy.

In this work, we focus on the semantic segmentation of food items to extract the region that each one covers in a given image, allowing us to accurately estimate the nutritional contribution of each dish. The relevance of this analysis lies in its potential uses, such as a mobile application that helps individuals better control and monitor food consumption for health-related issues (e.g., diet, allergy, intolerance, organ transplantation) or to maintain a healthy lifestyle.

Although several food image analysis methods have been proposed, implementing them on mobile devices without relying on external resources remains challenging. In most cases, a client-server architecture is adopted [20], which increases execution costs and also makes it dependent on the Internet connection.

For all these reasons, we have explored the generation of a lightweight DNN method for semantic food segmentation. More specifically, we have addressed these limitations by proposing a method that is both parameter-efficient and computationally effective, making it suitable for deployment on mobile devices. Our main contributions are as follows: 1) We evaluate the performance of several lightweight Convolutional Neural Networks (CNNs) as the backbone of DeepLabv3+ [8] for semantic food segmentation; 2) We propose a lightweight semantic food segmentation method by modifying the backbone and neck in the encoder of the standard DeepLabv3+; 3) We set the lightweight semantic food segmentation baseline on two popular public food datasets, UNIMIB2016 [10] and UECFoodPixComplete [29].

2 Related Work

2.1 Lightweight CNNs

Research in the field of CNNs has evolved towards the development of lightweight architectures, which is crucial for applications on mobile and embedded devices.

Reducing the required resources is important because it shortens training times and subsequently lowers the carbon footprint. On the other hand, implementing and running natively on mobile devices is more feasible, as the client-server approach is unsustainable due to the maintenance costs of servers and the need for a stable Internet connection. This dependency excludes areas far from cities, such as rural regions, where coverage is often poor or non-existent [33]. Notable examples of lightweight CNNs include MobileNet [15, 16, 34], EfficientNet [40, 41], and ShuffleNet [27, 48], known for their computational efficiency and low memory demands without sacrificing prediction quality. The strategies focus on optimized convolutional blocks, complexity reduction through factorization, and parameter tuning to achieve compact models with high performance.

In recent years, the use of lightweight models in general applications has been a growing trend in the field of computer vision. Examples of this include the work in [47], which proposes a vehicle color recognition method using a lightweight CNN. This approach uses feature maps of intermediate convolutional layers and the Spatial Pyramid Matching strategy for efficient representation and classification. Another example is LiTM-Net [43], a lightweight CNN designed to efficiently generate high dynamic range images from low dynamic range images on mobile devices. LiTM-Net uses a lightweight encoder for efficient feature extraction and upsampling blocks in the decoder to improve reconstruction quality. Additionally, IIRNet [23] focuses on a lightweight and efficient CNN for resource-constrained systems. This architecture uses Intensely Inverted Residual (IIR) blocks to reduce redundancy and model size while maintaining classification accuracy. Furthermore, this study [22] introduces a target detection approach in Synthetic Aperture Radar (SAR) images using a lightweight version of R-CNN. The detection method proposed improves vessel detection accuracy without increasing complexity and reduces detection speed.

These recent advances in the development of lightweight models have also improved the feasibility of using the technology for food recognition and calorie estimation in devices with few computational resources. Haque et al. [13] addressed the need for fast and reliable calorie estimation systems by proposing a parameter-optimized CNN that operates in real-time on handheld devices without relying on external servers. Sajith et al. [32] demonstrated the application of MobileNetV2 for nutrient estimation, achieving high accuracy and practical usability in a mobile app designed for diet management. Sheng et al. [38] introduced an approach that integrates transformer grouping and token shuffling techniques, balancing computational efficiency and recognition performance. More recently, Sheng et al. [37] refined this by developing the Efficient Hybrid Food Recognition Net (EHFR-Net), which combines CNNs with vision transformers to maintain spatial information, thereby enhancing recognition accuracy.

2.2 Semantic Segmentation

Image segmentation is a fundamental task in computer vision that divides an image into meaningful regions for analysis and understanding. The main types of segmentation include (1) semantic segmentation, which assigns class labels to each pixel to identify objects and areas [8,25]; (2) instance segmentation, which distinguishes individual objects even of the same class [6,14]; (3) edge-based segmentation, which highlights the contours and boundaries of objects [17,26]; (4) panoptic segmentation, which combines semantic segmentation and instance segmentation in a single task [9,18]; and (5) promptable segmentation, or segmentation bases on promising regions, which focuses on identifying and extracting significant regions of the interest in an image. These regions typically have distinctive features relevant to a specific task, such as object detection or semantic segmentation [19].

In this context, it is worth mentioning the recent advances in the design of efficient and effective lightweight networks for real-time semantic segmentation tasks. LEDNet [42] employs an asymmetric encoder-decoder architecture enhanced by channel splitting and shuffling operations within a ResNet backbone, alongside an attention pyramid network in the decoder, achieving a state-of-the-art balance between speed and accuracy with less than 1M parameters. Similarly, CGNet [45] introduces the context-guided block to efficiently integrate local and global context, achieving high segmentation accuracy while maintaining a small memory footprint with less than 0.5M parameters. LAANet [49] is based on an asymmetric encoder-decoder framework with innovative modules such as the efficient asymmetric bottleneck and attention-guided dilated pyramid pooling, achieving an optimal balance between model size, speed, and accuracy, with 0.67M parameters. Lastly, LMFFNet [39] proposes a balanced architecture that incorporates split-extract-merge bottleneck blocks and multiscale attention decoders, offering superior accuracy and speed with 1.4M parameters.

Research on segmentation models for food has experienced notable advances in computer vision, focusing on tasks such as dietary analysis and automatic recognition of foods in images. Several proposals include adaptations of popular CNN-based architectures, such as U-Net [31] or DeepLabv3 [7,8], specifically designed for accurate segmentation of ingredients and dishes in food images. This is the case for the work of Aslan et al. [3,4], who have extensively investigated automatic diet monitoring by applying networks such as DeepLab and SegNet to food and non-food segmentation. DeepLabV3 was also utilized in [29] with default parameter settings; a choice also explored in [5], alongside an adapted version of YOLACT, which was initially designed for instance segmentation. Incorporating uncertainty estimation into the segmentation process carried out by these popular architectures has been shown to add robustness to decision-making tools applied to nutrition monitoring and calorie estimation [1]. Novel models for food segmentation have also been proposed. GourmetNet [35] integrates spatial and channel attention mechanisms into an extended multi-scale feature representation. CANet [11] incorporates two distinctive modules: a cross-spatial attention module to enhance feature extraction efficiency and a channel attention module to improve feature representation. FoodSAM [21] explores zero-shot capability for food image segmentation and extends to instance, panoptic, and promptable segmentation, providing a comprehensive solution for food segmentation tasks. Finally, Zhu and Dai [50] introduced a framework for ingredient

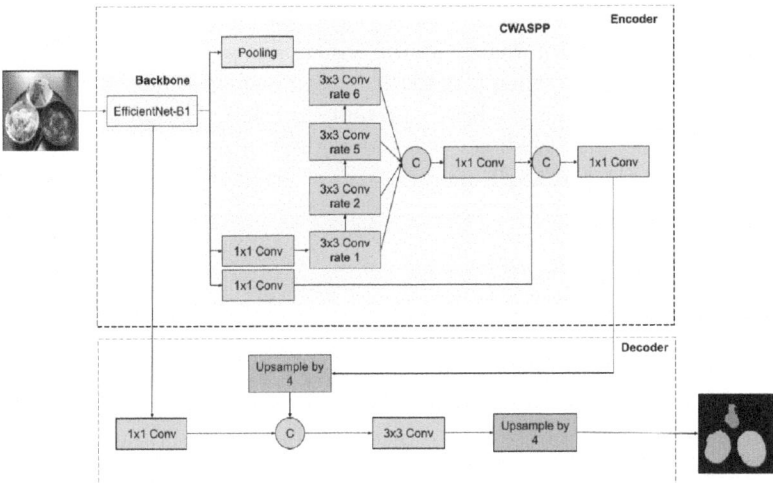

Fig. 1. Pipeline of the proposed Lightweight DeepLabv3+ method for semantic food segmentation. **C** (purple circle) stands for the concatenation operation and **rate** is the dilation factor of the convolutional filter. (Color figure online)

segmentation that uses the feature maps extracted by a CNN-based classification model. All these models have been recognized for their effectiveness in identifying and delineating regions of interest in food photos, making them valuable tools for dietary analysis and automatic food recognition applications. However, to the best of our knowledge, the use of lightweight architectures in semantic food segmentation has not been explored so far.

3 Lightweight DeepLabv3+ for Semantic Segmentation

The proposed lightweight DeepLabv3+ (LW-DeepLabv3+) method for semantic food segmentation modifies the standard DeepLabv3+ architecture by optimizing the encoder of the network, using EfficientNet-B1 as the backbone and replacing the Atrous Spatial Pyramid Pooling (ASPP) with the Cascade Waterfall Atrous Spatial Pyramid Pooling (CWASPP) [44]. The general schematic of LW-DeepLabv3+ is shown in Fig. 1, where a gray background has been placed to highlight the changes with respect to the original implementation. These changes are motivated by the high performance documented in the literature for methods based on DeepLabv3+ [44,46], which have confirmed the benefits of using EfficientNet-B1 or CWASPP, separately, in domains beyond food segmentation. Specifically for backbone selection, we selected EfficientNet-B1 after a thorough analysis of a comparative study on lightweight CNNs [24]. This decision was based on two main criteria: (1) its demonstrated performance, and (2) its compact architecture with less than 10M parameters. The maximum number of parameters was selected taking into account the feasibility of using the CNN architecture on mobile devices [33]. Regarding the replacement of the ASPP module, the decision was due to the remarkable performance improvement achieved in DeepLabv3+ when CWASPP is used instead [44].

In the following sections, we briefly present the original DeepLabv3+ method. Then, we describe the backbone selected for our proposal (EfficientNet-B1). Finally, we detail the CWASPP module.

3.1 DeepLabv3+

DeepLabv3+ [8], which is part of the DeepLab series, significantly outperforms its predecessor by incorporating a more sophisticated encoder-decoder structure to refine model outputs and improve segmentation accuracy. The encoder component captures rich semantic multi-scale information by using atrous convolutions. The decoder focuses on recovering object segmentation details (e.g., object boundary). This combination allows the model to maintain an optimal balance between accuracy and computational efficiency.

3.2 EfficientNet-B1

EfficientNet-B1 [40] is part of a family of models that uses a composite scaling method to optimally balance depth, width, and resolution dimensions. This allows for achieving high precision with efficient use of computational resources. This architecture offers an excellent accuracy-efficiency ratio, being suitable for resource-constrained applications.

3.3 Cascade Waterfall Atrous Spatial Pyramid Pooling

Cascade Waterfall Atrous Pyramid Pooling (CWASPP) [44] is an efficient and accurate alternative to the ASPP block, originally used in DeepLabv3+. Its name comes from the resulting architecture that remembers a waterfall [2] and atrous convolutions, which allow better capture of details at different spatial resolutions. The structure of the CWASPP module includes several layers of atrous convolutions in a cascade format. Unlike the ASPP standard, which applies atrous convolutions in parallel, each layer of CWASPP receives the output of the immediately preceding layer, allowing for a more efficient and progressive combination of features. A 1×1 convolution is used on the input to reduce the number of parameters. CWASPP replaces ASPP with 4 cascaded atrous convolutions, while maintaining the 1×1 convolution and pooling layer.

4 Experiment Setup

This section presents the datasets used, the details of the implementations, and the performance metrics used to evaluate the proposed method.

4.1 Datasets

Two public datasets for food segmentation are used in this research to evaluate the models' accuracy: UECFoodPixComplete and UNIMIB2016. The number of images, categories, and other key characteristics of both are detailed below.

UNIMIB2016 is an Italian food dataset that contains 1,027 images and 73 categories. The images have been manually segmented using carefully drawn polygonal boundaries. Captured on canteen trays under realistic but semi-controlled conditions, this dataset provides a valuable resource for training and evaluating semantic segmentation models, ensuring accuracy and relevance in real-world applications. This dataset includes training and test subsets, where 650 images are used for training and the remaining 360 for testing.

UECFoodPixComplete is a large dataset of Japanese food consisting of 10,000 images and 103 categories. Its authors divided the dataset into training and test sets, with 9,000 and 1,000 images, respectively. A manual and rigorous process was carried out for the segmentation of these images, resulting in smaller errors in the ground-truth masks compared to its predecessor, UECFoodPixComplete [12], which contains the same number of images and categories but with masks automatically built from the bounding boxes as a reference.

4.2 Implementation Details

Following the training protocol described in [8], except for the learning rate (LR) which is set empirically (see Sect. 5.4), all methods were initialized with ImageNet weights and trained on each dataset for 100 epochs, using a batch size of 16, minimizing cross-entropy loss using the SGD optimizer with an initial LR of 0.02 (the LR is divided by 10 for the encoder), a momentum of 0.9, and a weight decay of 1e−4. The learning policy adopts a polynomial LR with a power of 0.9, a maximum iteration of 56,000, and a minimum LR of 1e−6. The input size used for all methods is 512 × 512, so images from both datasets were resized to this dimension.

The experiments were conducted using the PyTorch framework, on a server with an NVIDIA GeForce RTX 4090-24GB GPU, a 13th generation Intel(R) Core(TM) i9-13900K CPU, and 62 GB of RAM.

4.3 Performance Metrics

The mean Intersection over Union (mIoU) was used in this research as the main evaluation metric. This metric is commonly used to assess the performance of semantic segmentation models in general objects and within the food domain. The mIoU metric is defined as follows:

$$mIoU = \frac{1}{N}\sum_{i=1}^{N}\frac{1}{K}\sum_{k=1}^{K}\frac{|y_{i,k} \cap \hat{y}_{i,k}|}{|y_{i,k} \cup \hat{y}_{i,k}|} \qquad (1)$$

Table 1. Comparison of DeepLabv3+ with different backbones in UECFoodPixComplete

Backbone	mIoU	Parameters	FLOPS (G)	Size (Mb)
MobileNetV3-S	56.44%	**2.19M**	3.66	**275.00**
MobileNetV3-L	61.24%	4.74M	5.35	434.78
EfficientNet-B1	**64.29%**	7.44M	**3.13**	671.37
ResNet18	59.65%	12.36M	19.10	464.53
ResNet50	65.88%	25.70M	37.62	1381.76

Table 2. Comparison of DeepLabv3+ with different backbones in UNIMIB2016

Backbone	mIoU	Parameters	FLOPS (G)	Size (Mb)
MobileNetV3-S	61.06%	**2.18M**	3.46	**271.17**
MobileNetV3-L	65.26%	4.73M	5.15	430.95
EfficientNet-B1	**69.12%**	7.43M	**2.93**	667.53
ResNet18	68.41%	12.35M	18.90	464.53
ResNet50	70.34%	26.70M	37.42	1381.76

where y_k and $\hat{y}_k$ are two sets with the labeled and predicted pixels for the k-th class and the i-th image, respectively. The operator $|.|$ represents the cardinality of the set and K is the total number of classes.

5 Results

This section presents the results obtained from extensive testing of the proposed model on the UECFoodPixComplete and UNIMIB2016 datasets, aimed at assessing its efficiency and accuracy in the task of semantic food segmentation. The model performance is evaluated using the mIoU metric, and comparisons are also made in terms of the number of parameters, FLOPS, and model sizes.

5.1 Performance When Changing the DeepLabv3+ Backbone

Table 1 and Table 2 detail the results of the DeepLabv3+ architecture on the UECFoodPixComplete and UNIMIB2016 datasets, respectively, using five different backbones: MobileNetV3-S, MobileNetV3-L, EfficientNet-B1, ResNet18, and ResNet50. Both tables show the performance of DeepLabv3+ for each backbone in terms of mIoU, parameters, FLOPS, and size. The results show how each backbone influences the performance achieved in food segmentation. It is observed that the heavier backbones, such as ResNet50 and ResNet18, have mIoU values similar to those of lightweight models such as MobileNetV3 and EfficientNet-B1. The latter, moreover, achieves a notable reduction in the number of parameters, FLOPS, and size, without a significant decrease in performance. This balance

Table 3. Benchmark results for semantic food segmentation on UECFoodPixComplete. * means that the reported value only takes into account the backbone.

Methods	Backbone	mIoU	Parameters	FLOPS (G)	Size (Mb)
DeepLabv3+ [8,29]	ResNet101	55.50%	45.70M	57.12	1,885.54
GourmetNet [35]	ResNet101	65.13%	48.80M	46.20	1,274.08
BayesianGourmetNet [1]	ResNet101	66.16%	48.80M	46.20	1,274.08
CANet [11]	ResNet50	68.90%	30.00M	196.69	-
YOLACT [5]	ResNext101	54.85%	78.44M*	29.14*	-
Zhu and Dai [50]	SLM-AttNet [50]	60.18%	-	-	-
FoodSam [21]	ViT-h	66.14%	636M*	1,016.72*	2,416*
DeepLabv3+	ResNet101	66.84%	45.70M	57.12	1,885.54
LW-DeepLabv3+ (Ours)	EfficientNet-B1	65.22%	6.79M	2.54	639.16

Table 4. Benchmark results for semantic food segmentation on UNIMIB2016. * means that the reported value only takes into account the backbone.

Methods	Backbone	mIoU	Parameters	FLOPS(G)	Size(Mb)
DeepLabv2 [3]	ResNet101	43.00%	42.5M*	52.19*	1,662.12*
SegNet [4]	ResNet101	44.00%	42.5M*	52.19*	1,662.12*
GourmetNet [35]	ResNet101	71.79%	48.80M	46.20	1,274.08
BayesianGourmetNet [1]	ResNet101	80.76%	48.80M	46.20	-
DeepLabv3+	ResNet101	72.65%	45.70M	57.12	1,885.54
LW-DeepLabv3+ (Ours)	EfficientNet-B1	74.49%	6.79M	2.54	639.16

between efficiency and accuracy highlights the effectiveness of lightweight models for practical applications that require optimized use of computing resources without sacrificing segmentation performance.

5.2 Benchmark Analysis with State-of-the-Art Methods

The performance of the SoA and the proposed method for semantic food segmentation on the two datasets considered are presented in Table 3 and Table 4, respectively. From the results presented in Table 3, it is clear that the different methods show varying levels of performance. Notably, the proposed LW-DeepLabv3+ achieves a competitive mIoU of 65.22% while significantly reducing the number of parameters and FLOPS compared to the other methods. We emphasize the contrast between lightweight models like LW-DeepLabv3+ with EfficientNet-B1 with only 6.79M parameters, and heavier counterparts such as DeepLabv3+ with ResNet101, which requires 45.70M. Additionally, the tradeoff can be observed and exemplified by CANet with ResNet50, which exhibits a higher mIoU of 68.90% but with a substantial computational cost of over 196.69 FLOPS. This underscores the importance of considering not just mIoU but also

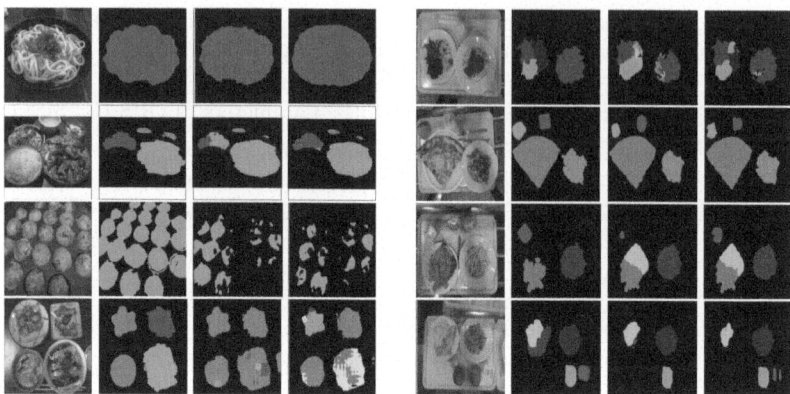

Fig. 2. Image samples for UECFoodPixComplete (left) and UNIMIB2016 (right) (a) original image, (b) ground truth (GT), (c) segmentation obtained with LW-DeepLabv3+ without the CWASPP modification, and (d) segmentation obtained with LW-DeepLabv3+. EfficientNet-B1 was used as the backbone in both cases.

computational efficiency in model selection. In the UNIMIB2016 dataset case, as shown in Table 4, LW-DeepLabv3+ achieves a superior mIoU of 74.49% compared to methods such as DeepLabv2 and SegNet, and demonstrates a significant reduction in the number of parameters and FLOPS. Although LW-DeepLabv3+ does not reach the best performance in terms of mIoU, having 6.67% less than BayesianGourmetNet, this difference is amply justified due to the notable reduction in the number of parameters and FLOPS.

Therefore, LW-DeepLabv3+ model achieves competitive performance in semantic food segmentation while maintaining a low parameter count and computational cost. This makes it a suitable candidate for deployment in resource-constrained environments without compromising on segmentation accuracy.

5.3 Qualitative Analysis of Results

Figure 2 (left) shows the results obtained for four representative sample images from the UECFoodPixComplete dataset to analyze the behavior of the proposed method. From the results, it can be seen that one of the biggest challenges identified for segmentation occurs when the same food class appears in several different regions. For example, in the third row, the method can correctly predict the class but has trouble recognizing the entire regions where the food appears. On the other hand, very good results are obtained when there is only one dish in the image, as exemplified in the first row. Regarding the images with several foods present, as in the second and fourth rows, it is observed that the segmentation regions do not cover the entire food. In the specific case of the second row, LW-DeepLabv3+ correctly segments the classes; however, without CWASPP (third column), it has problems segmenting to the correct class of certain dishes. Regarding the image in the fourth row, it shows that the proposed method with-

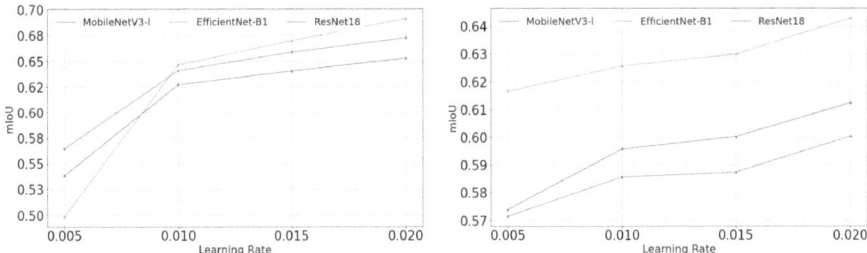

Fig. 3. Behavior of the backbones varying the LR in UNIMIB2016 (left) and UEC-FoodPixComplete (right).

out CWASPP miss-segmented everything with the same class, whereas including CWASPP allows for successful segmentation of some foods while having problems with others.

Figure 2 (right) illustrates some examples of success and failure in the segmentation results for four test images from the UNIMIB2016 dataset. It can be seen, in the second row, that when foods are well-separated and not mixed, the model can accurately segment the area and identify the respective class of the food item. On the other hand, when dishes contain overlapping mixtures of foods, the model often has problems accurately segmenting the correct region for each food class (e.g., in the first, third, and fourth rows). Specifically, for LW-DeepLabv3+ without the CWASPP, it can be observed that there are small regions of incorrectly predicted class pixels. In contrast, the implementation with CWASPP, despite having difficulties in fully capturing the food regions in the combo-plates, does not exhibit significant problems in class identification.

5.4 Ablation Study

Different LRs were evaluated to see how they affected model performance when the original DeepLabv3+ backbone was replaced with a lightweight CNN (ResNet18, MobileNetV3-L, and EfficientNet-B1). As shown in Fig. 3, the LR selected does not affect all backbones similarly. In the UNIMIB2016 case, it is visualized that EfficientNet-B1 with a LR of 0.005 achieves a performance below the other models. However, as the LR is increased, this model manages to surpass them by an increased difference. Regarding MobileNetV3-L and ResNet18, it is observed that the performance difference between these networks is maintained with the different LRs evaluated. In the UECFoodPixComplete case, we can see that with an LR of 0.005 both MobileNetV3-L and ResNet18 perform very similarly, unlike EfficientNet-B1, which outperforms them by a large difference. As the LR increases, MobileNetV3-L improves compared to ResNet18, while EfficientNet-B1 is still far better than the rest. Based on the results, a LR of 0.2 was set for the rest of the experiments.

On the other hand, Table 5 shows the effect of each proposed modification to the original DeepLabv3+ to make it lightweight. As can be observed, the modification of the backbone from ResNet101 to EfficientNet-B1 contributed to a huge reduction of parameters (6 times smaller) and FLOPS (≈95% less) with a loss of only about 3% of mIoU in both datasets. In addition, it can be seen that when the CWASPP is used with EfficientNet-B1 or ResNet101 there is a decrease in parameters and FLOPS to a lesser extent than changing of backbone. CWASPP produces a slight increase in mIoU in UNIMIB2016 and a more significant reduction in UECFoodPixComplete, which can be explained by the complexity of the data itself in each dataset. Regarding the use of both modifications (EfficientNet-B1 and CWASPP), the two datasets showed an increase of mIoU and a reduction of parameters and FLOPS with respect to their separate application. Specifically in UNIMIB2016 the proposed method LW-DeepLabv3+ increases mIoU by 1.84%, whereas in UECFoodPixComplete decreases by 1.62%, with a much smaller number of parameters and FLOPS than the original DeepLabv3+.

Table 5. DeepLabv3+ performance by varying backbone and pyramid pooling.

Dataset	Backbone	Pyramid Pooling	mIoU	Parameters	FLOPS(G)
UNIMIB2016	ResNet101	ASPP	72.65%	45.70M	57.12
	EfficientNet-B1	ASPP	69.12%	7.43M	2.93
	ResNet101	CWASPP	73.23%	43.06M	54.66
	EfficientNet-B1	CWASPP	74.49%	6.79M	2.54
UECFoodPixComplete	ResNet101	ASPP	66.84%	45.70M	57.12
	EfficientNet-B1	ASPP	64.29%	7.44M	3.13
	ResNet101	CWASPP	61.60%	43.06M	54.86
	EfficientNet-B1	CWASPP	65.22%	6.79M	2.54

6 Conclusions

This paper proposes a lightweight semantic food segmentation method based on DeepLabv3+. Initially, we explore several lightweight CNNs that could enhance computational efficiency (in terms of parameters and FLOPS) without significantly compromising segmentation performance, thereby facilitating deployment on mobile devices. After analyzing the results, EfficientNet-B1 was selected as the backbone, highlighting that accurate segmentation can be achieved with lightweight CNNs while optimizing parameter count, FLOPS, and size. In addition to the backbone, we propose another enhancement to DeepLabV3+ by replacing ASPP with a lighter version CWASPP, thus improving the mIoU and further reducing parameters and FLOPS. The proposed LW-DeepLabv3+ method was compared with SoA and showed comparable, and even better, mIoU values. Additionally, it showed significant improvements in terms of both parameters and FLOPS. In future work we consider exploring attention mechanisms (channel or spatial) to improve the definition of the predicted food regions, maintaining the limit of parameters that allow deployment on mobile devices.

Acknowledgments. This work was supported by the Agencia Nacional de Investigación y Desarrollo de Chile (ANID) (Grant No. FONDECYT INICIACIÓN 11230262).

Disclosure of Interests. The authors have no competing interests to declare that are relevant to the content of this article.

References

1. Aguilar, E., Nagarajan, B., Remeseiro, B., Radeva, P.: Bayesian deep learning for semantic segmentation of food images. Comput. Electr. Eng. **103**, 108380 (2022)
2. Artacho, B., Savakis, A.: Waterfall atrous spatial pooling architecture for efficient semantic segmentation. Sensors **19**(24), 5361 (2019)
3. Aslan, S., Ciocca, G., Schettini, R.: Semantic food segmentation for automatic dietary monitoring. In: IEEE 8th International Conference on Consumer Electronics-Berlin, pp. 1–6 (2018)
4. Aslan, S., Ciocca, G., Schettini, R.: Semantic segmentation of food images for automatic dietary monitoring. In: 26th Signal Processing and Communications Applications Conference, pp. 1–4 (2018)
5. Battini Sönmez, E., Memiş, S., Arslan, B., Batur, O.Z.: The segmented UEC food-100 dataset with benchmark experiment on food detection. Multimedia Syst. **29**(4), 2049–2057 (2023)
6. Bolya, D., Zhou, C., Xiao, F., Lee, Y.J.: YOLACT: real-time instance segmentation. In: Proceedings of the IEEE/CVF International Conference on Computer Vision, pp. 9157–9166 (2019)
7. Chen, L.C., Papandreou, G., Schroff, F., Adam, H.: Rethinking atrous convolution for semantic image segmentation. arXiv preprint arXiv:1706.05587 (2017)
8. Chen, L.-C., Zhu, Y., Papandreou, G., Schroff, F., Adam, H.: Encoder-decoder with atrous separable convolution for semantic image segmentation. In: Ferrari, V., Hebert, M., Sminchisescu, C., Weiss, Y. (eds.) ECCV 2018. LNCS, vol. 11211, pp. 833–851. Springer, Cham (2018). https://doi.org/10.1007/978-3-030-01234-2_49
9. Cheng, B., et al.: Panoptic-deeplab: a simple, strong, and fast baseline for bottom-up panoptic segmentation. In: Proceedings of the IEEE/CVF Conference on Computer Vision and Pattern Recognition, pp. 12475–12485 (2020)
10. Ciocca, G., Napoletano, P., Schettini, R.: Food recognition: a new dataset, experiments, and results. IEEE J. Biomed. Health Inform. **21**(3), 588–598 (2016)
11. Dong, X., Li, H., Wang, X., Wang, W., Du, J.: Canet: cross attention network for food image segmentation. Multimedia Tools Appl. 1–20 (2023)
12. Ege, T., Shimoda, W., Yanai, K.: A new large-scale food image segmentation dataset and its application to food calorie estimation based on grains of rice. In: Proceedings of the 5th International Workshop on Multimedia Assisted Dietary Management, pp. 82–87 (2019)
13. Haque, R.U., Khan, R.H., Shihavuddin, A., Syeed, M.M., Uddin, M.F.: Lightweight and parameter-optimized real-time food calorie estimation from images using CNN-based approach. Appl. Sci. **12**(19), 9733 (2022)
14. He, K., Gkioxari, G., Dollár, P., Girshick, R.: Mask R-CNN. In: Proceedings of the IEEE International Conference on Computer Vision, pp. 2961–2969 (2017)
15. Howard, A., et al.: Searching for mobilenetv3. In: Proceedings of the IEEE/CVF International Conference on Computer Vision, pp. 1314–1324 (2019)

16. Howard, A.G., et al.: Mobilenets: efficient convolutional neural networks for mobile vision applications. arXiv preprint arXiv:1704.04861 (2017)
17. Iannizzotto, G., Vita, L.: Fast and accurate edge-based segmentation with no contour smoothing in 2-D real images. IEEE Trans. Image Process. **9**(7), 1232–1237 (2000)
18. Kirillov, A., He, K., Girshick, R., Rother, C., Dollár, P.: Panoptic segmentation. In: Proceedings of the IEEE/CVF Conference on Computer Vision and Pattern Recognition, pp. 9404–9413 (2019)
19. Kirillov, A., et al.: Segment anything. In: Proceedings of the IEEE/CVF International Conference on Computer Vision, pp. 4015–4026 (2023)
20. Knez, S., Šajn, L.: Food object recognition using a mobile device: evaluation of currently implemented systems. Trends Food Sci. Technol. **99**, 460–471 (2020)
21. Lan, X., et al.: FoodSAM: any food segmentation. IEEE Trans. Multimedia (2023)
22. Li, Y., Zhang, S., Wang, W.Q.: A lightweight faster R-CNN for ship detection in SAR images. IEEE Geosci. Remote Sens. Lett. **19**, 1–5 (2020)
23. Li, Y., Zhang, D., Lee, D.J.: Iirnet: a lightweight deep neural network using intensely inverted residuals for image recognition. Image Vis. Comput. **92**, 103819 (2019)
24. Liu, H.I., et al.: Lightweight deep learning for resource-constrained environments: a survey. ACM Comput. Surv. (2024)
25. Long, J., Shelhamer, E., Darrell, T.: Fully convolutional networks for semantic segmentation. In: Proceedings of the IEEE Conference on Computer Vision and Pattern Recognition, pp. 3431–3440 (2015)
26. Lyu, H., Fu, H., Hu, X., Liu, L.: Esnet: edge-based segmentation network for real-time semantic segmentation in traffic scenes. In: 2019 IEEE International Conference on Image Processing (ICIP), pp. 1855–1859. IEEE (2019)
27. Ma, N., Zhang, X., Zheng, H.-T., Sun, J.: ShuffleNet V2: practical guidelines for efficient CNN architecture design. In: Ferrari, V., Hebert, M., Sminchisescu, C., Weiss, Y. (eds.) Computer Vision – ECCV 2018. LNCS, vol. 11218, pp. 122–138. Springer, Cham (2018). https://doi.org/10.1007/978-3-030-01264-9_8
28. Mo, Y., Wu, Y., Yang, X., Liu, F., Liao, Y.: Review the state-of-the-art technologies of semantic segmentation based on deep learning. Neurocomputing **493**, 626–646 (2022). https://doi.org/10.1016/j.neucom.2022.01.005, https://www.sciencedirect.com/science/article/pii/S0925231222000054
29. Okamoto, K., Yanai, K.: UEC-FoodPix complete: a large-scale food image segmentation dataset. In: Del Bimbo, A., et al. (eds.) ICPR 2021. LNCS, vol. 12665, pp. 647–659. Springer, Cham (2021). https://doi.org/10.1007/978-3-030-68821-9_51
30. Patterson, D., et al.: Carbon emissions and large neural network training. arxiv 2021. arXiv preprint arXiv:2104.10350
31. Ronneberger, O., Fischer, P., Brox, T.: U-net: convolutional networks for biomedical image segmentation. In: Navab, N., Hornegger, J., Wells, W.M., Frangi, A.F. (eds.) MICCAI 2015, Part III. LNCS, vol. 9351, pp. 234–241. Springer, Cham (2015). https://doi.org/10.1007/978-3-319-24574-4_28
32. Sajith, R., Khatua, C., Kalita, D., Mirza, K.B.: Nutrient estimation from images of food for diet management in diabetic patients. In: 2023 World Conference on Communication and Computing (WCONF), pp. 1–6. IEEE (2023)
33. San Woo, Y., Buayai, P., Nishizaki, H., Makino, K., Kamarudin, L.M., Mao, X.: End-to-end lightweight berry number prediction for supporting table grape cultivation. Comput. Electron. Agric. **213**, 108203 (2023)

34. Sandler, M., Howard, A., Zhu, M., Zhmoginov, A., Chen, L.C.: Mobilenetv2: inverted residuals and linear bottlenecks. In: Proceedings of the IEEE Conference on Computer Vision and Pattern Recognition, pp. 4510–4520 (2018)
35. Sharma, U., Artacho, B., Savakis, A.: Gourmetnet: food segmentation using multi-scale waterfall features with spatial and channel attention. Sensors **21**(22), 7504 (2021)
36. Sheng, G., et al.: Lightweight food image recognition with global shuffle convolution. IEEE Trans. AgriFood Electron. (2024)
37. Sheng, G., et al.: A lightweight hybrid model with location-preserving ViT for efficient food recognition. Nutrients **16**(2), 200 (2024)
38. Sheng, G., Sun, S., Liu, C., Yang, Y.: Food recognition via an efficient neural network with transformer grouping. Int. J. Intell. Syst. **37**(12), 11465–11481 (2022)
39. Shi, M., et al.: LMFFNet: a well-balanced lightweight network for fast and accurate semantic segmentation. IEEE Trans. Neural Netw. Learn. Syst. (2022)
40. Tan, M., Le, Q.: Efficientnet: rethinking model scaling for convolutional neural networks. In: International Conference on Machine Learning, pp. 6105–6114. PMLR (2019)
41. Tan, M., Le, Q.: Efficientnetv2: smaller models and faster training. In: International Conference on Machine Learning, pp. 10096–10106. PMLR (2021)
42. Wang, Y., et al.: Lednet: a lightweight encoder-decoder network for real-time semantic segmentation. In: IEEE International Conference on Image Processing, pp. 1860–1864 (2019)
43. Wu, G., Song, R., Zhang, M., Li, X., Rosin, P.L.: LiTMNET: a deep CNN for efficient HDR image reconstruction from a single LDR image. Pattern Recogn. **127**, 108620 (2022)
44. Wu, L., Xiao, J., Zhang, Z.: Improved lightweight deeplabv3+ algorithm based on attention mechanism. In: 2022 14th International Conference on Advanced Computational Intelligence (ICACI), pp. 314–319. IEEE (2022)
45. Wu, T., Tang, S., Zhang, R., Cao, J., Zhang, Y.: CGNet: a light-weight context guided network for semantic segmentation. IEEE Trans. Image Process. **30**, 1169–1179 (2020)
46. Yan, T., et al.: Semantic segmentation of gastric polyps in endoscopic images based on convolutional neural networks and an integrated evaluation approach. Bioengineering **10**(7), 806 (2023)
47. Zhang, Q., Zhuo, L., Li, J., Zhang, J., Zhang, H., Li, X.: Vehicle color recognition using multiple-layer feature representations of lightweight convolutional neural network. Sig. Process. **147**, 146–153 (2018)
48. Zhang, X., Zhou, X., Lin, M., Sun, J.: Shufflenet: an extremely efficient convolutional neural network for mobile devices. In: Proceedings of the IEEE Conference on Computer Vision and Pattern Recognition, pp. 6848–6856 (2018)
49. Zhang, X., Du, B., Wu, Z., Wan, T.: LAANet: lightweight attention-guided asymmetric network for real-time semantic segmentation. Neural Comput. Appl. **34**(5), 3573–3587 (2022)
50. Zhu, Z., Dai, Y.: A new CNN-based single-ingredient classification model and its application in food image segmentation. J. Imaging **9**(10), 205 (2023)

Ensemble Approach to Adaptable Behavior Cloning for a Fighting Game AI

José García[1][✉], Carlos Castro[1], and Carlos Valle[2]

[1] Departamento de Informática, Universidad Técnica Federico Santa María, Valparaíso, Chile
jose.garcia.14@sansano.usm.cl, Carlos.Castro@usm.cl
[2] Departamento de Ciencia de Datos e Informática, Facultad de Ingeniería, Universidad de Playa Ancha, Valparaíso, Chile
carlos.valle@upla.cl

Abstract. This work is centered around advanced Imitation Learning (IL) techniques to develop an agent capable of imitating human behavior in a videogame, where behavior changes with the mastery of the game. A common problem in supervised approaches is the acquisition of labeled data, which is often costly and can also be subject to concept drift. This work explores improvements in the adaptability and effectiveness of learning an expert policy in dynamic and complex environments such as video games using Behavioral Cloning (BC) and Meta-learning. We propose a learner based on a hybrid approach, utilizing both BC and a blending ensemble of deep models to enhance the agent's generalization capability on unseen trajectories with minimal human intervention. Results indicate that this approach improves training effectiveness and achieves comparable classification performance to a traditional DAgger approach with less data training, improving the base learner's training effectiveness and adaptation to new trajectories.

Keywords: Behavioral Cloning · Meta learning · Videogame agents · Convolutional Neural Networks · Long Short-Term Memory

1 Introduction

The challenge of developing agents that can act and react similarly to humans has been an area of interest in artificial intelligence research, especially in applications related to video games, as imitating human behavior can maximize user enjoyment and fun [20]. Humans modify their behavior and learn by interacting with a video game, improving and adapting their play style.

Data-related difficulties for the supervised learning techniques, such as obtaining labeled data and concept drift when the experts show new patterns are common [24]. Also, to imitate human behavior in faithful conditions implies working with limited information, in our case, video data only, so that the learner will not use any additional information in the environment not available to the user, like hidden position values. We work with heavy video gameplay that requires powerful hardware to create a faithful imitation. Working with systems or simulators like [3] usually avoids this problem by including extra information.

Given these difficulties, the main objective of this work is to develop an Imitation Learning (IL) agent that not only imitates human actions but also effectively represents these human strategies and adaptations in an end-to-end scenario. Using data from an actual video game and imitating the expert with the same limited information, with the tradeoff of being simple to deploy and apply directly without much extra work. We leverage Behavioral Cloning (BC) and Meta-learning in a combined proposal to achieve this objective. The source code and dataset created for this proposal are made available within this work.

BC is a traditional approach that uses supervised learning techniques to model the agent's policy based on explicit demonstrations of human behavior from experts. The simplicity of BC approaches lets us work with video data in favorable conditions. Although effective for replicating specific behaviors, it often faces limitations in generalization and adaptability to new situations. In complement, Meta-learning improves adaptability and knowledge transfer, enabling the agent to learn how to learn and adapt efficiently across multiple tasks.

This work proposes a hybrid model that combines the strengths of both approaches to create a more effective and adaptable system. This approach aims to capture the accuracy of human behavior through BC and overcome certain limitations of BC, such as adaptability and learning effectiveness, through meta-learning. To achieve this, we propose learners that combine Convolutional Neural Networks (CNN) and Long Short-Term Memory (LSTM) to process user gameplay. We compared this proposal with a DAgger [16] approach as a reference.

Section 2 details the state of the art in IL, describing existing techniques and some relevant applications, as well as investigation gaps to which this work tries to contribute. Section 3 describes the proposal of this work, how this proposal solves the problem theoretically, how we design and implement the learners, and gives details about the dataset and the learners. Section 4 describes the learner's training, how we evaluate our proposal, and the results yielded by the training schema. Section 5 finishes this paper with the conclusions of this work results and some open problems identified for future work.

2 State of the Art

2.1 Imitation Learning

IL focuses on behavior imitation by extracting relevant patterns to apply them in an environment [26]. It shares many components with Reinforcement Learning (RL) problems, such as the definition of states, agents, and an environment with which to interact. Unlike RL, which attempts to find an optimal policy according to a specific reward function of the agent over the environment [19], IL focuses on seeking a policy that maximizes the agent's similarity to the expert user in the environment. Both IL and RL address the general problem of solving

Markov Decision Processes (MDPs), finding a model composed of a set of transitions between states probabilistically conditioned on possible actions and an optimization function for the model. While both techniques provide solutions to the MDP, the main advantage of IL lies in the simplicity of obtaining a solution due to a smaller search space and simple problem design [21], as it does not require explicitly defining a reward function, a process that is non-trivial in RL.

There are two historical approaches in IL. First, BC focuses on using supervised learning algorithms to approximate the agent's policy, using demonstrations from the expert user and optimizing the similarity between the algorithm's inference and the model. [1] coins the first formal definition, proposing that BC consists of an algorithm inductively learning a set of rules to map the demonstrations of an expert user through state-action pairs. BC is straightforward to implement and understand. It can be efficient in scenarios where data is readily available. It has issues with generalization to new, unseen states because it strictly learns from the provided demonstrations, making it less adaptable to dynamic or varied environments. It also suffers from supervised learning issues such as the accumulation of errors and dependency on expert data quality.

Second, Inverse Reinforcement Learning (IRL) proposes an alternative approach to BC by using techniques that seek an appropriate reward function [17] for the imitation problem iteratively. The solution to the MDP proposed by an IRL algorithm derives from this estimation. By inferring the reward function, IRL can generalize to new situations not explicitly covered in the expert demonstrations while providing a deeper understanding of the expert's behavior by revealing the reward structure. IRL is typically more computationally expensive due to the iterative process of estimating the reward function and solving the RL problem. Also, defining and inferring an appropriate reward function can be complex, with extensive tuning and expert knowledge possibly required, even struggling with scalable search spaces.

Recent approaches in IL include Generative Adversarial Imitation Learning (GAIL), which uses adversarial techniques to train a discriminator and a generator model, proving particularly effective in complex tasks with high-dimensional action spaces [6], allowing for imitation without specifying an explicit reward function. However, GAIL inherits disadvantages from adversarial models, such as high computational cost when training the generator and discriminator and high sensitivity and instability due to hyperparameters. Another approach is Imitation from Observation (IfO), allowing an agent to learn behaviors solely by observing the actions of an expert without access to specific information about the actions, taking advantage in scenarios where it is not possible or practical to obtain direct state-action pairs from the expert [23]. It improves simplicity by not requiring the expert to interact actively for data collection. However, methods in IfO yield less precise policies due to the lack of explicit samples of the expert's original policy and heavily depend on data quality.

2.2 Behavioral Cloning and Meta-learning

BC methods allow for simple training with direct and explicit behavior demonstrations through supervised learning approaches, a desirable trait of algorithms that work with heavy data such as video data. These approaches inherit the classical disadvantages of supervised methods, such as error compounding and concept drift, which occur in continuous learning environments. [16] proposes Dataset Aggregation (DAgger) to solve those problems through interaction between the agent and the expert in the learning process, updating the available demonstrations to train robust policies and minimize compounding errors encountered in unseen states. Subsequently, advances in meta-learning, where techniques focus on estimating the policy through the intelligent use of various base models as predictors, are being studied despite it being under-explored in IL [13]. The meta-learning approach has demonstrated advantages such as improving policy robustness, knowledge transfer between base learners, and data efficiency [7]. This last point is essential as it contrasts with DAgger-based approaches when generating high-quality policies using only the available data.

Meta-learning application in IL [9,25] shows the system adapts quickly to new tasks using knowledge acquired from similar tasks, especially in environments where conditions or objectives change frequently. [5] proposes one of the first approaches based on neural networks, showing how meta-learning allows for rapid adaptation of learners. Meta-learning models applied to IL quickly adjust to new tasks with minimal training examples, optimizing generalization from multiple learning tasks. [11] simplifies adaptation to new demonstrations by tuning the model's weights directly towards those that perform best in similar trajectories. Unlike these approaches, this work focuses on segmenting datasets according to the user's domain over the system, allowing for heterogeneity among the patterns learned by the base learners in meta-learning.

2.3 Imitation and Video Games

[22] is one of the first to address the use of human-like AIs in videogames, highlighting the main advantage of reducing the solution search space. [2] proposes a convolutional model to control a character in a fighting game as an initial approach but does not consider temporal dependency relationships in the states of the MDP. [8] firstly groups strategies by similarity using clustering techniques to secondly learn a policy, providing a similar approach to ours by suggesting play-style-centric agents instead of focusing on the mastery of the expert. [12] presents a learning approach where collaboration with a human is needed, similar to DAgger, detailing the main problem is working with sparse data, an issue this proposal tries to solve by making the training process more effective and data efficient. [15] implements a BC algorithm for video games, concluding that the expected performance of this technique is acceptable for most basic patterns. [18] researches visual encoders for video game imitation, reviewing different architectures used as feature extractors for modern games (Minecraft, Minecraft Dungeons, and CSGO) with more complex patterns than

older games. Our proposal accounts for visual encoding by including different distributed CNN architectures for each base learner evaluated on our dataset obtained from Street Fighter IV, a game in a different genre. [14] implements an agent on a massive dataset for a modern game in the First Person Shooter genre, while our proposal does it on the fighting genre and with less data.

Previous works show multiple research gaps in this area, motivated by difficulties related to supervised learning, including data availability from modern video games in different genres, independent and identically distributed (IID) data patterns, and label imbalance, leading to poor performance in low-frequency classes. The research community has given more attention to solutions based on GAIL and IfO in the last years [26], leaving BC solutions under-explored. We try to fill this gap by proposing a BC learning schema that considers these problems. We propose a method to obtain a dataset to account for data availability and publish our dataset for public use. A LSTM architecture in our base learners recognizes IID patterns, and a meta-learning schema increases training effectiveness by using our data more efficiently. We evaluate our proposal using frequency-related metrics such as the weighted f1-score. Meta-learning also tackles the problem of adapting to new trajectories in this proposal.

3 Proposal

3.1 Solving an MDP

The basis of this proposal lies in solving an MDP by finding a policy that imitates the expert. We aim to solve the MDP with a focus on the adaptation and effectiveness of the policy, i.e., the policy should be robust to new trajectories and should not need more data to achieve high-performing results, reducing overall costs for creating the agent.

Definition 1. *We have an MDP, modified with memory, with the following components: a set of states $S = \{s_0, \ldots, s_n\}$, a set of actions $A = \{a_0, \ldots, a_n\}$, a set of window of ordered states $W = \{w_0, \ldots, w_n\}$ where $w_i = \{s_{i-k}, \ldots, s_{i-1}\}$ of length k, the probability $P(a_i|s_i, w_i)$ of choosing action a_i in state s_i with a memory of states w_i, and the reward function $R(a_i|s_i, w_i)$ for the transition.*

Proposition 1. *BC solves the MDP estimating P with a learned policy π obtained after training, minimizing the cross-entropy loss via backpropagation.*

Proof. We can treat the imitation problem as an MDP by defining the following components. A trajectory of states and actions from an expert τ^* is given by:

$$\tau : \{\langle (a_i, s_i, w_i) \rangle : i \in [k, \ldots, n]\}. \tag{1}$$

The transition probability function $P(a_i|s_i, w_i) = \pi(a_i|s_i, w_i; \theta)$ is estimated by a function π parameterized by θ. The reward function $R(a_i|s_i, w_i) = \mathbb{L}(\theta) =$

$\sum_{s_i,a_i,w_i \in \tau} \log(\pi(a_i|s_i, w_i; \theta))$, is the log-likelihood between π and τ. Maximizing the log-likelihood leads us to optimize the cross-entropy function:

$$\max_\theta (\mathbb{L}(\theta)) = \min_\theta \left(- \sum_{a_i,s_i,w_i \in \tau} \log(\pi(a_i|s_i, w_i; \theta)) \right). \quad (2)$$

Any policy π obtained by optimizing the cross-entropy function regarding τ will maximize the likelihood and thus solve the MDP by BC. As demonstrated in [10], a deep neural network using backpropagation for classification also optimizes the cross-entropy. Because of this, any deep neural network that optimizes the cross-entropy function is a solution for the MDP defined previously.

Solving the MDP can be turned into an optimization problem, with challenges such as the dependency on tuning the hyperparameters involved in the process, not tuned by an intelligent algorithm. The main task of inferring a good policy π, which is critical for BC, also has human error to account as a factor for a suboptimal policy. Another interesting challenge is the design of the problem as a multi-class classification, which is a design decision critical for any supervised algorithm, determining the loss function to optimize. Considering these challenges, our proposal includes a grid parameter search to reduce the chances of obtaining suboptimal results. We also defined this problem as a multi-category problem, leveraging the multi-class problem by using combined categories.

3.2 Proposed Method

The proposed solution consists of an end-to-end algorithm with three steps. First, a dataset is created by recording expert gameplay. Second, an agent is trained with this dataset to estimate the policy π using an ensemble learning technique. Third, the agent predicts actions in real-time to play the game.

Our dataset, obtained and processed to be published with this work, contributes to the scarce IL datasets focused on real video games, where copyright issues for older games and labor-intensive generation remain problems. It is processed with a stable FPS rate to divide the clips into frames, and the inputs are recorded synchronously with every frame. The process involves recording the screen of an expert playing the game, showing the video game and the inputs on the screen simultaneously. Then, the video session is divided into multiple clips, one for every fight of 2 to 3 rounds. The focus of every clip recorded is the mastery of the expert playing, as every clip shows different gameplay patterns every time the user adapts and executes more optimal strategies and combos.

We trained multiple base learners on the trajectories provided as demonstrations. These trajectories are segmented into groups according to a blending training scheme. We train each base learner on a subset of the total dataset. Subsequently, a meta-learner is trained on another subset of the dataset, ensuring no overlap in the subsets used by the base learners and the meta-learner. With the agent trained, we can establish an end-to-end process for the agent by running it against a fast screen-capture library to play the game.

The novelty of this approach lies in training the agent using a meta-learning strategy with a complex data type and base learner architecture in an under-explored body of knowledge such as meta-learning applied in BC [13]. Heterogeneity in feature extraction is crucial for our proposal to improve effectiveness in training stages and adaptability in unseen trajectories. We dive more deeply into the first two stages of the proposal to evaluate the performance objectively. We compare this method to an agent trained using the DAgger technique. This technique has become a relevant focus of research in the BC area over the last decade. A training scheme using DAgger establishes an iterative training process to correct failed trajectories, as shown in Algorithm 1. Since both meta-learning and DAgger do not specify particular learning architectures, the same base learning architecture is used for both approaches.

Algorithm 1: DAgger

Data: Initial dataset T_1 and expert policy τ^*, I iterations.
Result: Trained policy π.
Initialization:
The initial dataset $T_1 = \{(s, \tau^*(s)) \mid s \in \text{trajectories generated by } \tau^*\}$, is defined as the result of executing the expert policy τ^*.
The initial policy $\pi_1(\theta) = \min_\theta \sum_{\tau \in T_1} L(\pi(\tau, \theta), a)$, is trained using supervised learning on T_1.
Iterative Loop:
foreach $i \leq I$ **do**
    Generate trajectory data $\pi_i(x) = \{s \mid s \text{ is a state visited by } \pi_i \text{ in a run } x\}$.
    Obtain expert labels $\tau_i = \{(s, \tau^*(s)) \mid s \in \pi_i(x)\}$, for the states in $\pi_i(x)$.
    Add the new data to the aggregated dataset $T_{i+1} = T_i \cup \tau_i$.
    Train the policy $\pi_{i+1} = \min_\theta \sum_{\tau \in T_{i+1}} L(\pi(\tau, \theta), a)$, on the aggregated dataset.
end

3.3 Obtaining a Dataset

The data used in this work are gameplay recordings from the video game Street Fighter IV, where a human controls a Playable Character (PC) against a Non-playable Character (NPC). This dataset is applied to Definition 1 as described in Algorithm 2. The state s_i is a frame video capture of dimensions $128 \times 128 \times 3$. The action a_i is the expert input recorded for s_{i+d}, where d is the stride time between each frame used. The state also includes a memory window $w_i = s_{i-k \cdot d}, \ldots, s_{i-d}$, a window of gameplay frames with fixed length k and stride d. The expert has a set of possible actions to interact with the game. These actions are determined by all the possible combinations of inputs recorded. For this proposal, we use the 80% most frequent combinations of actions, resulting in a total of 34 labels:

$$\text{labels} = \{a^p, p \in [1, \ldots, 34]\}. \tag{3}$$

Algorithm 2: Dataset generation

Data: List of gameplay clips and inputs C, I.
Result: A list of state-action sequences τ as defined in (1).
Iterative Loop:
for $(clip, inputs)$ in (C, I) **do**
$\quad$ **for** $each$ $(window) \leq length(clip) - k \cdot d$ **do**
$\quad\quad s_i = clip[window + k]$
$\quad\quad a_i = inputs[window + k + d]$
$\quad\quad w_i = clip[window : window + k : d]$
$\quad$ **end**
$\quad \tau_{clip} = (S_{clip}, A_{clip}, W_{clip})]$
end
$\tau = [\tau_m], m \in C$

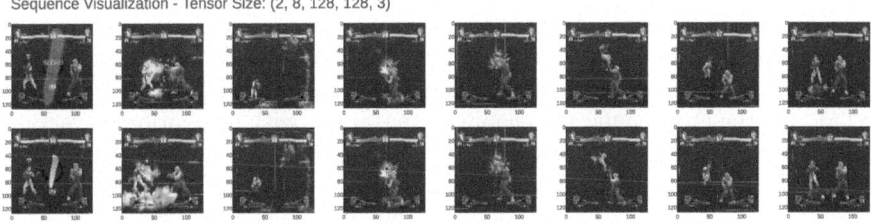

Sequence Visualization - Tensor Size: (2, 8, 128, 128, 3)

Fig. 1. Batch of 2 frame sequences. The dimensions of the tensor are (batch size, sequence length k, frame height, frame length, color channels)

To train with the window of gameplay shown in Fig. 1, the trajectory videos are transformed into a set of sequences of frames and labels by concatenating the current frame with the memory window and forming a sequence:

$$\text{base data} : \{(\text{sequence} : \text{concatenate}(w_i, s_i), \text{label} : a_i) : i \in [k \cdot d, \ldots, n]\}. \quad (4)$$

3.4 Training an Agent

The chosen strategy for the agent proposed is:

1. Base learner: A CNN extracts features along the sequence of frames from the gameplay. A LSTM receives this sequence of features to predict the action considering the window of previous states.
2. Meta-Learner: A shallow network trained with a blending approach.

Using different CNN architectures, heterogeneity is introduced in the first training stage in the feature extraction layer, thus leveraging meta-learning to select the best base learner according to the situation.

Base Learner Architecture. The base learner aims to process a sequence of frames as input to estimate the policy as a parametric probabilistic function for the action. It defines a parameterized policy $\pi(x, \theta)$ as the objective proposed in (2) by optimizing the categorical cross-entropy. To this purpose, the architecture proposed in Fig. 2 considers a feature extraction layer pre-trained in object recognition. We distribute this layer across the sequence, using the same weights for all the frames since every feature extractor receives the same frame sequence subset as training progresses. Once we obtain the feature maps, we apply a global max pooling to maintain important features such as border detections and heavy contrast, both key to our learner with a fixed monocolored background dataset. Consequently, we train the LSTM layer to process these vectors, taking advantage of temporal dependencies in the memory window. Finally, the learner models the probability of all possible actions using a fully connected layer with a softmax activation. The feature extraction layer in Fig. 2 is interchangeable with any other CNN. We chose ResNet50, MobileNetV2, and InceptionV3 as possible layers for feature extraction. All these CNN are pre-trained on Imagenet [4] to accelerate the object detection training. The feature extraction layer, the LSTM, and the Fully connected classifier are trained, with the feature extraction layer sharing weights over the sequence. After a grid search on all three validation dataset splits, a good set of hyperparameters was sequence length $k = 8$, sequence stride $d = 8$, and LSTM units $= 1000$ for the base learner.

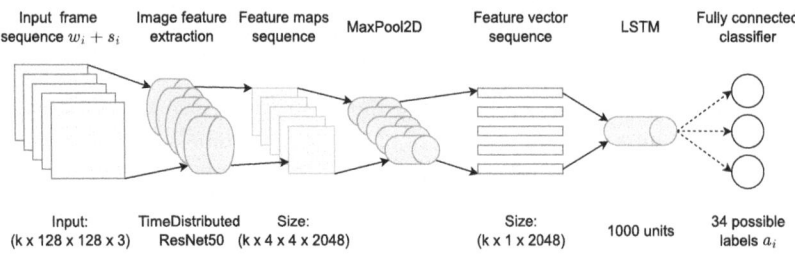

Fig. 2. Base learner architecture using ResNet50 as the feature extraction layer. From left to right, the frame sequence is processed by the feature extractors and turned into a sequence of feature maps, which then are pooled into a sequence of feature vectors, which are fed into an LSTM network that has a classifier as output.

Blending. Blending is a variation of stacking where the dataset is split for the base learners and the meta-learner, typically leaving a smaller portion of data for the meta-learner. Having better-trained base learners with an intelligent meta-learner trained avoids overfitting in unseen data. An ensemble learning algorithm $\Pi(\Theta) = M(\pi_1(\theta_1), \pi_2(\theta_2), \ldots, \pi_K(\theta_K), \theta_0)$, is defined as the solution, where π_k are the base learners, M is the meta-learner, and Π represents the complete ensemble with parameters $\Theta = \bigcup_{k=0}^{K} \theta_k$. Both π_k and Π solve the MDP in a supervised manner. We implement Π as a shallow network with only one Fully connected layer of 128 units.

We use the dataset defined in (4) for the ensemble, but the meta-learner receives a sequence of actions $\{\pi_j(a_i|s_i, w_i; \theta) : i \in [k \cdot s, \ldots, n], j \in [1, \ldots, K]\}$, instead of frames. We define the problem as multi category, so the action $a'_{i,j} = \pi_j(a|s_i, w_i; \theta) = \text{softmax}(\pi_j)$, obtained by the base learner is a sequence of actions received by the meta-learner, where every action is a vector with the probability of every possible action (3), giving the meta-learner more information as input from the base learners. As shown in Fig. 3, the proposed solution

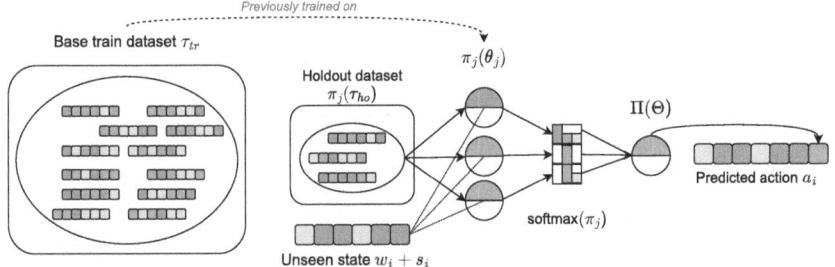

Fig. 3. Proposed blending scheme. The base learners (previously trained) identify different actions (shown as colors) probabilities from the holdout dataset, which a meta-learner uses to train. Real data follows the same flux to predict new actions. (Color figure online)

is trained as a two-step process:

1. Data split and base training: As defined in Algorithm 2, the dataset is a set of expert trajectories τ, which is divided randomly into two disjointed sets $T = T_{tr} \cup T_{ho}$, $T_{tr} \cap T_{ho} \in \emptyset$. Then, each base learner π_j is trained in $\tau_{tr} \in T_{tr}$ by optimizing (2).
2. Base inference and meta training: a new Holdout dataset $\{\pi_j(\tau_{ho}), j \in [1, \ldots, K]\} = \{a'_{i,j}, i \in \tau_{ho}\}$, is constructed using τ_{ho}, with the predictions $a'_{i,j}$ of each base learner over the holdout trajectories $\tau_{ho} \in T_{ho}$. Then, we train the meta-learner using this new dataset, optimizing (2) with our new policy $\Pi(\Theta)$. The meta learner Π receives sequence of actions instead of a sequence of frames, which are transformed into actions by the base learners.

4 Experiments and Results

4.1 Training

Our proposal includes three datasets, consisting of 135 gameplay clips divided into 823160 sequences of frames, which amounts to around 4.5 h of recorded gameplay. We set the background to a low-quality stage included in the game settings. This black background provides a good contrast with the characters. Each dataset considers a unique combination of playable characters, with Dataset 1 having the PC as Cammy and the NPC as Ken, Dataset 2 having the PC as Chun-Li and the NPC as Guile, and Dataset 3 having the PC as Sagat and the NPC as Ryu. Since the problem consists of imitating the expert without hidden information, we purposely avoid applying image augmentation to the dataset to perform as faithful imitation as possible. We created the dataset with an original copy of the game, which allows us to make it public without copyright issues for its study[1]. The base learner dataset is divided into train, validation, and testing subsets following a 73%, 10%, and 17% split, respectively. This percentage roughly allows an even split of clips for each dataset. We use the training dataset for all the experiments, the validation datasets for hyperparameter tuning, and the testing dataset for the results of this work.

Kaggle TPU VM v3-8 with 128 GB of RAM and Google Cloud Storage are used to handle a large amount of data within reasonable times by using Tensorflow data pipelines. We consider three training groups. First, two holdout sets, one for the base learners and one for the meta-learner, with one independent feature extraction layer for each base learner and a shallow network as the meta-learner. Next, a base learner trained with the DAgger scheme. This training is challenging as extra trajectories for correction are needed, which we obtain as described in Algorithm 1 . Lastly, we consider a baseline learner trained on all available data with a frozen ResNet as a feature extraction layer. This baseline will let us evaluate the importance of the feature extraction layer. All the code for this section is public[2].

We compare the proposed policy $\Pi(\Theta)$ to the performance of a base learner $\pi(\theta)$ using DAgger. This learner π will have access to more demonstrations than Π, which shows that reaching a more data-efficient approach with this proposal.

4.2 Evaluation

As defined in (3), this imitation problem has a finite and discrete set of actions, which makes this a classification problem. Classification metrics are widely used in video action recognition problems [27] like this proposal. The performance evaluation will consider mainly the Weighted F1-Score to account for label imbalance. This indicator will show the algorithm's ability to learn patterns with the 34 possible input combinations. Metrics like Precision and Recall are obtained as an average of the classes, measuring the trade-off between false positives and false negatives. Accuracy will serve as a reference metric, indicating the overall correctness of the model by dividing the number of correct predictions by the total number of predictions. Top10-Accuracy measures the accuracy of the most frequent actions. The first ten actions include the agent's basic movements plus one basic attack. This metric also indicates how well the policies learn simple patterns. Cross-entropy loss is the target of optimization by the learners.

---

[1] https://www.kaggle.com/datasets/gardethk/sfiv-v2.
[2] https://www.kaggle.com/code/gardethk/base-models.

It shows the capability of optimization of any particular learner. Since we aim to improve effectiveness, the proposed learner trains on fewer clips than any other learner in this work. The evaluating sets include new trajectories to measure the improvement of adaptation. We track the number of trajectories and sequences of frames to measure the efficiency of our proposal while training with less data. We replicate this process for each dataset defined in Sect. 3.3.

4.3 Metric Results

Table 1 shows the performance over the testing dataset. All learners suffered from overfitting in all datasets, with the meta-learner mitigating this effect the most. The baseline approach obtains the worst results in training in Datasets 1 and 2, which tells us that the feature extraction layer benefits from fine-tuning, as the baseline uses a frozen pre-trained layer, showing that the LSTM cannot estimate a good policy on its own with all the data. The base learners obtain mixed results. There is no CNN that achieves better results for all datasets. ResNet DAgger has better classification performance on unseen trajectories in Dataset 1, while the Meta-learner obtained the best results for Datasets 2. Dataset 1 is obtained with a PC highly dominated by the expert, while Datasets 2 and 3 include gameplay of new PCs for the expert, leading to longer trajectories for Datasets 2 and 3 as most gameplay rounds were longer. Specifically, Dataset 2 includes new gameplay elements like charged inputs, leading to different trajectories of actions that remain active during the memory window. Achieving better Accuracy but similar Top10-Accuracy indicates that the learner achieved better performance in low-frequency actions as the Accuracy gives us a non-weighted reference for all actions. All learners achieved high values of Top10-Accuracy, even the baseline, without any measures to account for imbalance in training, meaning the basic patterns associated with the movements of the character, staying still, jumping, and basic strikes are easy to learn. Combined inputs of directional strikes, grabs, and power moves are hard to imitate. Table 2 shows how many trajectories and sequences the training process considers for every learner schema, and the time spent in minutes training each learner. Dagger learners achieve these results with seven new correction trajectories in Dataset 1 that amount to an increase of 24% of the sequences available for training. In Datasets 2 and 3, an increase of 15% was not enough for DAgger learners to generalize better. Datasets 2 and 3 show that DAgger learners do not achieve the expected results, leading us to conclude that even more iterations with new data are needed. Our proposal achieves comparable results with fewer training data while improving the base learner's generalization performance. Also, the Dagger approach re-processes the incremental dataset in every iteration, giving the learner repeated exposure to the starting sequence dataset with improved new trajectories in between. Our proposal mimics this behavior by exposing every base learner to the starting sequence dataset. Our proposal takes more time to train in Dataset 1 and less time in Dataset 2 and 3, improving the effectiveness of the process for the training process in both this datasets. The ResNet feature extractor, which in the base learners achieves a lower cross-entropy value than

Table 1. Performance metrics across datasets is evaluated for the baseline, base learners, DAgger variants, and the meta-learner. These metrics include F1-Score, Precision, Recall, Accuracy, Top10-Accuracy, and Cross-entropy.

Metric	Dataset 1						Dataset 2						Dataset 3					
	F1 Score	Precision	Recall	Accuracy	Top10 Accuracy	Cross entropy	F1 Score	Precision	Recall	Accuracy	Top10 Accuracy	Cross entropy	F1 Score	Precision	Recall	Accuracy	Top10 Accuracy	Cross entropy
Baseline	0.071	1e−5	1e−5	0.114	0.694	3.11	0.298	0.613	0.223	0.379	0.895	2.134	0.448	0.719	0.487	**0.581**	0.920	1.622
Base learners Resnet50	0.523	0.541	0.490	0.534	0.988	0.874	0.358	0.675	0.274	0.431	0.898	1.981	0.513	0.725	0.431	0.579	0.927	1.509
Base learners MobileNetV2	0.230	0.390	0.157	0.259	0.857	2.464	0.411	0.618	0.314	0.454	0.931	1.905	0.487	0.697	0.466	**0.581**	**0.941**	1.541
Base learners InceptionV3	0.427	0.452	0.384	0.423	0.972	2.468	0.409	0.580	0.329	0.459	0.902	2.094	0.459	0.808	0.256	0.545	0.919	1.604
DAgger Resnet50	**0.694**	**0.685**	**0.652**	**0.670**	0.997	**0.783**	0.407	0.537	0.350	0.424	0.909	2.139	0.490	0.656	0.470	0.530	0.909	1.825
DAgger MobileNetV2	0.420	0.432	0.408	0.424	0.976	2.195	0.457	0.627	**0.402**	0.496	0.934	1.703	0.431	0.592	0.457	0.508	0.910	2.066
DAgger InceptionV3	0.554	0.543	0.518	0.539	**0.999**	0.871	0.408	0.495	0.365	0.411	0.822	3.127	**0.531**	0.629	**0.507**	0.544	0.900	2.147
Meta-learner	0.590	0.636	0.607	0.623	0.993	1.040	**0.509**	**0.702**	0.373	**0.548**	**0.935**	**1.489**	0.524	**0.809**	0.438	**0.581**	0.921	**1.481**

Table 2. Data used in training. The trajectory's length is different for every dataset. The meta-learner trains on fewer trajectories and fewer sequences than ResNet Dagger. The training time is measured in minutes. DAgger times are averaged considering the three prior variants.

Learner	Dataset 1			Dataset 2			Dataset 3		
	Trajectories τ	Sequences	Training time	Trajectories τ	Sequences	Training time	Trajectories τ	Sequences	Training time
Baseline	46	141495	91.2	33	230945	98.9	34	250385	64.4
Base Learners	30	75572	786.2	20	136793	168.3	20	146830	214.4
Meta-learner	9	31333	30.5	9	58111	24.7	9	64585	28.8
Meta-learner + Base Learners	39	106905	816.7	29	194904	193.0	29	211415	243.2
DAgger	46	141495	466.6	33	230945	407.9	34	250385	371.6

the meta-learner for Dataset 1, shows that it still can learn new features, which benefits DAgger by including unseen trajectories, but also causes overfitting to be more prevalent in Dataset 2 and 3. The meta-learner achieves higher metric values by combining only three base learners with different performances. While DAgger could benefit from running more iterations in Datasets 2 and 3, the meta-learner could benefit from including more heterogeneous base learners in the schema, a less expensive and low-effort way to improve the learner's performance, as more DAgger iterations are expensive in training time.

5 Conclusion

We propose a hybrid approach combining BC and Meta-learning to develop an agent for Street Fighter IV by estimating a policy Π. Our proposal improves the adaptability of the base learners and achieves comparable results in unseen trajectories to a DAgger approach while maintaining data efficiency by training with fewer data across three different datasets. By including two extra feature extraction layers in the base learners, such as MobileNetV2 and InceptionV3, the meta-learner improved the adaptability of the base learner, reducing the effect of overfitting. The meta-learner improves learning of frequent and infrequent actions, suggesting space for experimenting with larger datasets and more diverse base learners. Our proposal contributes to developing IL agents for video games that behave more like humans, improving fun in games.

6 Future Work

1. Agnostic feature extraction layers: A feature extractor ignoring the background to focus on the selected characters and in-game elements would make this approach work in gameplay with diverse backgrounds and characters.
2. Input to action: We designed the agent's actions to get better and more precise labels per frame. It has the disadvantage of not being able to perform combos and complicated inputs when running the model end-to-end.
3. Imitation from Observation: Since this approach processes video gameplay, proposing an IfO approach is only natural, improving the efficiency of the end-to-end process dramatically, as only video is needed.

Acknowledgments. There is no acknowledgments.

Disclosure of Interests. The authors have no competing interests to declare that are relevant to the content of this article.

References

1. Bain, M., Sammut, C.: A framework for behavioural cloning. In: Machine Intelligence, vol. 15, pp. 103–129 (1995)
2. Chen, Z., Yi, D.: The game imitation: deep supervised convolutional networks for quick video game AI. CoRR abs/1702.05663 (2017). http://arxiv.org/abs/1702.05663
3. Codevilla, F., Müller, M., López, A., Koltun, V., Dosovitskiy, A.: End-to-end driving via conditional imitation learning. In: International Conference on Robotics and Automation (ICRA) (2018)
4. Deng, J., Dong, W., Socher, R., Li, L.J., Li, K., Fei-Fei, L.: Imagenet: a large-scale hierarchical image database. In: 2009 IEEE Conference on Computer Vision and Pattern Recognition, pp. 248–255. IEEE (2009)
5. Finn, C., Abbeel, P., Levine, S.: Model-agnostic meta-learning for fast adaptation of deep networks. In: International Conference on Machine Learning, pp. 1126–1135. PMLR (2017)
6. Ho, J., Ermon, S.: Generative adversarial imitation learning. In: Advances in Neural Information Processing Systems, vol. 29 (2016)
7. Hospedales, T., Antoniou, A., Micaelli, P., Storkey, A.: Meta-learning in neural networks: a survey. IEEE Trans. Pattern Anal. Mach. Intell. **44**(9), 5149–5169 (2021)
8. Ingram, B., Van Alten, C., Klein, R., Rosman, B.: Creating diverse play-style-centric agents through behavioural cloning. In: Proceedings of the AAAI Conference on Artificial Intelligence and Interactive Digital Entertainment, vol. 19, pp. 255–265 (2023)
9. Menda, K., Driggs-Campbell, K., Kochenderfer, M.J.: Ensembledagger: a Bayesian approach to safe imitation learning. In: 2019 IEEE/RSJ International Conference on Intelligent Robots and Systems (IROS), pp. 5041–5048. IEEE (2019)
10. Murphy, K.P.: Machine Learning: A Probabilistic Perspective. MIT Press, Cambridge (2012)

11. Nichol, A., Achiam, J., Schulman, J.: On first-order meta-learning algorithms. CoRR abs/1803.02999 (2018). http://arxiv.org/abs/1803.02999
12. Niemantsverdriet, D.: Improvements in imitation learning for overcooked (2023)
13. Patacchiola, M., Sun, M., Hofmann, K., Turner, R.E.: Comparing the efficacy of fine-tuning and meta-learning for few-shot policy imitation. In: Conference on Lifelong Learning Agents, pp. 878–908. PMLR (2023)
14. Pearce, T., Zhu, J.: Counter-strike deathmatch with large-scale behavioural cloning. In: 2022 IEEE Conference on Games (CoG), pp. 104–111. IEEE (2022)
15. Pussinen, J.: Behavioural cloning in video games. Master's thesis, Itä-Suomen yliopisto (2021)
16. Ross, S., Gordon, G., Bagnell, D.: A reduction of imitation learning and structured prediction to no-regret online learning. In: Proceedings of the Fourteenth International Conference on Artificial Intelligence and Statistics, pp. 627–635. JMLR Workshop and Conference Proceedings (2011)
17. Russell, S.: Learning agents for uncertain environments. In: Proceedings of the Eleventh Annual Conference on Computational Learning Theory, pp. 101–103 (1998)
18. Schäfer, L., et al.: Visual encoders for data-efficient imitation learning in modern video games (2023)
19. Shakya, A.K., Pillai, G., Chakrabarty, S.: Reinforcement learning algorithms: a brief survey. Expert Syst. Appl. 120495 (2023)
20. Soni, B., Hingston, P.: Bots trained to play like a human are more fun. In: 2008 IEEE International Joint Conference on Neural Networks (IEEE World Congress on Computational Intelligence), pp. 363–369. IEEE (2008)
21. Tai, L., Zhang, J., Liu, M., Boedecker, J., Burgard, W.: A survey of deep network solutions for learning control in robotics: from reinforcement to imitation (2018)
22. Thurau, C., Bauckhage, C., Sagerer, G.: Imitation learning at all levels of game-AI. In: Proceedings of the International Conference on Computer Games, Artificial Intelligence, Design and Education, vol. 5 (2004)
23. Torabi, F., Warnell, G., Stone, P.: Behavioral cloning from observation. In: Proceedings of the 27th International Joint Conference on Artificial Intelligence. IJCAI'18, pp. 4950–4957. AAAI Press (2018)
24. Xiang, Q., Zi, L., Cong, X., Wang, Y.: Concept drift adaptation methods under the deep learning framework: a literature review. Applied Sciences **13**(11) (2023). https://doi.org/10.3390/app13116515, https://www.mdpi.com/2076-3417/13/11/6515
25. Zare, M., Kebria, P.M., Khosravi, A., Nahavandi, S.: A survey of imitation learning: algorithms, recent developments, and challenges (2023)
26. Zheng, B., Verma, S., Zhou, J., Tsang, I.W., Chen, F.: Imitation learning: progress, taxonomies and challenges. IEEE Trans. Neural Netw. Learn. Syst. (2022)
27. Zhu, Y., et al.: A comprehensive study of deep video action recognition. CoRR abs/2012.06567 (2020). https://arxiv.org/abs/2012.06567

TeleoWatch: Pose-Transformer-Based Advanced Action Recognition

Hanno Jacobs[(✉)] [iD] and Thambo Nyathi [iD]

Department of Computer Science, University of Pretoria, Pretoria, South Africa
hanno.jacobs.sa@gmail.com
https://www.up.ac.za/

Abstract. Recognition of human actions from videos is a valuable application for building management, security systems, accident prevention, accident intervention and several other applications. This study provides a framework for joint-based action recognition for human and non-human action recognition and common use cases for advanced intelligent closed circuit television management systems. This study offers a scaling and normalization algorithm that makes the model position agnostic for the person's position in the frame and allows for generalization for actions performed facing the camera at various angles. It also improves on previous pose-based action recognition systems (Bidirectional-LSTM) by using a transformer encoder architecture for the recognition task. The transformer encoder architecture uses the joint locations output from pose estimation computer vision models to train the model to recognise the common joint trajectories that constitute human actions. The proposed transformer encoder model provides better performance than bidirectional-LSTM models in both speed and accuracy and provides better generalisation to novel tasks on benchmark action recognition datasets such as the KTH dataset and the UR-Fall dataset.

Keywords: Computer Vision · Machine Learning · Action Recognition · Transformer · Bidirectional-LSTM · CCTV · Pose-Detection

1 Introduction

Human action recognition is valuable in building management, security systems, accident prevention, and intervention. This is essential for applications where recognizing human actions in real time can capture significant value for customers. The ability to recognise actions using purely vision-based approaches is essential for monitoring actions in the closed-circuit television (CCTV) context where unknown people perform actions in public spaces. Previous studies in computer vision-based approaches for action recognition largely fall into two categories: First, purely single-frame image-based object detection models [23] that have been trained to recognise a particular state in an action as valuable and second, video-based methods that take a certain number of frames in a rolling window to detect the temporal context of an action. A good example of singleframe-based methods would be fall detection which is based on categorising all people in a frame into either the state of "standing" or "fallen" (and sometimes even "falling" in between the two). This approach often performs worse than recurrent-based models that have access to the temporal information which is lost in single-frame methods [1, 14, 17]. Some of the reasons for these incorrect detections are primarily because the temporal context of why the person is in the "falling" or "fallen" state is lost. Therefore, someone who is lying down on a couch or sitting on the floor is often automatically categorised as "fallen" when the context of a few frames could reveal the truth. Another problem with this approach is that a purely Convolutional Neural Network (CNN) vision-based approach struggles to generalise to different people with different appearances in either the "standing" or "fallen" positions [16]. Another problem is that a person is essentially the same "object" for CNN networks whether they are standing or lying down. Therefore class categorisation mistakes are prevalent.

The problem of generalization to people in different positions is somewhat addressed in methods that use pose models to extract the skeletal joint locations of people in the different positions [1, 14] and then train another machine learning model to recognise the relative joint locations that constitute fallen or standing. This approach can generalize better than the purely CNN-based approach but still suffers from the lack of context which is inherent in recognising an action from a single frame that inherently has no temporal context.

Other commonly used methods are dynamic image networks [2], optical flow (moving images) [22] or texture descriptors [10] that blend multiple images into one image and then perform object detection on the blended image. This method allows for temporal context that allows for better recognition of the context of the action. However, the image then still suffers from the problem of CNN generalization to different people with different appearances performing the actions.

Current state-of-the-art methods for video action recognition rely on the use of pose-based computer vision models to extract the skeletal joints of a person performing a given action and feed them into a neural network that has a form of recurrence (LSTM/RNN/GRU) [4, 15, 20] or attention (Transformer) [6]. One of the tasks of this paper is to compare the performance of the recurrent BiLSTM [4] to the attention-based transformer method and evaluate their ability to generalize when adding new features (such as bounding box speed). We compare the two approaches' robustness for generalizing in CCTV environments where camera frame rates often vary due to dropped frames or varying camera frame rates and finally, we also evaluate our Transformer model on fall detection after training on our own custom dataset that contains falls and four other adversarial classes.

The goals of our action recognition system are: First, be able to recognise arbitrary actions that can generalize to work for humans and non-humans. Second, be able to generalize to work and recognize actions performed from a variety of different camera angles and people facing different directions (away from the camera, towards the camera and all the angles in between). Third, performs better than another state-of-the-art Bi-LSTM model. Fourth, ensure that the action recognition model works for fall detection such that it can work on unseen fall datasets and fifth, be able to work well for CCTV-specific situations and deployments (AI-on-camera deployment, multi-person batching, and be MLOps [12] pipeline friendly).

Human action recognition use-cases: Fall detection as well as identification of real vs fake falls, running detection (useful for indoors), aggression detection, ensuring factory tasks such as assembly of products are done correctly, ensuring that cleaning and janitorial tasks are done correctly, identifying poor lifting posture for manual labourers, recognition of exercises done in gyms for workout tracking, recognition of various actions taken on sports fields for automated statistics capture, recognition of poor throw/lift/run technique on sports field for team management and injury assessments, recognition of anomalous actions in public spaces, recognition of poor hand washing technique for surgeons or other high stakes environments, patient getting into or out of bed incorrectly, human-re-identification through gait.

Non-human action recognition: The modular approach of our approach which uses an easily retrainable pose detection model such as the YOLOv8 pose allows for this action recognition model to be used to identify non-human actions if the pose model is trained to place key points on non-human objects. This approach can be used to perform action recognition on any non-human object that can either be manipulated in space by an agent or on the non-human object itself as long as it has "joints" that can perform relative motion with other parts of itself. Some examples of use cases for non-human action recognition are recognition of different robot arm motions in a factory (recognition of faulty motion), recognition of non-round ball motion (tumbling vs spiralling), recognition of boom-gate opening and closing for facilities management, recognition of animal-actions for conservation.

2 Related Work

2.1 Action Recognition for Intelligent CCTV Systems

Intelligent CCTV systems are becoming ubiquitous in the area of building and facilities management and safety and security systems. The area of intelligent CCTV systems incorporating action recognition systems that go far beyond basic action recognition is necessary to enable insights into facility usage, security and accident prevention/response. An action recognition model that can be trained on any arbitrary set of actions and then deployed on a large scale for several edge devices is something that is all but unknown in the industry. The goal of this work is to create a model that can be trained on an arbitrary set of actions, and deployed at a large scale on several cameras with different camera angles, lighting conditions and volumes of people walking through the view of the camera. A successful action recognition model should be created with parallelization and batching strategies in mind, given that a large number of people should be able to be analysed at once. With this in mind, speed, accuracy, and model sizes must be able to work on varying image sizes, camera angles, camera frame rates, edge devices and lighting conditions. The ideal model should be able to be retrained on labelled false-positive and false-negative data to aid the improvement that MLOps [12] processes pipelines provide in optimizing models for the particular cameras that they are run on. To aid in the MLOps process there should be the ability to adjust model sizes for deployment on low-powered edge devices as well as the ability to scale up the model architecture to build a large model that can run on a powerful GPU cluster to be used to relabel false-positive detections for adding to the training dataset (as proposed in [26]). Therefore, models like Transformers that can be more easily scaled up to create large models like those used in large language models are better for the task than recurrent models like the LSTM which often suffer from the vanishing or exploding gradient problem for long sequences. Transformers further benefit from being more easily parallelizable than LSTMs or other recurrent networks due to the lack of recurrence which is intrinsically harder to parallelize than Transformers.

2.2 Fall Detection in Public Spaces

The purpose of fall detection systems is to automatically identify falls, prompt appropriate responses, facilitate quicker medical attention and reduce the delay in accessing help after a fall. Traditional monitoring methods often depend on manual intervention such as emergency buttons [9], which may not be readily available after a fall, or wearable sensors such as watches that are required to be worn by each person for whom falls are wished to be detected. These methods are not useful when posed in the context of CCTV or building management systems. The ideal fall detection system for CCTV-based systems should be able to recognise falls, in public spaces, and have a near-zero false-negative as well as a low false positive rate. The vision-based solution allows for falls to be detected in public spaces, even in persons who are not expected to be fall risks.

2.3 LSTM Action Recognition Model

Previous work in pose-based action recognition done by Ramirez et al. [20] illustrates how a successful application of skeleton-based action recognition can be performed to recognise different actions, including falls, based on the temporal positions of human joints. This paper provides the state-of-the-art for fall detection for joint locations fed into a recurrent network (LSTM). The work by Chen et al. [4] shows how an attention-guided Bi-LSTM can detect actions without using the skeleton points. Current state of the art methods use pose-skeleton LSTM/Bi-LSTM models for recognition tasks (such as fall detection) as shown by Inturi et al. [8]. Therefore this will be the benchmark that our Transformer model should beat. Therefore the rest of our work will compare Bi-LSTM model to our proposed Transformer model.

The Bidirectional-LSTM action recognition model enables the recurrence relationship of the joints to be captured for the actions to be recognised. The bi-directionality of the model allows the model to learn recurrence relationships in both the forward and reverse directions. The problem with using a recurrence model such as an LSTM is the common vanishing [7] and exploding gradient problem [18] that hampers the model's ability to generalize for both long and short sequences of data with the same model architecture (number of layers and layer sizes). This problem is avoided by using the attention-based Transformer model. The LSTM model's sensitivity to sequence lengths means it has tremendous trouble generalising to actions performed over varying frames per second with the same models. This makes it unsuitable for most CCTV corner cases where frame rates can vary due to dropped frames and different default camera frame rates that are common for large buildings with cameras that are often supplied by several different vendors.

2.4 Transformers in Action Recognition

The Transformer model makes use of the multi-head attention mechanism that is introduced by Vaswani et al. [25]. Previous work for action recognition using Transformer models [6, 11, 19] demonstrate that Transformers show promising results for action recognition when combined with temporal joint information and the powerful attention mechanism. The usefulness of Transformers in action recognition tasks for CCTV systems extends beyond the promised accuracy and speed increases suggested [6, 11, 19]. The Transformer-based implementation by Do and Kim [6] relies on grouping human joints for optimization which results in high accuracy on human action recognition at the cost of generalizability to non-human action recognition tasks. The transformer architecture with its self-attention mechanism can tend to different sequence lengths which provide promise for CCTV systems which commonly drop frames and have varying frame rates for different cameras that are deployed on the same site.

3 Proposed Approach

3.1 Pose Detection Model

The proposed pose detection model that is used to extract key points (joints) is a YOLOv8x-pose model [24] which is among the state of the art for pose detection models. The model returns a bounding box along with coordinates for the following key points (referred to as joints) on the people in the frame: nose, eyes, ears, shoulders, elbows, wrists, hips, knees and ankles. These key points are fed into the Transformer model one frame at a time after they are scaled and normalized to remove positional information. This model can be swapped out for any other pose model that returns key points.

3.2 Scaling and Normalization Algorithm

An important pre-processing step that is imposed on the key points generated by the pose detection model is the scaling and normalization of the data. This is important seeing that the data should be represented in as uniform a format as possible given the model will try to learn the underlying patterns and trends in the pose estimation key points. The method taken to ensure that the data for actions is represented constantly is described in the scaling and normalization algorithm in Algorithm 1. This algorithm and pre-processing step are essential for making sure that the location of the person in the frame and the relative orientation of the person acting are not taken into account to aid in the generalization of the model over the same action done in other places in the frame and also being done in other directions concerning the camera angle (front on vs facing away from the camera and all the other angles in between). This scaling and normalization algorithm is designed to be pose model agnostic and work on non-human objects, unlike the setup by Do and Kim [6] which relies on grouping human joints for optimization.

Algorithm 1 Normalize and Scale Keypoints

```
 1: norm_val_x, norm_val_y ← joints_coordinates[T=0][norm_index]
 2: for each time_step in joints_coordinates do
 3:     for each keypoint in time_step do
 4:         keypoint_x ← keypoint_x - norm_val_x
 5:         keypoint_y ← keypoint_y - norm_val_y
 6:     end for
 7: end for
 8: norm_val_x, norm_val_y ← joints_coordinates[T=end][norm_index]
 9: for each time_step in joints_coordinates do
10:     for each keypoint in time_step do
11:         keypoint_x ← keypoint_x / norm_val_x
12:         keypoint_y ← keypoint_y / norm_val_y
13:     end for
14: end for
15: signed_max_val ← max(abs(joints_coordinates))
16: for each time_step in joints_coordinates do
17:     for each keypoint in time_step do
18:         keypoint_x ← keypoint_x / signed_max_val
19:         keypoint_y ← keypoint_y / signed_max_val
20:     end for
21: end for
22: return joints_coordinates
```

3.3 Examples of the Scaling and Normalization Algorithm in Action

The selected normalization keypoint should ideally be one of the shoulders since the shoulders' keypoints are the most commonly visible keypoints when taking videos of people whether from the front or the back. When a key point such as the nose or ear is selected then the pose models often output no point and then give a poor output guess when the person is looking away from the camera and these facial features are not visible. This would result in poor performance and generalization on the scaling and normalization algorithm. Using a shoulder joint location is almost always output by the model (as long as the shoulders are framed) since the training data that pose models are trained on has labelled joint data for people when they are looking away of the camera as well. This is not the case for facial features such as nose or ear locations.

An example of the scaling and normalization algorithm in action in Fig. 1 conveys both relative position information as well as velocity information to the model that receives the keypoint information as input. The velocity information is carried over by the relative distance between the key points for motions that are performed since the frame rate is constant.

Note that the action looks upside down when due to the relative distance of the ankle's key points being far away from the normalization joint which is selected (left shoulder) thus they dominate the third portion of the scaling and normalization algorithm that normalizes the points to be kept between -1 and + 1 for x and y coordinates. If the normalization keypoint selected for normalization were in the bottom half of the body then the plots would look right-side up.

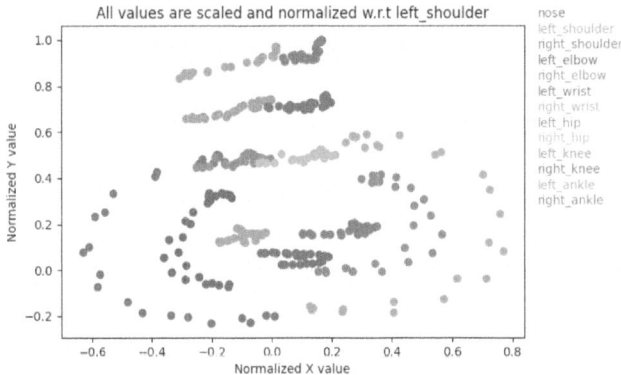

Fig. 1. Scaled and normalized jumping jack keypoints

3.4 Transformer Action Recognition Model

The Transformer model proposed makes use of the same architecture as the pose-based Bi-LSTM but with the Bi-LSTM model swapped out by Transformer encoder layers. The encoder section of the transformer model is the necessary section of the model that enables using multi-head attention on a recognition task such as the task of action recognition in the problem at hand. The architecture used for the action recognition application is shown in Fig. 2. This architecture can effectively attend to vastly varying sequence lengths as long as the model is at least large enough to handle the longest sequence.

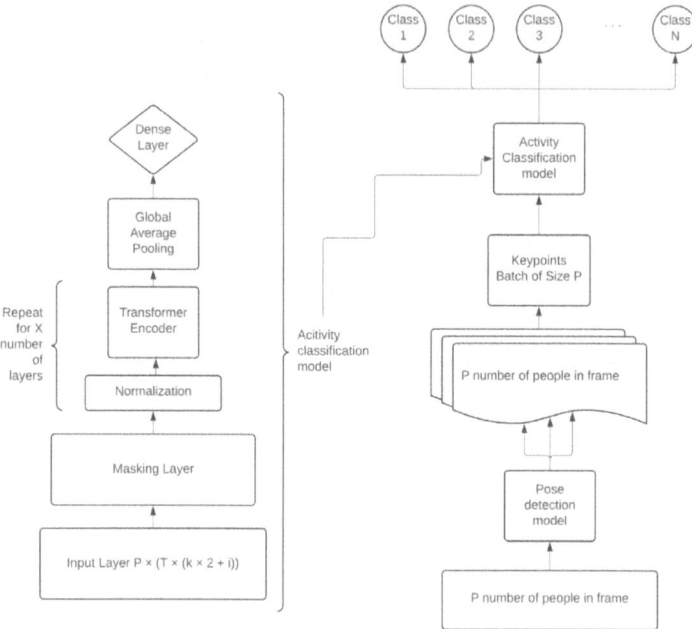

Fig. 2. Transformer encoder model used

4 Experimental Setup

Each of the clips in all of the datasets is limited in duration to an arbitrary max duration of 5 seconds per repetition. This can be adjusted to any length that makes sense for the actions that are to be recognised, however, keeping the max length of actions at 3 to 5 seconds provides good performance. A max duration of 5 seconds is set, given that most actions that humans perform which provide value to action recognition systems take less than 5 seconds per repetition. Training data for our custom dataset is captured such that each training clip constitutes exactly one repetition of the action. Training data is purposely done at different speeds to have a wide variety of durations between 0 and 5 seconds. This aids the model to recognise the same action regardless of how long it takes to complete a repetition of the action. The sequences are all right-zero-padded so as to allow for seamless masking using the Keras input masking library [5].

4.1 Own Custom Dataset

For the development of a fall detection model, a custom dataset is created of the following actions, performed by one person from one camera angle: fall, squat, punching (a "1–2-punch combination"), jumping jacks, and kicking. The purpose of this dataset is twofold, firstly, create a comprehensive set of examples for true falls as well as a set of adversarial actions. This ensures that the model is able do differentiate falls from other actions and secondly see if the actions performed by one person from one camera angle can generalize to the same actions performed by other people from other camera angles. The actions are performed with the person facing towards as well as away from the camera and all the angles in between. This is done so that the action can be recognised regardless of which direction the person is facing when they act.

The primary adversarial action that is of the most value is the squat class. Squats mimic falls in the sense that they both have a strong downward motion by the person acting, but the fall is more noisy in terms of the joints of the person acting and squats have the upward motion during the concentric portion of the repetition. In CCTV systems people who squat down to pick something up off the floor or people bending down to tie shoes are commonly mistaken for falls, therefore adding the squat class as an adversarial class is a smart choice. The custom dataset consists of example videos of each of the five categories (100–150 examples per action) done by one person from one angle. The goal is to see if a single person performing different falls, from a single camera angle can create a model that can generalize to the UR-Fall dataset which consists of multiple people doing different types of falls from a different camera angle than that used for recording the falls on our custom dataset that include falls. The models are trained on the custom dataset with an 80/20 train-test split. The 80% that is used for training is then further split up to an 80/20 train-validation split (resulting in a final split of 64/16/20 train, val, test split).

4.2 KTH Dataset

The KTH dataset created by Schüldt et al. [21] is a dataset that consists of 6 different actions performed by 25 different people taken from several different angles: boxing, handclapping, handwaving, walking, jogging, and running. These actions allow for discrimination between the actions that the hands make for the first 3 classes and discrimination between bipedal motion done at different speeds and intensities for the last 3 classes. Good performance on this benchmarking dataset suggests a model that is good at discriminating between similar actions done by different people and therefore will be able to generalize well to arbitrary actions done by different people from different camera angles.

4.3 UR-Fall Dataset

The UR-Fall dataset [13] is a dataset of 30 falls that are performed by 5 different people from one camera angle. The falls are performed from standing/walking to falling position and also from seated to fallen position. Good performance on this dataset indicates that the model can identify falls that occur from different initial positions.

4.4 Hardware

The experiments were developed on a computer with the following specifications: MacBook Pro (14-inch, 2021) with an Apple M1 Pro chip, which includes an 8core CPU (6 performance cores and 2 efficiency cores) and a 14-core GPU. The system has 16 GB of unified memory (RAM) and a 512 GB SSD for storage, running macOS Monterey. The approach was developed using Python 3.10 in Visual Studio Code (VSCode).

5 Results

5.1 Inference Speed vs Accuracy Comparisons

A summary of the model's performance for different training batch sizes is shown in Fig. 3 for testing done on the KTH dataset (64/16/20 train, val, test split). From this figure, it is clear that the performance of the Transformer model is more accurate than that of the Bi-LSTM model when tuned for accuracy and similar speeds and also faster when tuned for similar accuracy and better speeds (The models are trained on a 14" M1-pro MacBook pro, the absolute speeds are not important, note the relative speeds differences). The figure shows that the transformer can be tuned to either be of relatively similar accuracy as the Bi-LSTM with much better speeds or similar speeds with much better accuracy. It is worth noting that the Transformer model performs relatively flat and predictably for both accuracy and speed when trained on different batch sizes whilst the LSTM is very sensitive to batch sizes (and sequence lengths) and has poor accuracy when trained at batch sizes below 32 as well as at a batch size of 90. This is indicative of training dynamics that are very sensitive to gradient updates which the transformer model is not.

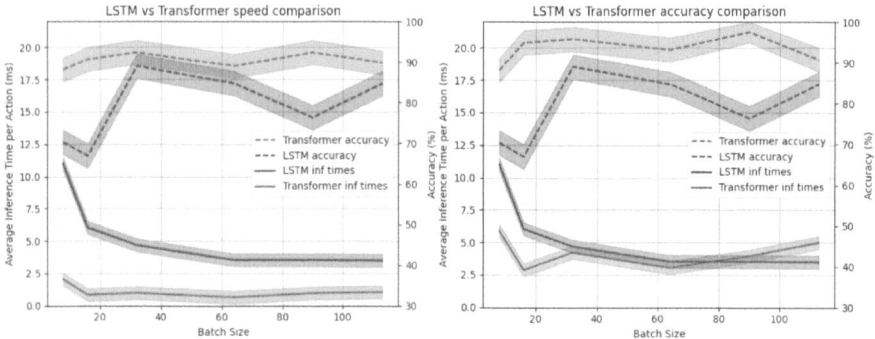

Fig. 3. Bi-LSTM vs Transformer

The results of testing the Bi-LSTM and Transformer models at their optimal batching numbers (batch size is 32 for Bi-LSTM and 90 for Transformer) for 5 training runs each yields average accuracies that are shown in Table 1. This table suggests that the Transformer is the better model overall when comparing accuracy for all the tested datasets, not just the KTH dataset as displayed in Fig. 3. Note that the UR-Fall dataset is not trained on at all, the models are trained on our dataset with the specified training split (64/16/20 train, val, test split) and then the UR-Fall dataset is used completely as an unseen test dataset. Therefore the ability of the Bi-LSTM and the Transformer models to perform well on the UR-Fall dataset at a high confidence level (set high 85% to prevent false positives) suggests that the scaling and normalization algorithm allows for the action recognition models to generalize to never before seen actions done from different camera and body angles, by different people, as long as they perform the same action.

The Transformer model's results on our own dataset, KTH dataset and URFall datasets are shown in confusion matrices and Metric-Confidence level curves in Figs. 4, 5, and 6.

Table 1. Accuracies of Bi-LSTM vs Transformer models on the tested datasets

	Bi-LSTM	Transformer
Our own dataset	77.0%	95.5%
KTH dataset	88.1%	97.5%
UR-Fall dataset (at 85% confidence)	83%	97%

5.2 Model Size Comparisons

When running a model on the edge for CCTV applications model size is a very important factor to consider. The advancements of AI-on-camera devices, such as the Sony IMX500 [3], allow for lightweight machine learning models to be run directly on the camera without streaming the video to a more powerful device for inference. These cameras often have a maximum model size limit that is 10 MB [3]. The size of the Bi-LSTM model is 16.45 MB compared to the Transformer model which is 1.77 MB. This means that the Transformer model is not only faster, and more accurate but also an order of magnitude smaller, enabling edge deployment of several different Transformer action recognition models directly on a single camera.

6 Limitations

The classification model has decreased precision when a high percentage of the keypoints obtained from the pose model are occluded. This can be solved by only making predictions when a high enough percentage of keypoints are present over the rolling window. The classification model can be trained to work well for extreme camera angles (very high/low w.r.t the subject) but this will require extra representative data in the training set. The model is limited by the performance of the pose model. This limitation can be mitigated by retraining the pose model on more data that is representative of difficult cases.

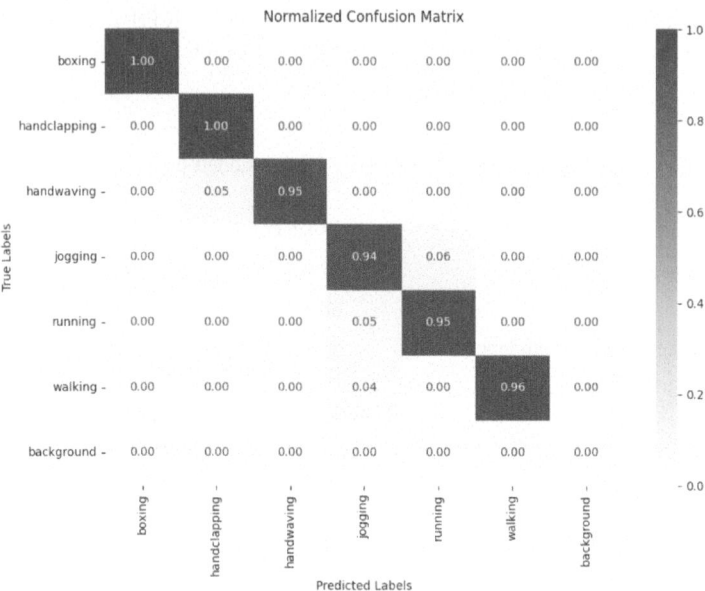

Fig. 4. Transformer confusion matrix on the KTH dataset for a 40% confidence level

7 Conclusion

The Transformer model implemented in this paper accomplishes all the goals of the paper as stated in Sect. 1. The model can generalise to non-human applications due to its pose model agnostic approach that is not specially tuned for humans, seeing as the scaling and normalization algorithm is made to be pose model agnostic. The model can generalize well to arbitrary actions as demonstrated by its performance on our dataset as well as the KTH dataset [21] and UR-Fall dataset [13] as is demonstrated by the model's performance on different camera angles, body angles and people. This is shown by its ability to perform well on our dataset (95.5% test set accuracy) which contains multiple body angles, the KTH dataset (97.5% test set accuracy) which contains multiple people and camera angles and the 97% accuracy on the unseen UR-Fall dataset which has multiple people and different fall types that were not trained on. The model has better accuracy than the Bi-LSTM model on all of the tested datasets as shown in Table 1, better batching performance as shown in Fig. 3 and smaller model size as discussed in Sect. 5.2. The model is also designed with CCTV deployments in mind and allows for easy scaling up of the model for MLOps relabelling. The model is easily parallizable which aids in better batch processing and faster inference than the recurrent LSTM models due to the Transformer architecture.

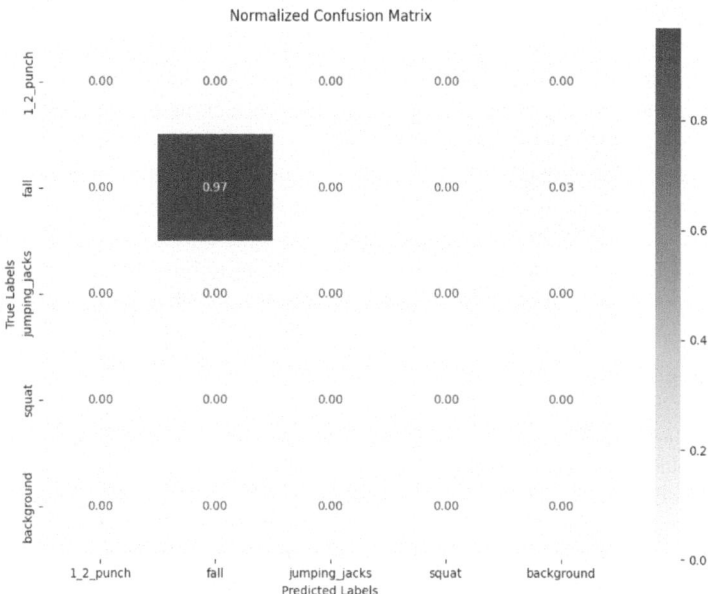

Fig. 5. Transformer confusion matrix on the UR-Fall dataset for a 85% confidence level

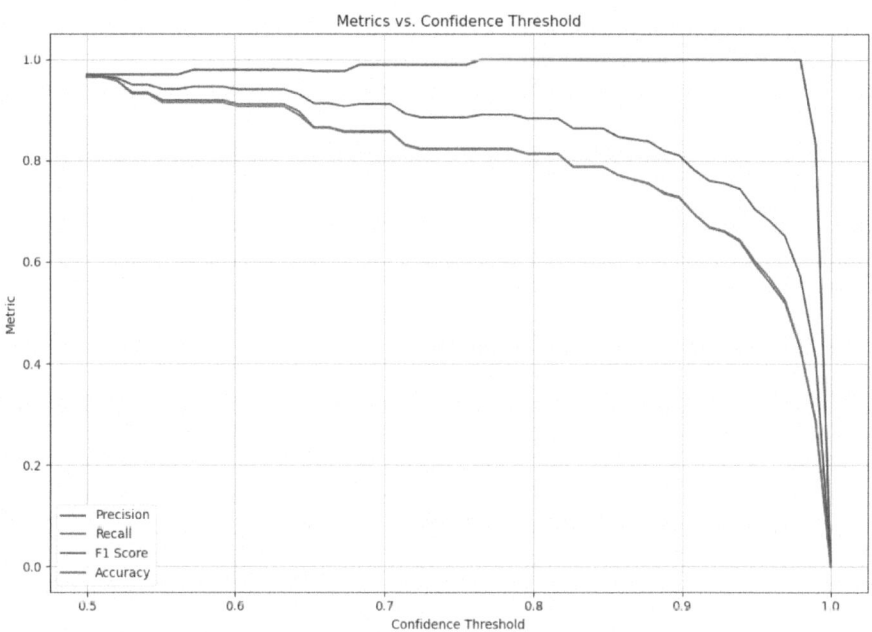

Fig. 6. Transformer metrics on the KTH dataset

8 Future Work

Further work can be done on hyperparameter tuning and model architecture tuning which may involve adding residual layers to the transformer model. Further work can be done on creating an augmentation scaling and normalization algorithm that can translate training pose data from one angle to another to simulate different camera angles. This algorithm could also include further augmentation that could be used to "speed up" or "slow down" actions, drop certain joints add noise to the joints to create augmented datasets and even create fake adversarial joints to actions to force the model to learn the underlying joints and motions that are most important for each actions.

References

1. Ali, A., Pinyonatpong, E., Wang, P., Dorodchi, M.: Skeleton-based human action recognition via convolutional neural networks (CNN). arXiv preprint arXiv:2301.13360 (2023)
2. Bilen, H., Fernando, B., Gavves, E., Vedaldi, A., Gould, S.: Dynamic image networks for action recognition. In: 2016 IEEE Conference on Computer Vision and Pattern Recognition (CVPR), pp. 3034–3042. IEEE (2016). https://doi.org/10.1109/CVPR.2016.331
3. Bonazzi, P., Rüegg, T., Bian, S., Li, Y., Magno, M.: Tinytracker: ultrafast and ultra-low-power edge vision in-sensor for gaze estimation (2023). https://doi.org/10.1109/SENSORS56945.2023.10325167
4. Chen, Y., Li, W., Wang, L., Hu, J., Ye, M.: Vision-based fall event detection incomplex background using attention guided bi-directional LSTM. IEEE Access **8**, 161133–161147 (2020). https://doi.org/10.1109/ACCESS.2020.3021795
5. Chollet, F., et al.: Keras (2015). https://keras.io
6. Do, J., Kim, M.: Skateformer: skeletal-temporal transformer for human action recognition. arXiv preprint arXiv:2403.09508 (2024)
7. Hochreiter, S.: The vanishing gradient problem during learning recurrent neural nets and problem solutions. Int. J. Uncertain. Fuzz. Knowl.-Based Syst. **6**, 107–116 (1998). https://doi.org/10.1142/S0218488598000094
8. Inturi, A.R., Manikandan, V.M., Garrapally, V.: A novel vision-based fall detection scheme using keypoints of human skeleton with long short-term memory network. Arab. J. Sci. Eng. **48**, 1143–1155 (2023). https://doi.org/10.1007/s13369-022-06684-x
9. Karar, M., Shehata, H., Reyad, O.: A survey of iot-based fall detection for aidingelderly care: sensors, methods, challenges and future trends. Appl. Sci. **12**, 3276 (2022). https://doi.org/10.3390/app12073276
10. Kellokumpu, V., Zhao, G., Pietikäinen, M.: Recognition of human actions using texture descriptors. Mach. Vis. Appl. **22**(5), 767–780 (2009). https://doi.org/10.1007/s00138-009-0233-8
11. Khan, S., Naseer, M., Hayat, M., Zamir, S.W., Khan, F.S., Shah, M.: Transformers in vision: a survey. arXiv preprint arXiv:2101.01169 (2022)
12. Kreuzberger, D., Kühl, N., Hirschl, S.: Machine learning operations (mlops): overview, definition, and architecture. arXiv preprint arXiv:2205.02302 (2022)
13. Kwolek, B., Kepski, M.: Human fall detection on embedded platform using depthmaps and wireless accelerometer. Comput. Methods Prog. Biomed. **117**(3), 489–501 (2014). https://home.agh.edu.pl/~bkw/research/pdf/2014/KwolekKepski_CMBP2014.pdf
14. Li, C., Zhong, Q., Xie, D., Pu, S.: Skeleton-based action recognition with convolutional neural networks. In: 2017 IEEE International Conference on Multimedia Expo Workshops (ICMEW), pp. 597–600 (2017). https://doi.org/10.1109/ICMEW.2017.8026285

15. Lin, C.B., Dong, Z., Kuan, W.K., Huang, Y.F.: A framework for fall detection based on openpose skeleton and lstm/gru models. Appl. Sci. **11**(1), 329 (2021). https://doi.org/10.3390/app11010329
16. Long, K., Haron, H., Ibrahim, M., Eri, Z.: An image-based fall detection system using you only look once (yolo) algorithm to monitor elders' fall events (2021)
17. Orozco, C.I., Xamena, E., Buemi, M.E., Berlles, J.J.: Human action recognition in videos using a robust cnn lstm approach. Cienc. Tecn. **20**, 23–36 (2020)
18. Philipp, G., Song, D., Carbonell, J.G.: The exploding gradient problem demystified-definition, prevalence, impact, origin, tradeoffs, and solutions (2018)
19. Plizzari, C., Cannici, M., Matteucci, M.: Spatial temporal transformer network for skeleton-based action recognition. arXiv preprint arXiv:2012.06399 (2020)
20. Ramirez, H., Velastin, S.A., Aguayo, P., Fabregas, E., Farias, G.: Human activity recognition by sequences of skeleton features. Sensors **22**(22), 3991 (2022). https://doi.org/10.3390/s22113991
21. Schüldt, C., Laptev, I., Caputo, B.: Recognizing human actions: A local svm approach. In: Proceedings of the ICPR (2004)
22. Sevilla-Lara, L., Liao, Y., Güney, F., Jampani, V., Geiger, A., Black, M.J.: On the integration of optical flow and action recognition. arXiv preprint arXiv:1712.08416 (2017)
23. Shinde, S., Kothari, A., Gupta, V.: Yolo based human action recognition and localization. In: Procedia Computer Science, International Conference on Robotics and Smart Manufacturing (RoSMA2018), pp. 831–838. Elsevier (2018). https://doi.org/10.1016/j.procs.2018.07.112
24. Ultralytics, et al.: ultralytics/yolov8: Yolov8 - new state-of-the-art yolo model (2023). https://github.com/ultralytics/ultralytics. Accessed 31 May 2024
25. Vaswani, A., et al.: Attention is all you need (2023)
26. Zhang, X., Zhou, Z., Chen, D., Wang, Y.E.: Autodistill: an end-to-end framework to explore and distill hardware-efficient language models. arXiv preprint arXiv:2201.08539 (2022)

Fruit Deformity Classification Through Single-Input and Multi-input Architectures Based on CNN Models Using Real and Synthetic Images

Tommy D. Beltran[1], Raul J. Villao[1], Luis E. Chuquimarca[1,2](✉), Boris X. Vintimilla[1], and Sergio A. Velastin[3,4]

[1] ESPOL, CIDIS, ESPOL Polytechnic University, Guayaquil, Ecuador
{tomdbelt,rjvillao,lchuquim,boris.vintimilla}@espol.edu.ec
[2] UPSE, FACSISTEL, UPSE Santa Elena Peninsula State University, La Libertad, Ecuador
[3] School of EECS, Queen Mary University of London, London, UK
sergio.velastin@ieee.org
[4] Department of Computer Science, University Carlos III, Madrid, Spain

Abstract. The present study focuses on detecting the degree of deformity in fruits such as apples, mangoes, and strawberries during the process of inspecting their external quality, employing Single-Input and Multi-Input architectures based on convolutional neural network (CNN) models using sets of real and synthetic images. The datasets are segmented using the Segment Anything Model (SAM), which provides the silhouette of the fruits. Regarding the single-input architecture, the evaluation of the CNN models is performed only with real images, but a methodology is proposed to improve these results using a pre-trained model with synthetic images. In the Multi-Input architecture, branches with RGB images and fruit silhouettes are implemented as inputs for evaluating CNN models such as VGG16, MobileNetV2, and CIDIS. However, the results revealed that the Multi-Input architecture with the MobileNetV2 model was the most effective in identifying deformities in the fruits, achieving accuracies of 90%, 94%, and 92% for apples, mangoes, and strawberries, respectively. In conclusion, the Multi-Input architecture with the MobileNetV2 model is the most accurate for classifying levels of deformity in fruits.

Keywords: Fruit · Deformity · Multi-Input · CNN models · Real and Synthetic data

1 Introduction

One of the most critical stages in the post-harvest process of fruits and vegetables is the selection and classification through quality inspection (parameters of ripeness, deformation, and defects) [1]. This process, typically meticulous and mechanized, is usually carried out manually by specialized personnel in the field. The significance of this stage lies in the commercial value attributed to the product once evaluated, as higher-quality products tend to command higher selling prices in the market. Additionally, both small-scale retailers and supermarkets, driven by their commercial needs, also consider various parameters when purchasing these foods [2]. Moreover, consumers often take into account the visual appeal of the fruit when making their purchases.

The issue lies in the fact that the quality assessment of fruits and vegetables is carried out through manual processes, turning it into a slow and inconsistent procedure due to the subjectivity associated with the evaluation criteria of the individuals responsible for inspection. This human factor diminishes the efficiency of the evaluation process and adversely impacts the commercial value of the fruits, affecting both consumers in general and farmers/suppliers, who may experience decreased profits due to the quality of their products. Therefore, the purpose of this study is to propose an improvement in the fruit classification process by developing deep learning models that enable the automatic classification of one of the main quality assessment parameters: the level of fruit deformation. In this work, apples, mangoes, and strawberries are analyzed as the fruits under study.

This article is structured as follows. Section 2 provides a review of the state of the art regarding CNN models applied to the identification of fruit deformities. Section 3 describes the contributions and the proposed methodology for conducting the work. In Sect. 4, the results of identifying levels of deformities in fruits using Single-Input and Multi-Input architectures with public CNN models are presented. Finally, the study's conclusions are provided in Sect. 5.

2 Literature Review

This section describes the state of the art concerning the analysis of deformities in fruits, specifically apples, mangoes, and strawberries. It also reviews the techniques and models that have been employed in previous studies for deformity analysis in these fruits.

2.1 Fruit Quality Inspection

Quality inspection is an indispensable aspect of the food industry. Numerous food items and pre-processed products rely on fruits in their production, prompting both retailers and wholesalers to seek fruits of the highest quality and appearance. This approach ensures consumer satisfaction and may result in higher remuneration for the seller when marketing their products.

To assess fruit quality, two methodological approaches are employed: invasive and non-invasive methods [3]. Invasive methods, also known as destructive methods, allow for physicochemical analysis by assessing nutritional parameters such as sugar content, acidity, and others through the manipulation of the fruit.

On the other hand, non-invasive methods enable the evaluation of external features such as deformities (shape, size), ripeness level [4], and defect detection without compromising the fruit's integrity [5,6].

International markets demand that fruits undergo rigorous quality control, leading to the establishment of international standards by organizations such as the Economic Co-operation and Development (OECD), United States Department of Agriculture (USDA), and the regulatory bodies of the People's Republic of China.

When it comes to deformities, it is crucial to assess both the shape and size of the fruit. While Chinese and American regulations address both characteristics, European standards focus solely on the fruit's shape. Incorporating the evaluation of both characteristics would ensure a more accurate prediction of deformities. However, considering size adds a level of complexity to data collection. Given the impracticability of establishing a controlled environment for image acquisition and the paucity of publicly available data concerning size evaluation, this project opted to focus solely on the parameter of shape rather than size. Consequently, only shape classes delimited by the standards set forth by the OECD for deformity assessment are taken into consideration. For this reason, the categories employed for each fruit were: Extra Class, First Class, Second Class, and Ungraded which will be the class where we will send all those images that are already outside of any limit allowed by the OECD.

2.2 Fruit Deformities

During the growth and development process of fruits, alterations in their shape and size can occur as a result of various factors, including genetic causes [7]. A deformed fruit exhibits an atypical shape that does not align with the characteristic features of its species. In this study, we investigate the detection of deformities in fruits using CNNs. It is essential to consider the nature of these deformities to achieve precise classification and enable CNNs to learn to recognize fruits with deformities.

When dealing with deformities, typically two aspects are evaluated: shape and size. International organizations responsible for establishing quality parameters assess these aspects. Some examples include the OECD [8], the USDA [9], and governmental entities that establish their criteria, such as the Republic of China [10]. Table 1 summarizes the primary parameters evaluated in most fruits.

In [11] and [12], both the shape and diameter of the apple are considered for categorization. However, the limitation of size as an evaluation parameter is that it requires the measurement of fruit diameters. Therefore, it is necessary to capture images under controlled conditions, taking into account parameters such as the working distance between the camera and the target to facilitate diameter calculation.

According to the OECD, for a fruit to be classified as Extra Class, Class I, or Class II, it must meet minimum requirements for each parameter in quality inspection such as ripeness, defects, and shape. If a fruit fails to meet the criteria for classification within these classes, it is deemed of low quality and categorized

Table 1. Parameters evaluated for classifying deformities according to various international organizations.

Organizations	Categories	Shape	Size
OECD	Extra-Class, Class 1, Class 2	✓ Typical Shape of fruit	✗ Does not assess
USDA	Extra-Fancy, Fancy, Utility	✓ Typical Shape of fruit	✓ Measurement of the fruit's major diameter.
Republic of China	Excellent, Class 1, Class 2, Substandard	✓ Typical Shape of fruit	✓ Measurement of the fruit's major diameter. Measurement of the fruit's circumference from the top view.

as Ungraded. The following sections elaborate on the permissible levels of deformities in apples, mangoes, and strawberries within international standards for commercialization.

Apple Deformities. According to [13], the classification of deformities in apples is reflected in the classes depicted in Fig. 1: Extra Class, Class 1, Class 2. The OECD stipulates that apples must have a symmetrical surface without evident deviations or bumps. This clearly describes the typical shape of an apple, which, during its development, should have a symmetrical form on both sides, as if it were a mirror image. When the apple lacks this symmetrical shape, it is considered deformed [14].

An apple whose distance between one end and its axis is similar to the distance between the axis and the other is considered symmetrical. As can be observed in Fig. 2, both sides of the "top" part of the apple grow similarly.

In the context of this research, all apples, regardless of their family, must have a symmetrical shape. Therefore, apples from different families can be included in the image dataset since they all contribute to the recognition of deformities in their symmetry.

Fig. 1. Apple Classification based on OECD standard.

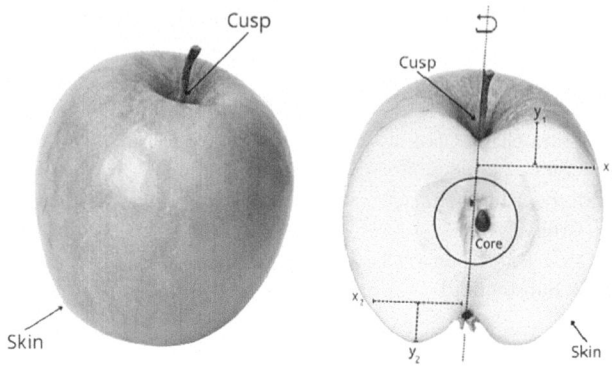

Fig. 2. Apple Symmetry.

Mangoes Deformities. According to [13], the classification of deformities in mangoes is reflected in the classes depicted in Fig. 3: Extra Class, Class 1, Class 2. According to the parameters set by the OECD, mangoes should have an oval shape. However, there are different mango families, so we cannot guarantee that every mango will have a similar shape. This study will focus on the properties of mangoes from Mangifera Indica, specifically the Tommy Atkins variety, and its traditional shape is ovoid (see Fig. 4).

Strawberries Deformities. According to [13], the classification of deformities in strawberries is reflected in the classes depicted in Fig. 5: Extra Class, Class 1, Class 2. Just like with apples and mangoes, there are various species and families of strawberries. Fortunately for this study, strawberry varieties are distinguished by their size and flavor characteristics, but the traditional conical shape is maintained [15]. As a result, any strawberry family will be included in the image collection (see Fig. 6). A strawberry is considered symmetrical if its shape from the calyx to the tip is like an inverted cone.

Fig. 3. Mangoes Classification based on OECD standard.

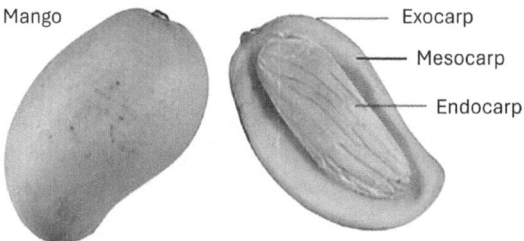

Fig. 4. Mango with elongated oval shape.

Fig. 5. Strawberries Classification based on OECD standard.

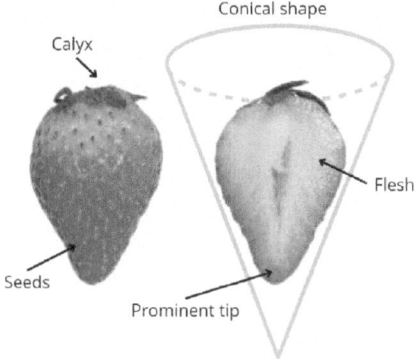

Fig. 6. Strawberry Symmetry.

2.3 Generation of Datasets

When utilizing CNN models, it is essential to have an extensive database of images from which the model can extract meaningful features for each classification. However, having a large quantity of training images does not conclusively equate to obtaining a model with high accuracy. In [16], it was concluded that inadequate preprocessing of input images resulted in a model with low learning levels. From this perspective, the need to refine the dataset to ensure the quality of the input data for CNN models becomes evident.

Studies such as [11,17], and [18] add value to their image datasets by employing preprocessing techniques, which involve segmenting the object of interest, applying enhancement processes, or using images in different visible spectra.

When acquiring images, it is common to turn to public platforms, although at times it is crucial to maintain strict control over the input data, as discussed in [11]. On the other hand, a proprietary acquisition process is often chosen, where the environment is prepared, and parameters such as luminosity and focus distance can be managed, among others. Furthermore, if the aim is to increase the number of training images, Data Augmentation techniques can be employed. The use of synthetic images is also an option. The work in [4] utilizes the Unreal Engine game engine to create realistic images of bananas at different stages of ripeness. In [19], on the other hand, DALL-E mini, a Text-to-Image artificial intelligence, was used to generate images of apples and mangoes with various types of defects such as rot, scabs, and blemishes. In both studies, the use of this type of image contributed to improving the model's performance.

2.4 Classification of Deformities Using CNN Models

In recent years, the application of CNN models to images in the agricultural sector has gained significant attention, particularly in the classification of fruits based on various external quality parameters. While research has been conducted on identifying ripeness levels in fruits and classifying healthy or defective fruits using CNN models, there is a notable scarcity in the literature regarding the classification of fruit deformities. This study specifically addresses this issue by focusing on the parameter of deformation, which is crucial for ensuring the quality, categorization, and marketability of fruits.

The study in [12] focused on identifying the shape and size of apples. A modified VGG16 model was used, which consisted of two fully connected layers (traditional in VGG architectures), and an additional four convolutional layers were added. As a result of the modification, the model achieved an accuracy of 99.04%.

The study in [20] proposes an apple size classification system using their own CNN, the LightNet model. The model's generalization capacity was verified as it was tested with images of zizanias, yielding similar results in model accuracy. The proposed method achieves an accuracy of 99.31%, surpassing that obtained by the AlexNet, VGG11, ResNet18, MobileNet, MobileNetV2, and ShuffleNetV2 models.

Currently, there is a noticeable scarcity of published works on the classification of deformities in fruits using CNN models. This research gap highlights a significant opportunity for future investigations in the field. The application of deep learning techniques, such as CNNs, in the detection and classification of deformities in fruits represents a promising and relevant area, both for the agricultural industry and for the automation of quality inspection processes. Therefore, researchers are encouraged to explore and develop innovative approaches in this direction, aiming to enhance the efficiency and accuracy of fruit deformity classification.

2.5 Architectures Topology

Single-Input Architectures process a single type of input data at a time. This is the most common architecture and is used in a wide range of applications. Multi-Input Architectures can process two or more types of input data simultaneously. This allows the network to learn complex relationships between different types of data, such as RGB images and images in other spectrums. The architecture has multiple branches, each processing a type of input data, and then concatenates the features extracted from each branch to perform the final classification or prediction task [21, 22].

3 Proposed Methodology

Throughout this section, the methodology employed in the development of this work is outlined. Additionally, it is important to highlight the contributions that this study offers to the scientific community:

- Three distinct datasets were developed for this study: Initially, a real dataset was compiled, sourced from public image repositories, and carefully refined according to international standards for shape classification. The second dataset, synthetic in nature, was generated using a blend of "text-to-image" and "image-to-image" tools, to produce high-fidelity images to enrich the training of CNN models. Lastly, a dataset comprising fruit silhouettes was created, wherein only the fruit shapes were meticulously extracted. Furthermore, the dataset generated by this study is made publicly available at the following link: (https://github.com/luischuquim/Deformed-fruits.git).
- The accuracies obtained by the three evaluated models: VGG16, MobileNetV2, and CIDIS are compared. It is important to highlight that the MobileNetV2 model has demonstrated superior performance compared to the other evaluated CNN models.
- The contribution of this study lies in the evaluation of two different training approaches: Single-Input architecture, which allows for progressive and linear learning in contrast to Multi-Input architecture training, involving the simultaneous training of two models with different input approaches. In the case of Multi-Input architecture training, two branches are trained separately, one with RGB images and the other with silhouette images.

3.1 Real Image Acquisition

As with any machine learning model, the availability of a large number of images is essential for CNN models to effectively learn and generalize results. In this regard, it was considered that the appropriate number of images for each deformity category would be around 5,000; meaning a total of 20,000 images per fruit type were required to classify the 4 types of deformities. To compile this dataset, various online sources were explored, such as Kaggle, Github, and Mendeley. Initially, datasets already classified similarly to the established categories (Extra,

Class I, Class II, and Unclassified) were sought. However, the search did not yield any public datasets with these specific classes, although datasets intended for other types of classifications were found. Faced with this situation, it was decided to adopt any available fruit image datasets, such as apples, mangoes, and strawberries, and subsequently classify the images according to the deformity class. Following this, the manual classification of each image obtained from public platforms was refined. Each image was categorized according to the accepted limits for each class. At the same time, the dataset was curated by removing fruit images that did not contribute significantly to the training process. This included images with non-uniform backgrounds that were difficult to remove, as well as images of fruits captured from the top or bottom perspectives.

At the end of the classification process, a significant imbalance was observed among the deformity categories, with the Extra Class having the highest number of images, in contrast to Class II and Ungraded. To address this issue and balance the image dataset, data augmentation techniques were applied, increasing the number of images by a factor of up to 8x for the classes that required it.

3.2 Synthetic Image Acquisition

The generation of synthetic images was carried out in multiple stages, aiming to reach 5,000 images per class for each type of fruit, resulting in a synthetic dataset of 20,000 images per fruit. The initial phase entailed generating near-perfect images for each fruit and category using Easy Diffusion tool version 2.5 [23]. This tool facilitated the creation of diverse image variants corresponding to the categories from texts. However, at a certain point, it ceased to provide significant variations, or the generated images diverged from the expected outcome. Consequently, the decision was made to transition to generating images from existing ones.

Following the approach outlined in [24], small sets of images (between 200 and 800) were created, depending on the different shape scales for each category (Extra, Class I, Class II, Ungraded) and fruit, especially for those classes that were poorer, such as Class II and Ungraded. Models were then trained using LORA with AUTOMATIC 1111 [25]. With the results obtained, the image generator Kohya [26] was used. Unlike Easy Diffusion, this tool allows us to load models trained and generated by AUTOMATIC 1111. This enables us to combine the trained image model with text and generate additional variations beyond what the model learned through LORA.

3.3 Model Pre-training Approach

The training started from the CIDIS model [4]. This model, being custom-made, does not have an initial weight matrix. Therefore, it was first trained with the synthetic image dataset. Once it learned the characteristics of the synthetic data, the learning was transferred to further training using the real image dataset. Figure 7 illustrates the process graphically. In the Model Pre-training approach, we utilized the VGG16 and MobileNetV2 architectures with pre-trained weights, prepared for classification. Thus, in the Model Pre-training process of these architectures, only real images were employed, unlike the CIDIS model, which was used for the initial synthetic phase and subsequent Model Pre-training with real images.

3.4 Multi-input Network Approach

In the analysis of implementing a Multi-Input architecture, the focus lies primarily on evaluating the differences between the input images of each branch [21,27,28]. Subsequently, the result obtained from each branch is unified through a feature fusion process, and classification is carried out after a Multi-layer Perceptron (MLP) layer. Given that this approach prioritizes the study of fruit shape, it was necessary to disregard features typically recognized by CNNs, such as color and texture in the images. Therefore, it was decided that the input images for one of the branches would be RGB images of the fruits (from the real and synthetic datasets), while the other branch would receive the silhouettes of the same fruits in those images, as depicted in Fig. 8.

For this approach, the CIDIS, MobileNetV2, and VGG16 models were used again, with the difference that the CIDIS model was trained without initial weights, while the MobileNetV2 and VGG16 models were initialized with the same weights found before training with the Model Pre-training approach.

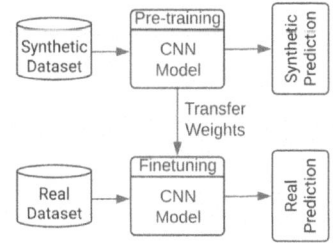

Fig. 7. Model Pre-training Approach

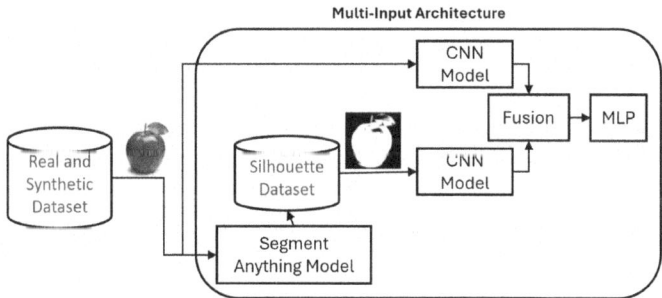

Fig. 8. Multi-Input Network.

4 Experimental Results

In this section, we elaborate on the datasets of real and synthetic images, which were both acquired and generated to ensure high-quality data for evaluating CNN models. Subsequently, we provide a detailed presentation of the results obtained for each training approach, taking into account the various CNN models utilized.

After gathering the minimum required set of real fruit images, data augmentation techniques, including rotations, horizontal flips, and vertical flips, were employed to achieve the agreed-upon number of images for training, validation, and testing (80/10/10). The resulting image size was fixed at 224 × 224 pixels. In summary, a total of 20,000 images per fruit were obtained, evenly distributed among the four deformation categories (see Table 2). On the other hand, given the scarcity of suitable real images for fruit deformity analysis, synthetic datasets were generated leveraging the capabilities provided by the Stable Diffusion tool, as well as the technologies mentioned in Sect. 3.2. These tools produced high-quality synthetic images with dimensions of 512 × 512 pixels. In many cases, images from the real dataset were used as a basis for generating similar synthetic images. Subsequently, data augmentation techniques were employed to rotate and flip the images, aiming to introduce variability into the dataset. Similar to the real image set, a total of 20,000 synthetic images per fruit were obtained, evenly distributed among the deformity categories.

Table 2. Number of real and synthetic images in the fruit dataset.

Fruits	Extra Class		First Class		Second Class		Out of Class	
	Real	Synthetic	Real	Synthetic	Real	Synthetic	Real	Synthetic
Apples	4050 (*950)	5000	1741 (*3259)	5000	651 (*4349)	5000	1961 (*3039)	5000
Strawberry	2782 (*2218)	5000	875 (*4125)	5000	759 (*4241)	5000	679 (*4321)	5000
Mango	4317 (*683)	5000	1889 (*3111)	5000	655 (*4345)	5000	734 (*4266)	5000

Note: The numbers marked with (*) represent augmented images.

Furthermore, silhouette images were prepared, which are closely linked to the shape of an object. In our case, fruit analysis revolves around its shape. Therefore, it is crucial to train the models using fruit silhouettes. For the automatic generation of these silhouette images, the Segment Anything Model (SAM) tool was utilized. The generation of fruit silhouette images was successful in most cases (see Fig. 9). For fruit images with incorrect silhouettes, individual processing was applied, or they were excluded entirely, to obtain a refined and high-quality dataset.

Once the datasets were prepared, hyperparameters were set for training each CNN model. Each training was conducted with a different combination based on

Fig. 9. Obtaining the Silhouette of Each Fruit

the obtained results. The use of various optimizers, such as Adam, Nadam, and RMSProp, along with variable learning rates, is noted. Typically, trainings were configured to run until epoch 30, and the learned weights were restored from the epoch with the best performance.

4.1 Results with Model Pre-training Approach

When training and evaluating the CNN models using only real data, the results shown in Table 3 were obtained. These results provide a detailed assessment of the models' performance, including validation accuracy, validation loss, precision, recall, F1-score, and test accuracy. Table 3 offers a thorough comparison of these performance metrics across different fruits and model architectures, providing a clear understanding of each approach's effectiveness in image classification. However, the results shown in Table 4 are those obtained with the Model Pre-training approach, which achieves better results than the models trained only with real images. We based our selection on the model with the highest accuracy and the lowest loss.

When training and evaluating the CNN models using only real data, as shown in Table 3, the results were suboptimal due to the limited amount of available data and the imbalance in the number of images across classes. This imbalance negatively impacts the models' ability to generalize and correctly classify images, resulting in lower performance metrics. However, by applying the data augmentation methodology through the inclusion of synthetic images, as detailed in Table 4, the models' performance was significantly improved. This approach allowed for the balancing of the datasets and provided the models with a greater diversity of examples during training, leading to increased accuracy and reduced loss. Therefore, the use of synthetic images not only mitigates the issues associated with small and imbalanced datasets but also optimizes the models' ability to achieve more precise classifications.

Table 3. Training Results with CIDIS, MobileNetV2, and VGG16, Model Training Approach using only Real Data.

Fruit	Model	Val Accuracy	Val Loss	Precision	Recall	F1-score	Test Accuracy
Apple	CIDIS	0.6815	0.8491	0.6100	0.5900	0.5500	0.5915
	MobileNetV2	**0.7848**	**0.5851**	**0.6100**	**0.6100**	**0.5700**	**0.6095**
	VGG16	0.7747	0.6073	0.5600	0.5700	0.5200	0.5685
Mango	CIDIS	0.5155	1.2260	0.4400	0.4100	0.4100	0.4400
	MobileNetV2	**0.5155**	**0.9612**	**0.6700**	**0.6700**	**0.6700**	**0.6710**
	VGG16	0.5746	1.0673	0.4900	0.4800	0.4800	0.4770
Strawberry	CIDIS	0.5601	1.3034	0.4900	0.4100	0.4000	0.4110
	MobileNetV2	**0.6971**	**1.2005**	**0.7000**	**0.7000**	**0.7000**	**0.6980**
	VGG16	0.6694	1.0141	0.6300	0.6100	0.6000	0.6100

Table 4. Training Results with CIDIS, MobileNetV2, and VGG16, Model Pre-training approach.

Fruit	Model	Val Accuracy	Val Loss	Precision	Recall	F1-score	Test Accuracy
Apple	CIDIS	0.8956	0.3645	0.8900	0.8800	0.8800	0.8806
	MobileNetV2	**0.9325**	**0.2087**	**0.9100**	**0.9000**	**0.9000**	**0.9044**
	VGG16	0.9050	0.2511	0.9000	0.8900	0.8900	0.8931
Mango	CIDIS	0.08679	0.3819	0.8700	0.8700	0.8700	0.8694
	MobileNetV2	**0.9420**	**0.1594**	**0.9400**	**0.9400**	**0.9400**	**0.9425**
	VGG16	0.9022	0.2290	0.9600	0.9600	0.9600	0.9604
Strawberry	CIDIS	0.8010	0.5258	0.8200	0.8100	0.8200	0.8135
	MobileNetV2	**0.9189**	**0.1945**	**0.9200**	**0.9200**	**0.9200**	**0.9220**
	VGG16	0.8962	0.2646	0.9100	0.9100	0.9100	0.9050

In this way, it is determined that the best architecture for this type of training has been MobileNetV2 for all fruits. It is worth mentioning that in the case of mango, VGG16 had the highest precision; however, the loss was very high. Therefore, we consider that a better result can be achieved with MobileNetV2.

4.2 Results with Multi-Input Network Approach

We relied on studies that validated the results of Multi-Input networks for extracting features between two similar images. In our approach, we feed one branch with an RGB image and the other with the silhouette of the same image. This process effectively identifies similarities or differences in the shape of the object in the image, in our case, the shape of the fruit.

The Multi-Input network approach was employed with the complete dataset of real and synthetic images, resulting in a larger volume of images than Model Pre-training. All models obtained with this approach exhibited an accuracy above 90%, with the MobileNetV2 model achieving the highest accuracy for apples and strawberries, whereas for mangoes, the VGG16 model was superior (see Table 5).

Based on the results obtained, it is evident that the MobileNetV2 model with the Multi-Input architecture achieved the best outcomes for apples and

Table 5. Training Results with CIDIS, MobileNetV2, and VGG16, Multi-Input Network Approach

Fruit	Model	Val Accuracy	Val Loss	Precision	Recall	F1-score	Test Accuracy
Apple	CIDIS	0.7626	1.6818	0.8639	0.8366	0.8107	0.8249
	MobileNetV2	**0.8068**	2.2276	**0.9257**	**0.9158**	**0.9161**	**0.9257**
	VGG16	0.8060	1.9643	0.8966	0.8920	0.8832	0.8844
Mango	CIDIS	0.9234	0.2577	0.9237	0.9232	0.9233	0.9232
	MobileNetV2	0.9440	0.1505	0.9394	0.9393	0.9390	0.9393
	VGG16	**0.9501**	**0.1696**	**0.9551**	**0.9539**	**0.9541**	**0.9551**
Strawberry	CIDIS	0.9188	0.2496	0.9145	0.9171	0.9154	0.9171
	MobileNetV2	**0.9197**	**0.2330**	**0.9455**	**0.9228**	**0.9213**	**0.9394**
	VGG16	0.8769	0.4013	0.9623	0.9204	0.9190	0.9204

strawberries, with an accuracy of 92% and 93%, respectively. However, the VGG16 model proved to be superior for mangoes, with an accuracy of 95%. Consequently, the most favorable results for the Multi-Input approach compared to the Single-Input approach are summarized in Table 6.

Table 6. Results Comparison

Fruits	Models	Accuracy
Apple	MobileNetV2 (Single-Input)	0.9044
	MobileNetV2 (Multi-Input)	**0.9257**
Mango	MobileNetV2 (Single-Input)	0.9425
	VGG16 (Multi-Input)	**0.9551**
Strawberry	MobileNetV2 (Single-Input)	0.9220
	MobileNetV2 (Multi-Input)	**0.9394**

The results obtained in this study have significant implications for both research and the agricultural industry. The high accuracy achieved by the MobileNetV2 and VGG16 models in classifying deformed fruits suggests that these architectures can be implemented in automated quality classification systems, significantly reducing the reliance on manual and subjective evaluations. This not only improves the consistency and speed of the inspection process but also optimizes the supply chain by ensuring that higher-quality products reach consumers, which, in turn, could increase profitability for farmers and suppliers. Additionally, the insights gained on the effectiveness of Multi-Input architectures compared to Single-Input provide a solid foundation for future research in automating quality assessment in other types of fruits and agricultural products.

The findings of this study, focused on the parameter of fruit deformation, enhance and complement the results obtained in previous research, such as the classification of healthy and defective fruits [19], and the assessment of banana ripeness levels [4]. These advancements contribute to the development of a comprehensive fruit categorization system based on international standards for external quality control, thereby strengthening the ability to ensure consistent standards in the classification and commercialization of agricultural products.

The increase in computational cost of the multi-input approach compared to the single-input method may seem like a challenge, especially in contexts with limited resources. However, it is crucial to consider the tangible benefits this approach offers in terms of precision and efficiency in fruit classification. By utilizing multiple inputs, the model can learn from different features and patterns present in both RGB and silhouette images, thereby enhancing its ability to differentiate the levels of deformities in fruits. This improvement in accuracy not only reduces the error rate in classification but also minimizes the risk of deformed products reaching the market, which can negatively impact both consumers and producers.

5 Conclusions

A public dataset has been generated, presenting a wide range of images, including both real and synthetic images of apples, mangoes, and strawberries, thus addressing their morphological diversities. This dataset is evenly divided for each fruit into 5,000 samples for each of the four defined classes: Extra Class, First Class, Second Class, and Out of Class. Furthermore, it is supplemented with an additional set of 20,000 silhouette images generated from each RGB image, providing a more comprehensive and detailed representation of the morphological characteristics of the apples in the dataset. For this work, a dataset was obtained that enabled successful training. AI data generation tools such as Stable Diffusion played a fundamental role in achieving this accomplishment. Analyzing the shape of fruits involves being very careful with the images used. The results obtained show that training with binary images where only shape details are present significantly aids learning, as is the case with the Multi-Input Network approach. To achieve the goal of having a good dataset, it is essential to obtain as many images as possible that are easy to segment, such as images with a consistent background. Evaluating different architectures such as MobileNetV2, VGG16, and the one proposed by CIDIS has helped us understand that for this type of task, it is more advantageous to use lightweight architectures like MobileNetV2, as it yielded the best accuracy results.

Both approaches, Model Pre-training, and the Multi-Input Network, have yielded good results in deformity classification. However, the Multi-Input Network approach stood out for its efficiency, grounded in the capability of the branches corresponding to RGB images and silhouette images to capture distinctive characteristics of fruit deformities. In the case of the branch processing RGB images, it benefited from detailed information regarding edges, textures, color patterns, and specific shapes, all of which are fundamental elements for the precise identification of deformities. Conversely, the branch associated with silhouette images allowed for the consideration of aspects such as shape, contours, edges, and pixel distribution within the silhouette, thereby enriching the representation of the structural characteristics of the fruits. For future work, a comparison with ViT and DINO ViT models is proposed, as they have demonstrated high efficacy in image classification tasks, to evaluate their performance in classifying fruit deformities compared to lightweight architectures like MobileNetV2. Additionally, it is essential to improve the quality of the dataset using advanced artificial intelligence tools to manage, clean, and select visual data quickly and at scale. These enhancements will ensure a more robust foundation for model training and evaluation, enhancing their ability to accurately and efficiently identify deformities.

Acknowledgment. This work has been partially supported by the ESPOL-CIDIS-11-2022 project.

References

1. Olorunfemi, B.J., Kayode, S.E.: Post-harvest loss and grain storage technology - a review. Turk. J. Agric.-Food Sci. Technol. **9**(1), 75–83 (2021)
2. Shewfelt, R.L., Prussia, S.E.: Challenges in handling fresh fruits and vegetables. In: Postharvest Handling, pp. 167–186. Elsevier (2022)
3. Vetrekar, N.T., et al.: Non-invasive hyperspectral imaging approach for fruit quality control application and classification: case study of Apple, Chikoo, Guava fruits. J. Food Sci. Technol. **52**(11), 6978–6989 (2015). https://doi.org/10.1007/s13197-015-1838-8
4. Chuquimarca, L.E., Vintimilla, B.X., Velastin, S.A.: Banana ripeness level classification using a simple CNN model trained with real and synthetic datasets. In: VISIGRAPP (5: VISAPP), pp. 536–543 (2023)
5. Chuquimarca, L., Vintimilla, B., Velastin, S.: Classifying healthy and defective fruits with a multi-input architecture and CNN models. In: 2024 14th International Conference on Pattern Recognition Systems (ICPRS), pp. 1–7. IEEE (2024)
6. Coello, O., Coronel, M., Carpio, D., Vintimilla, B., Chuquimarca, L.: Enhancing Apple's defect classification: insights from visible spectrum and narrow spectral band imaging. In: 2024 14th International Conference on Pattern Recognition Systems (ICPRS), pp. 1–6. IEEE (2024)
7. Behera, S.K., Rath, A.K., Mahapatra, A., Sethy, P.K.: Identification, classification & grading of fruits using machine learning & computer intelligence: a review. J. Ambient. Intell. Human. Comput. 1–11 (2020)
8. Lidror, A., Prussia, S.E.: Improving quality assurance techniques for producing and handling agricultural crops. J. Food Qual. **13**(3), 171–184 (1990)
9. Huang, K.M., Guan, Z., Hammami, A.M.: The US fresh fruit and vegetable industry: an overview of production and trade. Agriculture **12**(10), 1719 (2022)
10. Zhou, J.H., Kai, L., Liang, Q.: Food safety controls in different governance structures in China's vegetable and fruit industry. J. Integr. Agric. **14**(11), 2189–2202 (2015)

11. Wang, J., et al.: Grading detection of "Red Fuji" apple in Luochuan based on machine vision and near-infrared spectroscopy. PLoS ONE **17**(8), e0271352 (2022)
12. Hu, G., et al.: Infield apple detection and grading based on multi-feature fusion. Horticulturae **7**(9), 276 (2021)
13. Porat, R., Lichter, A., Terry, L.A., Harker, R., Buzby, J.: Postharvest losses of fruit and vegetables during retail and in consumers' homes: quantifications, causes, and means of prevention. Postharvest Biol. Technol. **139**, 135–149 (2018)
14. Chakrabarti, A., Michaels, T.C., Yin, S., Sun, E., Mahadevan, L.: The cusp of an apple. Nat. Phys. **17**(10), 1125–1129 (2021)
15. Liu, L., et al.: The flavor and nutritional characteristic of four strawberry varieties cultured in soilless system. Food Sci. Nutrition **4**(6), 858–868 (2016)
16. Vujović, Ž, et al.: Classification model evaluation metrics. Int. J. Adv. Comput. Sci. Appl. **12**(6), 599–606 (2021)
17. Sun, L., Liang, K., Song, Y., Wang, Y.: An improved CNN-based apple appearance quality classification method with small samples. IEEE Access **9**, 68054–68065 (2021)
18. Garillos-Manliguez, C.A., Chiang, J.Y.: Multimodal deep learning via late fusion for non-destructive papaya fruit maturity classification. In: 2021 18th International Conference on Electrical Engineering, Computing Science and Automatic Control (CCE), pp. 1–6. IEEE (2021)
19. Pacheco, R., González, P., Chuquimarca, L.E., Vintimilla, B.X., Velastin, S.A.: Fruit defect detection using cnn models with real and virtual data. In: VISIGRAPP (4: VISAPP), pp. 272–279 (2023)
20. Cao, J., et al.: An automated zizania quality grading method based on deep classification model. Comput. Electron. Agric. **183**, 106004 (2021)
21. Mesa, A.R., Chiang, J.Y.: Multi-input deep learning model with RGB and hyperspectral imaging for banana grading. Agriculture **11**(8), 687 (2021)
22. Pipitsunthonsan, P., et al.: Palm bunch grading technique using a multi-input and multi-label convolutional neural network. Comput. Electron. Agric. **210**, 107864 (2023)
23. Brade, S., Wang, B., Sousa, M., Oore, S., Grossman, T.: Promptify: text-to-image generation through interactive prompt exploration with large language models. In: Proceedings of the 36th Annual ACM Symposium on User Interface Software and Technology, pp. 1–14 (2023)
24. Somepalli, G., Singla, V., Goldblum, M., Geiping, J., Goldstein, T.: Diffusion art or digital forgery? Investigating data replication in diffusion models. In: Proceedings of the IEEE/CVF Conference on Computer Vision and Pattern Recognition, pp. 6048–6058 (2023)
25. Hidalgo, R., Salah, N., Chandra Jetty, R., Jetty, A., Varde, A.S.: Personalizing text-to-image diffusion models by fine-tuning classification for AI applications. In: Arai, K. (ed.) IntelliSys 2023. LNCS, vol. 822, pp. 642–658. Springer, Cham (2023). https://doi.org/10.1007/978-3-031-47721-8_44
26. Masrouri, M., Qin, Z.: Towards data-efficient mechanical design of bicontinuous composites using generative AI. Theor. Appl. Mech. Lett. **14**(1), 100492 (2024)
27. Dua, N., Singh, S.N., Semwal, V.B.: Multi-input CNN-GRU based human activity recognition using wearable sensors. Computing **103**(7), 1461–1478 (2021). https://doi.org/10.1007/s00607-021-00928-8
28. Choudhary, A., Mishra, R.K., Fatima, S., Panigrahi, B.K.: Multi-input CNN based vibro-acoustic fusion for accurate fault diagnosis of induction motor. Eng. Appl. Artif. Intell. **120**, 105872 (2023)

CNN Sensitivity Analysis for Land Cover Map Models Using Sparse and Heterogeneous Satellite Data

Sebastián Moreno[✉][iD], Javier Lopatin[iD], Diego Corvalán, and Alejandra Bravo-Diaz

Universidad Adolfo Ibáñez, Santiago, Chile
sebastian.moreno@uai.cl

Abstract. Land cover maps provide detailed information on the land use of territories, which is useful for public policy making. Constant changes in the landscape limit the usefulness of these maps over time, so they need to be constantly updated. In this context, remote sensing images combined with the use of deep neural networks can be used for this purpose. Although several models are trained on different datasets, we do not know their ability to transfer the learned patterns to new data. In this paper, we evaluate several pre-trained semantic segmentation models on deep convolutional neural networks (CNN) using freely available global RGB data from Sentinel-2. Four CNN models with 32 different architectures were evaluated on data from three continents, on seven different classes. The results show that the best model is the PSP-Net with seresnet18, obtaining a test macro F1 score of 0.4950 when the model is trained with data augmentation and fine-tuning.

Keywords: Land cover maps · deep neural networks · transfer learning

1 Introduction

Land cover maps are crucial for characterizing land use at a specific location and time, including information on water surfaces, vegetation, rocks, soil, roads, and urban areas [1]. This information is essential for public policy decisions, regional development and planning, and natural resource management [2]. Similarly, land cover maps can show changes over time and space, which is important because land use constantly changes at different scales. These changes, mainly caused by humans, can be drastic and significantly impact people, economies and the environment. These actions range from urban expansion, deforestation for agricultural, livestock, and forestry expansion, and land loss due to forest fires. Therefore, it is crucial to obtain updated, accurate, and high-resolution land cover maps through the use of remote sensing techniques.

Remote sensing is a technique used to obtain data from the Earth's surface at various scales by detecting and monitoring the physical and chemical characteristics. For this purpose, multi-sensor platforms located at a certain distance from the Earth's surface, such as aircraft, drones and satellites, mainly collect information from reflected or emitted radiation. The development of this technology has increased the number of high spatial resolution images, such as satellite images, which are ideal for land cover mapping.

Creating a land use model from high-resolution remote sensing imagery involves the classification of pixels, where each pixel is assigned a category label. Traditionally, this classification process is done manually, which can be time-consuming. However, this process can be automated using machine learning techniques. Machine learning techniques learn from the data and then apply that learning to make predictions or decisions without being specifically programmed to perform the task. In the context of remote sensing data, deep neural networks have proven to be highly effective at interpreting spatial patterns in remote sensing data, making them a valuable tool for generating land cover maps.

Deep learning models are increasingly being used to create land cover from remotely sensed imagery. Convolutional Neural Networks (CNNs) are a common alternative to incorporate the neighborhood context into the modeling. CNNs, especially Unet [3], PSPNet [4], LinkNet [5], and FPN [6], present a high classification performance for generating land cover maps. However, many models exist, and we still do not know their effectiveness using satellite imagery when sparsely labeled data is available.

This paper evaluates deep neural network models for predicting land cover classes using sparse and highly heterogeneous data. For this purpose, data from three continents, freely available through an open-source database, are used to train the models. We performed a sensitivity analysis to assess the suitability of the models under a wide range of geographic and climatic features, which is expected to improve their ability to generalize and predict on the test dataset.

2 Current Work

Deep learning models have been widely applied in remote sensing for land cover mapping, thanks to the increase in the number of satellite images. The acquisition of images from space-based platforms such as Sentinel-2, GeoEye, and Pleiades, among others, can explain this increase in high-quality images.

In general, the adoption of deep learning models has fundamentally changed the landscape of image processing, particularly with the emergence of convolutional neural networks (CNNs). CNNs represent a type of deep neural network architecture that learns directly from data without requiring manual convolutional feature extraction. These networks are mainly applied to image data, where we can generally observe four main tasks, as shown in Fig. 1: (a) image classification, which involves assigning a class to an entire image; (b) object detection, which aims to locate and identify classes within an image, usually by framing the target species with a box; (c) semantic segmentation, which delineates the explicit spatial extent of the target class; and (d) instance segmentation, which individually localizes each instance of the class in the image [7].

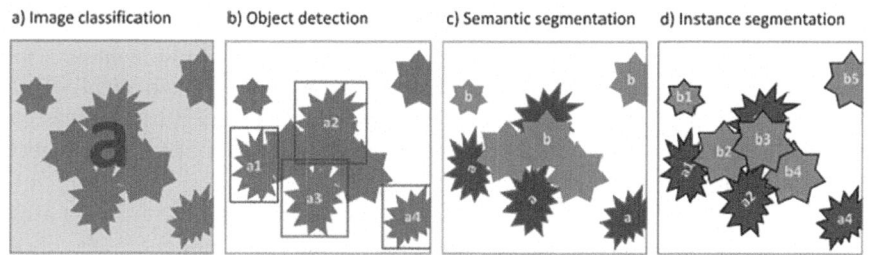

Fig. 1. Main tasks for CNNs. (a) Imagescene classification, where a single class is assigned to the entire image or part of it; (b) Object detection, bounding boxes are produced for each object in the image; (c) Semantic segmentation, a class is assigned to each pixel of the image; (d) Instance segmentation, bounding boxes are generated (like object detection), and it delineates the object boundary as in semantic segmentation [7].

For land cover mapping, image classification (Fig. 1a) and semantic segmentation models (Fig. 1c) are the most relevant CNN applications [8]. However, a drawback of the CNN image classification approach is that all pixels within the image share the same class label, resulting in a reduced resolution for a land cover map. Therefore, semantic segmentation, which provides pixel-level information, is preferred [9,10]. Given their high performance, semantic segmentation models are becoming the most used CNN models in the remote sensing community, especially those with high-resolution images.

Semantic segmentation has been used to solve various problems in different contexts. For example, it has been used in building footprint generation [11], road network mapping [12], human settlement mapping [13], and environmental monitoring [14,15]. However, one of its main applications is in element detection, such as water depth [16], plant species [17], tree species [18], plantation crop areas [19], crop types [20,21], and even building detection [22] and urban spaces [23,24].

For land cover mapping with CNNs, a mosaic-like approach is often used due to the large number of overlapping images. In this approach, images are divided into tiles, which may overlap, along with their respective labels. These labeled tiles are used to train the models; then, once applied to all sub-images, they are reassembled to generate the final land cover map [25–27]. This process has generally been applied considering two approaches: pre-trained models [28–32] and models created from scratch [33–37].

There are a large number of articles that focus on the creation of land cover maps using pre-trained models. For example, Dewangkoro and A. Arymurthy [28] performed land cover classification using CNNs: Where ResNet50 achieved 64.3%, InceptionV3 80.3%, and VGG16 94.5%, using EuroSat data. Marmaris et al. [29] used ImageNet to apply knowledge to a land classification problem, achieving 92.4% accuracy. Chen and Zhang [30] pre-trained a classifier on the VOC (Visual Object Classes) and PASCAL (Pattern Analysis, Statistical Modeling and Computational Learning) challenge datasets, which was then used to extract initial representations from GoogleMap images for remote sensing object detection. In their results, they achieved precision and recall above 90%. Finally, Zhang et al. [32] proposed new deep learning U-net models for urban land cover classification based on very high-resolution satellite imagery, achieving accuracy results above 80%.

There is also a considerable amount of work on models built from scratch. Wang et al. [33] proposed a lightweight multi-scale CNN model for classification of high-resolution remote sensing images in China, achieving an accuracy of 90.36% for Beijing and 89.64% for Qingdao. Li et al. [34] proposed a fully convolutional multi-scale convolutional neural network (MSFCN) for spatiotemporal images. They achieved an accuracy of about 87%. Tong et al. [35] proposed a scheme for training transferable deep models that are pre-trained on labeled land cover datasets and can be applied to unlabeled high-resolution images. Their results achieved an accuracy ranging from 80% to 96%, depending on the dataset. Hu et al. [36] used a novel deep convolutional neural network to extract land cover information in the city of Qinhuangdao, achieving an accuracy of 83%.

Finally, it is worth mentioning that the vast majority of pre-trained networks are built using RGB information, which represents the three basic information channels with wavelengths of the electromagnetic spectrum of blue, green, and red (visible spectrum). However, remote sensing data usually contains more information (e.g., infrared) that cannot be exploited when using pre-trained networks. While it is possible to use a pre-trained network with more than RGB information, this type of work is uncommon [38, 39].

3 Methodology

In this paper, we evaluate the performance of convolutional neural network (CNN) models for land cover prediction based on a four-step methodology: 1) data acquisition; 2) image preprocessing; 3) modeling; and 4) evaluation.

3.1 Data Acquisition

The data for this paper correspond to the LandCoverNet dataset of Radiant MLHub, which contains more than 8,934 images of 256 × 256 pixels with a resolution of 10 m. Each pixel in these images is labeled with one of seven possible land cover classes. The land cover classes include water, natural bare ground, artificial bare ground, woody vegetation, cultivated vegetation, (semi-)natural vegetation, and snow/ice.

The images were taken in 2018 with observations from the Sentinel-2 satellite. Sentinel-2 is a constellation of satellites developed by the European Space Agency (ESA) as part of the Copernicus program. These satellites have a multispectral sensor that captures images in different spectral bands. In the case of Sentinel-2, 13 spectral bands are depicted, covering a wide range of the electromagnetic

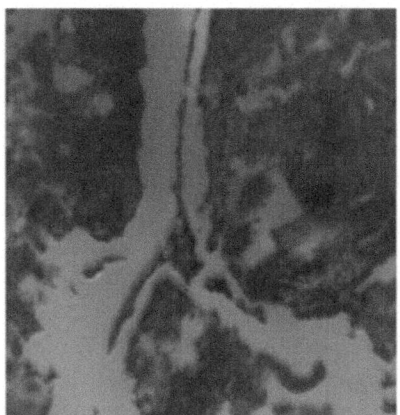 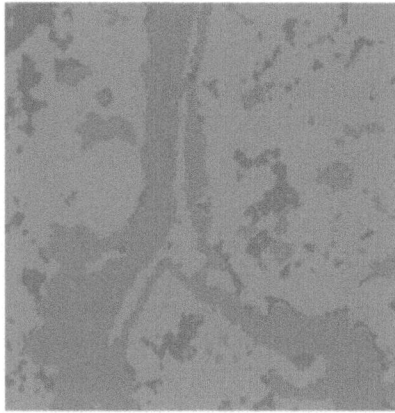

Fig. 2. Image data example. Left: the original RGB image. Right: the mask of the image, where each pixel is labeled with one of the seven possible classes.

spectrum from the visible (∼400 nm) to the shortwave infrared (∼2,500 nm). The spatial resolution of these images ranges from 10 m to 60 m, depending on the wavelength. However, we used the RGB bands (blue, green, and red; 10 m), which can be used with pre-trained neural networks.

3.2 Image Preprocessing

The RGB bands represent reflectance values, which are not the pixel values often used in pre-trained CNN models. Therefore, we normalized pixel values between 0 and 1 to train CNN models. In addition, to improve the quality of the training images, images with clouds were removed using the provided SCL band (sen2cor algorithms [40]).

Due to data availability on the platform, we reduced the data set to 200 images for each of the following continents: Africa, South America, and North America (other continents were not available), obtaining a total of 600 images. Each image has a 256 × 256x3 shape, and is accompanied by a label file that provides detailed information about the associated land cover classes. These labels are represented in grayscale with values ranging from 1 to 7 to indicate the seven land cover classes: C1-Water (4.69%), C2-Artificial bare soil (2.78%), C3-Natural bare soil (10.19%), C4-Snow/Ice (0.70%), C5-Woody vegetation (24.07%), C6-Cultivated vegetation (20.89%), C7-(Semi) Natural vegetation (36.68%). We applied a one-hot encoding to split these classes into single binary labels. An example of the RGB images with their corresponding class labels can be seen in Fig. 2.

3.3 Modeling

In this paper, we evaluated 128 different models for semantic segmentation. We evaluated four pre-trained convolutional neural network (CNN) models (Net [3],

Pyramid Scene Parsing Network (PSPNet) [4], Linknet [5], and Feature Pyramid Network (FPN) [6]) and 32 architectures per model, displayed at the bottom of the right plot Fig. 5.

We use a feature extraction approach in the first part of the evaluation. In this approach, we fix the pre-trained weights of the encoder and train only the decoder weights with the new data. After comparing all the models, we select one of the best models and perform fine-tuning.

Fine-tuning is an essential technique in machine learning used to adapt a pre-trained model to a new task [41]. In the context of this paper, this corresponds to unfreezing the decoder weights and including them in the training process. In this paper, we use a progressive fine-tuning approach. In each experiment, an increasing number of network layers were unfrozen, starting with the last encoder layers and progressing to the initial layers. The model retained and applied the more general features learned from the pre-trained model by keeping the initial layers. Then, at last, as the initial layers were unfrozen, the model could learn and adapt to more general features of the given task. Once this process was complete, we selected the best model and applied data augmentation.

Data augmentation is a commonly used technique to increase the amount of training data available to improve performance. In this paper, we implemented image rotation and flipping, which are geometric transformations [42]. These transformations allow the generation of new images without changing the pixel values, thus preserving the integrity of the original data. Specifically, we applied a 90-degree rotation to the right, yielding a second set of images, and then inverted the rotated images. This process tripled the number of training images.

In the training process, each of the models was trained with images from South America, North America, and Africa (the data from the other continents were not available). We split the data into 70% training, 15% validation, and 15% testing. This type of split was preferred over k-fold cross-validation because of its lower computational cost. The categorical cross-entropy loss function was used to train the models due to the multiclass nature of the problem. The Adam optimizer was used with a learning rate of 0.0001. To avoid overfitting, each model was trained for 30 epochs (most curves stabilized before the 30 epochs), and the epoch with the lowest loss function on the validation data was selected.

3.4 Evaluation

The following performance metrics were used to evaluate the models: F1-score, intersection over union (IoU), and training time. F1-Score and IoU use precision and recall information, where precision is the proportion of correctly classified values out of the total values predicted to be positive, and recall or sensitivity is the proportion of correctly classified values out of the total positive values.

The F1-score is a model evaluation metric equal to the harmonic mean of precision and recall. The formula for the F1-score is given by:

$$F_1 = 2 \cdot \frac{\text{precision} \cdot \text{recall}}{\text{precision} + \text{recall}} = \frac{2 \cdot TP}{2 \cdot TP + FP + FN}$$

where TP is true positive, FP is false positive, and FN is false negative. For the model evaluation, we consider the simple average F1 score for each class.

Intersection-Over-Union (IoU) is a widely used metric in computer vision and image processing to evaluate the accuracy of object detection. It is calculated as the ratio of the intersection between the predicted region and the reference region of an object, divided by the union of these two regions. Higher IoU values indicate more accurate detection, with 1.0 representing a perfect match, while values close to 0.0 indicate a significant lack of overlap between the detection and the actual object. The IoU is a measure similar to the F1-score, also known as the F0-score, since its formula is $IoU = \frac{TP}{TP+FP+FN}$, giving less importance to the correct values.

Finally, we also evaluated the models' training time due to the complexity of these networks. Training time generally depends on the size of the dataset, network complexity, and the processing capacity of the hardware used.

4 Results

In this section, we present the results of the 128 trained models. Recall that we use four different models: U-net, PSPNet, Linknet, and FPN, and they were tested with 32 different architectures. After the first set of results, we selected the best model and applied fine-tuning and data augmentation to further improve its performance. Finally, we analyzed performance by class.

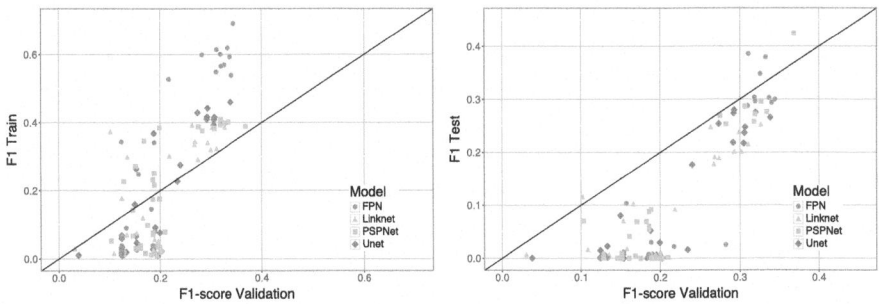

Fig. 3. Comparison of the 128 trained models for land map generation. Left: Training versus validation data. Right: Test versus Validation data. Note that plots have different scales to improve the visualization of the data points.

4.1 Feature Extraction Results

Figure 3 shows the F1-score results of the 128 trained models. The left plot shows the training versus the evaluation data, while the right plot shows the test versus the evaluation data. As can be seen, although we tried to control for overfitting as much as possible, several models from the FPN architecture were overfitted. In contrast, PSPNet and LinkNet appear to be less overfitted, with PSPNet performing better.

To analyze the difference between the different architectures used for the PSPNet models, we plotted the same results only for the PSPNet model. Figure 4 shows the F1-score results of the 32 architectures of PSPNet. In the results, we can see that seresnet18 has the highest F1-score in the validation data with 0.3752, and similar values for training (0.3889) and test (0.4247).

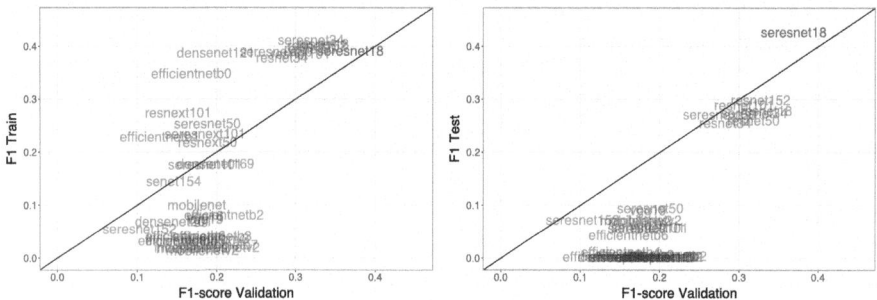

Fig. 4. F1-scores results for the 32 architectures of the PSPNet model. Left: Training versus validation data. Right: Validation versus Test data.

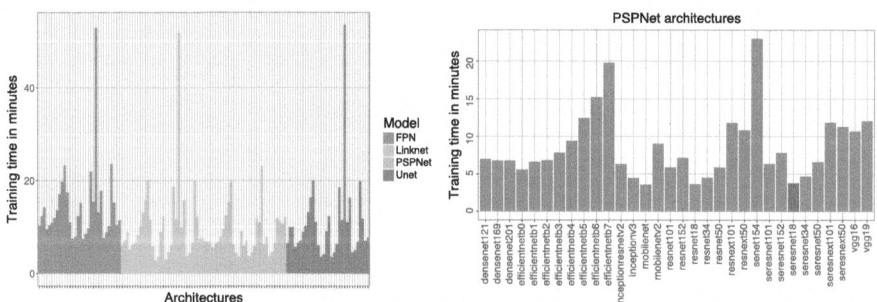

Fig. 5. Training times in minutes for the 128 models. Left: Training time, we get rid of architecture names to avoid cluttering. Right: training time of the 32 PSPNet architectures.

Finally, Fig. 5 shows the training time of all models (left plot) and PSPNet architectures (right plot). In the left plot, we show the training in minutes of the 32 architectures for each model (FPN in red, LinkNet in green, PSPNet in cyan, and Unet in purple). We can observe a clear pattern where the training time depends mainly on the architectures and not on the model. So, considering the importance of the PSPNet architectures, we analyzed their training times in minutes. As can be seen in the right plot, the seresnet18 (red bar), the architecture with the best F1-score performance, is also one of the fastest models to train. For these reasons, we chose seresnet18 for the next experiments, where we evaluated soft tuning and data augmentation. Note that similar results were obtained when analyzing the IoU, so they were removed due to space.

4.2 Fine-Tuning and Data Augmentation Results

Since the seresnet18 showed one of the best results, we proceeded to analyze the impact of training the model using fine-tuning and data augmentation. The experiment consisted of unfreezing the last encoder layers and gradually increasing the number of unfrozen layers up to 100 layers. Specifically, tests were conducted by unfreezing the last 5, 10, 20, 40, 60, 80, and 100 layers of the encoder. This approach allowed us to evaluate how the number of frozen layers affects the model's performance in the semantic segmentation task. In addition to progressively unfreezing the layers of the model, the difference in the loss function between training and test was also analyzed to avoid over-fitting models.

Table 1. Fine-tuning results for the PSPNet model with the seresnet18 architecture. The first column shows the number of unfrozen layers from 0 to 100 (0 being the base model). The second column shows the training time, and the subsequent columns show the performance measures of the F1-score for training, validation, and test data.

Unfrozen layers	Training time (mins)	F1-score Train	F1-score Validation	F1-score Test
0	3.701	0.3889	0.3752	0.4247
5	5.506	0.4944	0.4281	0.4266
10	5.194	0.5187	0.4439	0.4472
20	5.025	0.5391	0.4332	0.4303
40	4.683	0.5049	0.4607	0.4551
60	4.326	0.5366	0.4583	0.4550
80	**4.094**	**0.5253**	**0.4798**	**0.4786**
100	2.927	0.4752	0.4537	0.4541

The results from Table 1 show an increase in the performance of the model when only some of the layers are unfrozen. This is to be expected, since fine-tuning increases the number of parameters of the model to be trained, allowing a better adaptation of the model to the data. However, the highest F1-score for the validation and test data was obtained with 80 unfrozen layers. Thus, we believe that increasing the number of unfrozen layers led to an overfitted model due to the high number of new parameters to train and the lack of new data. Finally, as expected, the training time increases in proportion to the number of unfrozen layers. However, a high number of unfrozen layers (100 layers) had even less time than most of the models. This could be explained because the overfitted model could take less time to stop.

Table 2. Data augmentation and fine-tuning results for the PSPNet with the seresnet18 architecture. The presented models are Original, Augmentation (Original + data augmentation), Fine-tuning (Original with fine-tuning), and Augmentation + Fine-tuning.

Model	Training time (mins)	F1-score Train	F1-score Validation	F1-score Test
Original	3.701	0.3889	0.3650	0.4247
Augmentation	9.920	0.3728	0.3552	0.4620
Fine-tuning	4.094	0.5253	0.4798	0.4786
Augmentation + Fine-tuning	**9.023**	**0.5314**	**0.5157**	**0.4950**

Table 2 shows the results for data augmentation using the original and fine-tuned PSPNet with seresnet18 architecture. As expected, data augmentation increases the training time, doubling the corresponding time. However, it improved the performance of both models. Thus, a combination of fine-tuning and data augmentation could be considered for this type of problem, especially when the number of data points is reduced and the model has a large number of parameters.

4.3 Class Analysis

In the last part of the experiment, we analyzed the behavior of the model for each class. Figure 6 shows the confusion matrix, where each cell shows the percentage of data that the model predicts for each of the classes (note that each row adds up to 100%). Recall, the label and its percentage distribution based on the number of pixels are: "C1-Water" (4.69%), "C2-Artificial bare soil" (2.78%), "C3-Natural bare soil" (10.19%), "C4-Snow/Ice" (0.70%), "C5-Woody vegetation" (24.07%), "C6-Cultivated vegetation" (20.89%), and "C7-(Semi) Natural vegetation" (36.68%).

It can be observed that the model has different performance in different classes. "C1-Water" and "C5-Woody vegetation" show good performance with accuracies (percentage of correctly classified data points for that class) of 67.38% and 73.55%, respectively. However, "C2-Artificial bare soil" and "C4-Snow/Ice" present difficulties with accuracies of only 29.14% and 0.33%, respectively. In addition, there is considerable confusion between certain classes. For example, "C1-Water" is predicted to be "C5-Woody vegetation" 28.20% of the time, and something similar happens with "C2-Artificial bare soil", which is predicted to be "C7-(Semi) Natural vegetation".

These results suggest that while the model can effectively discriminate some classes, it struggles with others. This could be due to inherent similarities between certain classes, or uneven representation of classes in the training data. For example, "C4-Snow/Ice", represents only 0.70% of the pixels of the total data, and "C2-Artificial bare soil", represents only 2.78% of the data, and is thus underrepresented.

On the other hand, "C7-Natural vegetation", which has the largest number of pixels with 36.68% of the data, also has a poor performance. In particular, it is correctly predicted only 49.42% of the time, and 29.71% is predicted as "C5-Woody vegetation". These two classes have similarities in the visible spectrum. For example, "C7-Natural vegetation" includes meadows, grasslands, and natural areas or spaces with vegetation, and "C5-Woody vegetation" includes forests and areas where trees and shrubs predominate. Finally, "C7-Natural vegetation" also shares similarities with "C6-Cultivated Vegetation," which includes agricultural activities and crops, which may justify the confusion between the classes.

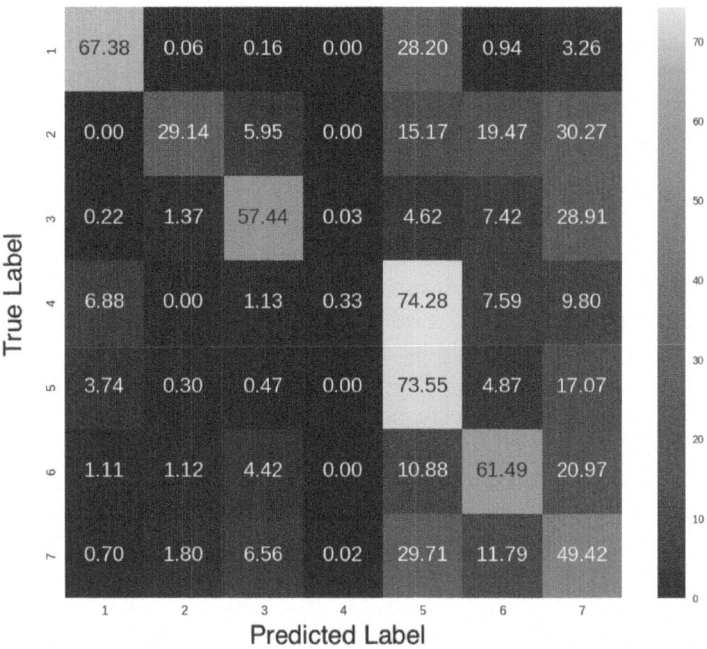

Fig. 6. Confusion matrix for the PSPNet model with seresnet18, using fine-tuning with 80 frozen layers and data augmentation. Each cell represents the percentage of predictions for a specific class in relation to the actual class. The classes in numerical order are: C1-Water, C2-Artificial bare soil, C3-Natural bare soil, C4-Snow/Ice, C5-Woody vegetation, C6-Cultivated vegetation, and C7-(Semi) Natural vegetation.

An example of the model's behavior can be seen in Fig. 7, where we compare the original image (left plot) and mask (middle plot) with the model's prediction (right plot). The figure show that the model correctly classifies most of the water and snow/ice (dark green and pink pixels), but fails to classify the many small patches of natural vegetation (gray pixels).

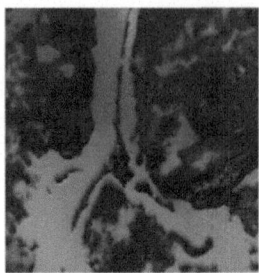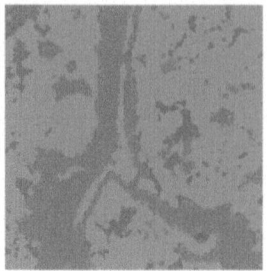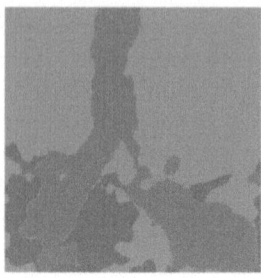

Fig. 7. Classification example. Left: the original RGB image. Middle: the mask of the image, where each pixel is labeled with one of the seven possible classes. Right: the final classification made by the PSPNet model with Seresnet18.

5 Conclusions

This paper investigates the sensitivity of land cover models using semantic segmentation and global satellite data from Sentinel-2. A total of 128 pre-trained CNN models were trained and evaluated using four different models: U-net, PSPNet, Linknet, and FPN, and 32 architectures per model.

The results show that the PSPNet model with seresnet18 outperformed the other models in terms of F1 score performance. This model also proved to be robust, as its performance was similar in both the training and validation sets. Fine-tuning and data augmentation further improved the performance of the PSPNet with seresnet18. In particular, unfreezing the last 80 layers provided a good balance between retaining general features learned from the pre-trained model and adapting to more specific features for the semantic segmentation task. Data augmentation also helped the model to generalize better to the test data, despite the increase in training time.

Class-wise analysis revealed that the model's performance varied across different classes. Underrepresented classes in the data, such as "snow/ice" and "artificial bare soil", caused difficulties for the model. On the other hand, the overrepresented class "Semi-natural vegetation" also presented learning difficulties, which could be explained by its similarity to other classes.

These results underperform other Sentinel-2-based land cover analyses using semantic segmentation techniques. We suggest that this could be due to two main reasons: First, the LandCoverNet dataset from Radiant ML-Hub may have major problems with its labeled data, adding additional noise to the models. We cannot be sure about this hypothesis because the dataset has not been used extensively for experiments, so we cannot compare our results with other work. Second, CNN models may have consistent problems when dealing with very heterogeneous datasets, like here, where we have data from three continents, especially when using small samples (600 observations). If this is the case, we could use dissimilarity metrics to pre-estimate the chances of successful learning transfer in such cases. Furthermore, one could consider including more channels to improve the model's performance by providing more spectral information. We believe that the use of global datasets such as LandCoverNet has great potential for revealing large-scale patterns and for experimenting with highly heterogeneous datasets, but as in this investigation, it may also introduce too much variation for direct use in local applications and decision-making. In such cases, it is often preferable to fine-tune a generic model with local data to accurately capture land use characteristics. Further work should also include model adaptation to landscape change and drought, as the browning of vegetation can significantly alter reflectance.

References

1. Lafontaine, J.H., Hay, L.E., Viger, R.J., Regan, R.S., Markstrom, S.L.: Effects of climate and land cover on hydrology in the southeastern U.S.: potential impacts on watershed planning. J. Am. Water Resour. Assoc. **51**(5), 1235–1261 (2015)
2. Herrera-Benavides, J., Pfeiffer, M., Galleguillos, M.: Land subdivision in the law's shadow: unraveling the drivers and spatial patterns of land subdivision with geospatial analysis and machine learning techniques in complex landscapes. Landsc. Urban Plan. **249**, 105106 (2024)
3. Ronneberger, O., Fischer, P., Brox, T.: U-net: convolutional networks for biomedical image segmentation. In: Navab, N., Hornegger, J., Wells, W.M., Frangi, A.F. (eds.) MICCAI 2015. LNCS, vol. 9351, pp. 234–241. Springer, Cham (2015). https://doi.org/10.1007/978-3-319-24574-4_28
4. Zhao, H., Shi, J., Qi, X., Wang, X., Jia, J.: Pyramid scene parsing network. In: IEEE Conference on Computer Vision and Pattern Recognition, pp. 6230–6239 (2017)
5. Chaurasia, A., Culurciello, E.: Linknet: exploiting encoder representations for efficient semantic segmentation. In: IEEE Visual Communications and Image Processing (VCIP), pp. 1–4 (2017)
6. Seferbekov, S., Iglovikov, V., Buslaev, A., Shvets, A.: Feature pyramid network for multi-class land segmentation. In: Proceedings of the IEEE Conference on Computer Vision and Pattern Recognition (CVPR) Workshops (2018)
7. Kattenborn, T., Leitloff, J., Schiefer, F., Hinz, S.: Review on convolutional neural networks (CNN) in vegetation remote sansón. ISPRS J. Photogramm. Remote. Sens. **173**, 24–49 (2021)
8. Qin, R., Liu, T.: A review of landcover classification with very-high resolution remotely sensed optical images—analysis unit, model scalability and transferability. Remote Sens. **14**(3) (2022)
9. Xue, Z., Li, J., Cheng, L., Du, P.: Spectral-spatial classification of hyperspectral data via morphological component analysis-based image separation. IEEE Trans. Geosci. Remote Sens. **53**(1), 70–84 (2015)
10. Liu, Y., Fan, B., Wang, L., Bai, J., Xiang, S., Pan, C.: Semantic labeling in very high resolution images via a self-cascaded convolutional neural network. ISPRS J. Photogramm. Remote. Sens. **145**, 78–95 (2018)
11. Li, W., He, C., Fang, J., Zheng, J., Fu, H., Yu, L.: Semantic segmentation-based building footprint extraction using very high-resolution satellite images and multi-source GIS data. Remote Sens. **11**(4), 403 (2019)

12. Abdollahi, A., Pradhan, B., Sharma, G., Maulud, K.N.A., Alamri, A.: Improving road semantic segmentation using generative adversarial network. IEEE Access **9**, 64381–64392 (2021)
13. Qiu, C., Schmitt, M., Geiß, C., Chen, T.-H.K., Zhu, X.X.: A framework for large-scale mapping of human settlement extent from sentinel-2 images via fully convolutional neural networks. ISPRS J. Photogramm. Remote. Sens. **163**, 152–170 (2020)
14. Blaschke, T., Lang, S., Lorup, E., Strobl, J., Zeil, P.: Object-oriented image processing in an integrated GIS/remote sensing environment and perspectives for environmental applications. Environ. Inf. Plan. Polit. Publ. **2**, 555–570 (2000)
15. Yuan, X., Sarma, V.: Automatic urban water-body detection and segmentation from sparse ALSM data via spatially constrained model-driven clustering. IEEE Geosci. Remote Sens. Lett. **8**(1), 73–77 (2011)
16. Muhadi, N.A., Abdullah, A.F., Bejo, S.K., Mahadi, M.R., Mijic, A.: Deep learning semantic segmentation for water level estimation using surveillance camera. Appl. Sci. (Switzerland) **11**(20) (2021)
17. Fricker, G.A., Ventura, J.D., Wolf, J.A., North, M.P., Davis, F.W., Franklin, J.: A convolutional neural network classifier identifies tree species in mixed-conifer forest from hyperspectral imagery. Remote Sens. **11**(19) (2019)
18. Dechesne, C., Mallet, C., Le Bris, A., Gouet-Brunet, V.: Semantic segmentation of forest stands of pure species as a global optimization problem. ISPRS Ann. Photogramm. Remote Sens. Spatial Inf. Sci. **IV-1/W1**, 141–148 (2017)
19. Oliveira, W.C.d.S., Braz Junior, G., Gomes Junior, D.L.: Semantic segmentation of the cultivated area of plantations with u-net. Commun. Comput. Inf. Sci. **1519**(CCIS), 3–14 (2022)
20. Yang, S., Chen, Q., Yuan, X., Liu, X.: Adaptive coherency matrix estimation for polarimetric SAR imagery based on local heterogeneity coefficients. IEEE Trans. Geosci. Remote Sens. **54**(11), 6732–6745 (2016)
21. Jadhav, J.K., Singh, R.P.: Automatic semantic segmentation and classification of remote sensing data for agriculture. Math. Models Eng. **4**(2), 112–137 (2018)
22. Rottensteiner, F., et al.: The ISPRS benchmark on urban object classification and 3d building reconstruction. ISPRS Ann. Photogramm. Remote Sens. Spatial Inf. Sci. **I-3**(1), **1**(1), 293–298 (2012)
23. Volpi, M., Ferrari, V.: Semantic segmentation of urban scenes by learning local class interactions. In: Proceedings of the IEEE Conference on Computer Vision and Pattern Recognition Workshops, pp. 1–9 (2015)
24. Fang, F., Yuan, X., Wang, L., Liu, Y., Luo, Z.: Urban land-use classification from photographs. IEEE Geosci. Remote Sens. Lett. **15**(12), 1927–1931 (2018)
25. Yang, H., Yu, B., Luo, J., Chen, F.: Semantic segmentation of high spatial resolution images with deep neural networks. GIScience Remote Sens. **56**(5), 749–768 (2019)
26. Chen, T.-H.K., Qiu, C., Schmitt, M., Zhu, X.X., Sabel, C.E., Prishchepov, A.V.: Mapping horizontal and vertical urban densification in Denmark with landsat time-series from: to 2018: a semantic segmentation solution. Remote Sens. Environ. **251**, 2020 (1985)
27. Wei, P., Chai, D., Lin, T., Tang, C., Du, M., Huang, J.: Large-scale rice mapping under different years based on time-series sentinel-1 images using deep semantic segmentation model. ISPRS J. Photogramm. Remote. Sens. **174**, 198–214 (2021)
28. Dewangkoro, H., Arymurthy, A.: Land use and land cover classification using CNN, SVM, and channel squeeze & spatial excitation block. IOP Conf. Ser. Earth Environ. Sci. **704**, 012048 (2021)

29. Marmanis, D., Datcu, M., Esch, T., Stilla, U.: Deep learning earth observation classification using imagenet pretrained networks. IEEE Geosci. Remote Sens. Lett. **13**(1), 105–109 (2016)
30. Chen, Z., Zhang, T., Ouyang, C.: End-to-end airplane detection using transfer learning in remote sensing images. Remote Sens. **10**(1) (2018)
31. Alem, A., Kumar, S.: Transfer learning models for land cover and land use classification in remote sensing image. Appl. Artif. Intell. **36**(1), 2014192 (2022)
32. Zhang, P., Ke, Y., Zhang, Z., Wang, M., Li, P., Zhang, S.: Urban land use and land cover classification using novel deep learning models based on high spatial resolution satellite imagery. Sensors **18**(11) (2018)
33. Wang, X., Zhao, Y., Liu, D., Sun, G., Zhang, A., Li, J.: A lightweight and multi-scale CNN model for land-cover classification with high-resolution remote sensing images. In: International Geoscience and Remote Sensing Symposium (IGARSS), vol. 2021-July, pp. 5989 – 5992 (2021)
34. Li, R., Zheng, S., Duan, C., Wang, L., Zhang, C.: Land cover classification from remote sensing images based on multi-scale fully convolutional network. Geo-Spatial Inf. Sci. **25**(2), 278–294 (2022)
35. Tong, X.-Y., et al.: Land-cover classification with high-resolution remote sensing images using transferable deep models. Remote Sens. Environ. **237** (2020)
36. Hu, Y., Zhang, Q., Zhang, Y., Yan, H.: A deep convolution neural network method for land cover mapping: a case study of Qinhuangdao, China. Remote Sens. **10**(12) (2018)
37. Saranya, K., Bhuvaneswari, K.S.: Semantic annotation of land cover remote sensing images using fuzzy CNN. Intell. Autom. Soft Comput. **33**(1), 399–414 (2022)
38. Dong, R., et al.: Improving 3-m resolution land cover mapping through efficient learning from an imperfect 10-m resolution map. Remote Sens. **12**(9) (2020)
39. Weikum, A., Robinson, C., Rose, S.: Torchgeo: a PyTorch domain library for geospatial data and models (2021)
40. Main-Knorn, M., Pflug, B., Louis, J., Debaecker, V., Müller-Wilm, U., Gascon, F.: Sen2Cor for sentinel-2. In: Image and Signal Processing for Remote Sensing XXIII, vol. 10427, pp. 37–48, SPIE (2017)
41. Razavian, A.S., Azizpour, H., Sullivan, J., Carlsson, S.: CNN features off-the-shelf: an astounding baseline for recognition. In: IEEE Computer Society Conference on Computer Vision and Pattern Recognition Workshops, pp. 512–519 (2014)
42. Hao, X., Liu, L., Yang, R., Yin, L., Zhang, L., Li, X.: A review of data augmentation methods of remote sensing image target recognition. Remote Sens. **15**(3) (2023)

Video Game Joystick by Recognizing Breathing Patterns

Diego Robles[1,2,3(✉)], Andrea Lira[4], Carla Taramasco[3], and Jorge Mauro[1]

[1] Escuela de Kinesiología, Universidad Diego Portales, Santiago, Chile
[2] Escuela de Ingeniería Informática, Universidad de Valparaíso, Valparaíso, Chile
diego.robles@postgrado.uv.cl
[3] Instituto de Tecnología para la Innovación en Salud y Bienestar, Valparaíso, Chile
[4] Departamento de Morfología, Facultad de Medicina, Universidad Andrés Bello, Santiago, Chile

Abstract. Chronic respiratory diseases are conditions that can severely affect lung function as patients grow. The gradual decline in lung function may increase the risk of developing chronic obstructive pulmonary disease (COPD) in adulthood. Therefore, it is important to incorporate pulmonary rehabilitation programs from an early age. The use of games has become an excellent tool to promote long-term treatment adherence. In this context, an innovative joystick design is presented for pediatric respiratory rehabilitation through the development of interactive games. The joystick integrates a differential pressure sensor to monitor the user's breathing patterns. The centerpiece of this system is the Arduino Nano 33 BLE Sense microcontroller, equipped with a preloaded machine learning model that accurately identifies four key respiratory patterns essential for therapeutic management. Through the control of an avatar, users can complete various challenges by executing these breathing patterns, making the joystick a novel and effective tool for respiratory rehabilitation. The experimental design for real-time identification and recognition of breathing patterns demonstrated a sensitivity of 96.25%, indicating the ability of the system to accurately detect the required breathing patterns, ensuring reliable feedback and improving the effectiveness of the rehabilitation process.

Keywords: Respiratory Pattern Detection · Rehabilitation Gaming · Machine Learning Integration

1 Introduction

1.1 Clinical Gaps in Children Respiratory Rehabilitation

Pulmonary rehabilitation is a personalized treatment that combines therapies such as muscular training, education, psychosocial intervention, and lifestyle changes. Its goal is to improve both the physical and mental condition of people suffering from chronic respiratory diseases. Additionally, it aims to promote adherence to healthy habits that can contribute to improving their quality of life in the long term [28]. Therefore, a rehabilitation program should consider various elements, not limited solely to aspects related to physical rehabilitation. The integration of these components can help increase the motivation of the children's families to ensure that they comply with the treatment properly [23].

Despite favorable results being observed in the health metrics of children who have initiated a rehabilitation process, adherence to this type of intervention is lower compared to pharmacological treatment. Furthermore, this adherence decreases as the patient's age and severity of the condition progress. However, this trend can be modified if the patient finds meaning in the treatment or if it has a positive impact on their quality of life. [18]. Other barriers that can affect treatment adherence include the perception of controlling adults, non-participation norms, and adult-centered medical care [17,27]. Most rehabilitation programs are based on physical exercise [11,14] and use respiratory muscle training devices that focus on technique and results [4]. However, they lack **experience in rehabilitation**, resulting in a focus not only on the adult, but also on the treatment process instead of the patient themselves. This factor could explain the gaps in treatment adherence, which can have direct repercussions on the final outcomes. In addition to program design issues, factors such as transportation costs and family or work responsibilities can reduce the likelihood of regular attendance to a rehabilitation program in low-income countries. On the other hand, institutions also face difficulties in providing services, which are related to the availability and experience of staff, as well as the economic resources necessary for staff and equipment [5]. In Chile, there are few institutions that offer these types of programs [28].

The limited availability of pulmonary rehabilitation programs for children, the lack of integration of a rehabilitation system into the childhood routine, highly medicalized programs geared towards adults, and an increase in symptoms in response to socio-environmental factors lead us to propose an innovative solution. This involves a pulmonary rehabilitation system based on a video game that is incorporated in a fun way into children's routine, allowing them to play together in a community.

1.2 Recognition of Respiratory Patterns as a Combination of Buttons on a Joystick

Rehabilitation based on training through video games has shown positive results in the clinical setting compared to standard programs [19,24,26]. Most of these experiences focus on physical training programs that aim to improve physical capacity, strength, and quality of life. In addition, some interventions use video games as a tool to educate about the disease and the use of medications [28].

Some research has used simple video games to improve lung function parameters through lung-machine interaction, controlling respiratory flow [31]. On the other hand, some studies have used video games as a stimulus for children to perform lung function tests [32]. Among the medical devices developed to monitor lung function in pediatric patients with chronic respiratory diseases is

Alunacare® (https://alunacare.com/), which integrates a video game to track patients at home and integrate health care into their daily lives.

Despite initial development, respiratory rehabilitation programs lack essential features to achieve effectiveness, such as training in specific breathing patterns. It is proposed to integrate these patterns into the video game as commands for the avatar's movement, on the condition that they are recognized by a model that classifies the breathing pattern as input and the avatar's movement as output (Fig. 1). In our research, we developed an experimental protocol for the real-time identification and recognition of respiratory patterns, which will allow us in the future to establish the respiratory commands with which the avatar moves in the game.

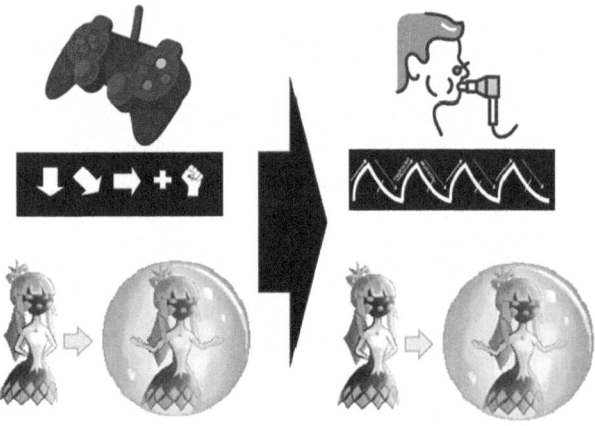

Fig. 1. Example of manual and breathing commands to use superpower

2 Method

2.1 Joystick Design

The joystick design resembles traditional video game controllers, with buttons and a lever for movement. However, this innovative design includes a differential pressure sensor to detect the user's breathing patterns. In addition, it uses an Arduino Nano 33 BLE Sense microcontroller with a built-in machine learning model in its memory. This model is specially designed to accurately identify four essential breathing patterns, crucial for respiratory disease rehabilitation (see Fig. 2). The MPXV7002DP pressure sensor, manufactured by NXP Semiconductors, is renowned for its high precision and sensitivity in measuring small pressure differences, making it ideal for applications that require that level of accuracy.

Technical Features: Measuring Range: ±2 kPa (±0.3 psi), Output: Analog voltage proportional to differential pressure, Power: 5V DC, Accuracy: Typical accuracy within 2.5% of reading. High sensitivity to small changes in pressure. Includes temperature compensation for stability in ranges from −10°C to 60 °C.

Operating Principle: Uses a piezoresistive transducer that converts differential pressure into an electrical signal. This transducer consists of a silicon membrane in which piezoresistive resistors are implanted. Measures the pressure difference between two ports. One of the ports (P1) receives the reference pressure (normally the environment), and the other port (P2) receives the pressure to be measured. Integrates, a circuit to convert the transducer signal into an analog output voltage that is proportional to the differential pressure.

Applications in Respiratory Patterns:

Airflow Measurement: is a useful tool in respiratory monitoring systems. The velocity of airflow can be calculated by using the differential pressure between two points in an airway.

Respiratory Volume Detection: The detection of respiratory volume is done by integrating the flow signal over time, which allows estimating tidal volume.

Continuous Monitoring: Can be used in continuous monitoring devices to capture respiratory patterns, including rate and depth of breathing.

The Arduino Nano 33 BLE Sense is a compact and powerful platform for deploying machine learning (ML) applications on embedded devices. By combining an advanced microcontroller, multiple embedded sensors, and connectivity, it allows you to create machine learning solutions right on the device without the need for cloud infrastructure. Its microcontroller corresponds to an nRF52840 from Nordic Semiconductor. Its architecture is a 32-bit ARM Cortex-M4. Clock frequency: 64 MHz and has 1 MB of Flash Memory and 256 KB of RAM.

2.2 Experimental Protocol

The test consisted of replicating four breathing patterns (Buteyko breathing, spontaneous breathing, pursed lips breathing, and breath stacking). Each subject was instructed by a clinician specialized in the area of respiratory rehabilitation, ensuring the quality of the respiratory pattern performed (See Fig. 3). Data were collected under laboratory conditions. The tests were performed in a sitting position with a face mask, T-connector with expiratory and inspiratory valve, and recording port to which the sensor was directly connected.

2.3 Respiratory Patterns

Tidal Volume at Rest (Spontaneous Ventilation): The tidal volume is the amount of air exchanged in each breath, following a regular pattern between inspiration and expiration (Fig. 3 a). This volume is sensitive to the body's metabolic demands, varying from a low tidal volume at rest to a higher one

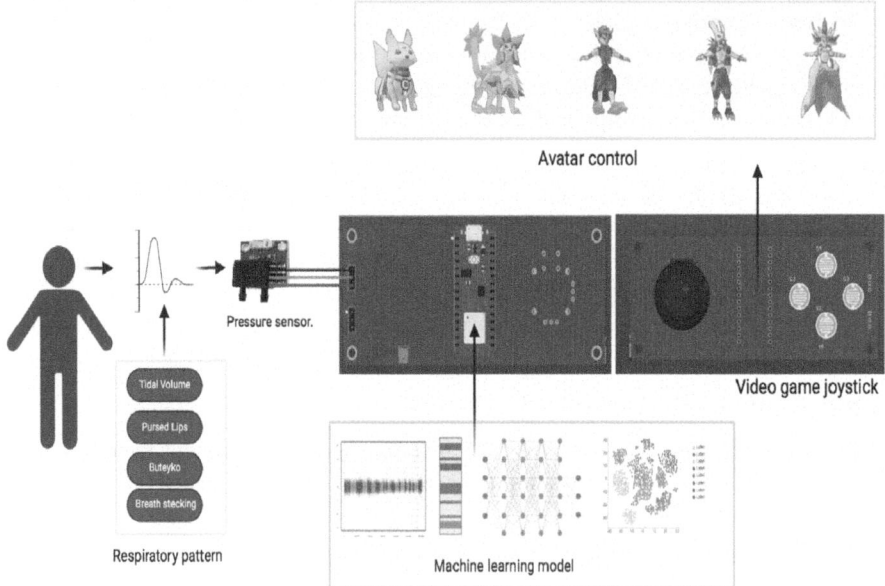

Fig. 2. Joystick operation diagram.

during exercise [22]. As the basic amount of air exchanged in each breath, its measurement at rest helps us establish a baseline for understanding a person's rest. Additionally, its relationship with total lung capacity allows us to detect the level of effort a person is exerting during an activity or at rest. When tidal volume approaches vital capacity, it indicates a higher energy expenditure by the individual.

Pursed Lip Breathing: The pursed lip technique involves taking a deep inhalation through the nose with the mouth closed, followed by a slow exhalation for 4 to 6 s with the lips in a whistling or kissing position (Fig. 3b) [30], similar to blowing out a candle. This type of breathing reduces respiratory rate and increases tidal volume, which in turn decreases the partial pressure of carbon dioxide in the blood and increases the partial pressure of oxygen. In addition, it helps reduce the sensation of breathlessness, improving exercise tolerance and reducing limitations in daily life [12]. These benefits are explained by the decrease in pulmonary hyperinflation, a common symptom in chronic respiratory diseases such as asthma, COPD, cystic fibrosis, and others, which can cause several of the symptoms reported by patients.

Buteyko: The Buteyko breathing pattern consists of periods of controlled reduction of breathing, known as 'slow breathing' or 'reduced breathing', accompanied by periods of breath-holding, called 'control pauses' and 'extended pauses' [8] (Fig. 3c). Studies that have used the Buteyko method have observed a decrease in medication use and an improvement in symptom control, as well as an increase in lung function values and the well-being of individuals [7,15,29]. However, there is still no definitive evidence on the reduction of medication use [29]. The effectiveness of this technique lies in controlling the pathophysiological mechanism of hyperventilation, which can lead to alkalosis and a lower release of oxygen to the tissues. According to the theory of the method, alterations in pH and tissue hypoxia could be related to the classic symptoms of asthma, such as cough, dyspnea, and wheezing [8]. Although this theory is not fully proven, there is evidence supporting its application as a beneficial breathing pattern for pulmonary rehabilitation.

Breath Stacking: Similar to the sniffing pattern, also called stacked sniffing technique or voluntary breath stacking, consists of performing a series of slow and staggered inhalations ranging from functional residual capacity to total lung capacity, including a pause at the end of inhalation [6] (Fig. 3d). While this type of respiratory exercise is recommended to increase lung volume in certain pathologies [3], its effects on respiratory mechanics depend mainly on the patient's muscle contraction to generate tidal volume [1]. Therefore, it is used in rehabilitation therapies with the purpose of improving diaphragmatic function and strengthening respiratory muscles [2].

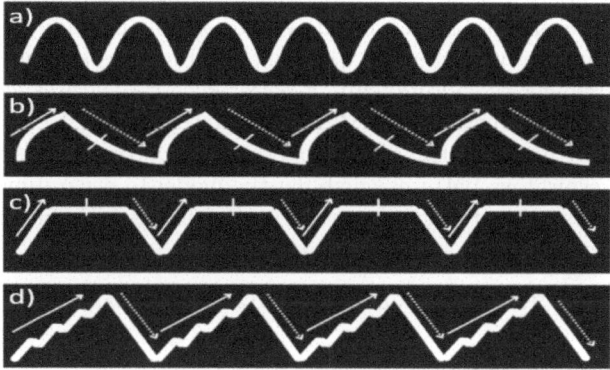

Fig. 3. Respiratory patterns.

2.4 Dataset

A total of 10 healthy young university students between 21 and 27 years of age, 5 women and 5 men (See Table 1) without respiratory disorders were evaluated. Each of them performed the different breathing patterns in a controlled environment supervised by an expert in respiratory rehabilitation, teaching them in detail how to perform each pattern studied before taking the measurement. Each subject performed each breathing pattern 10 times during 20 s of recording, obtaining a total of 100 recordings for each breathing pattern. Being a total of 400 records.

Table 1. Description of Sample.

	Age (years)		Size (cm)		Weight (kg)	
	Female	Male	Female	Male	Female	Male
Mean	22	24	160.8	174.1	62.4	74.6
Std. Deviation	1.2	2.7	6.2	0.7	10.9	8.7

2.5 Signal Processing

The feature extraction process from the respiratory signals was carried out using the Edge Impulse software, which allows for the precise handling and analysis of raw data. Initially, three key signals were derived from the raw data: Differential Pressure (measured in Pascals), Volumetric Flow (measured in cubic meters per second), and Flow Velocity (measured in meters per second). To ensure the quality and accuracy of these signals, a sixth-order Butterworth low-pass filter with a cutoff frequency of 30 Hz was applied, effectively minimizing any high-frequency noise.

For the feature extraction itself, a window of 1000 milliseconds was used with a step size of 1000 milliseconds, ensuring that each window captures significant portions of the respiratory signal for analysis. From these windows, 33 features in the time domain were extracted, which included metrics such as RMS, peak heights, and frequencies. Additionally, 5 power spectral features were derived, focusing on specific frequency bands (0.1–0.5 Hz, 0.5–1.0 Hz, 1.0–2.0 Hz, and 2.0–5.0 Hz). These features, as outlined in Table 2, provide a comprehensive representation of the respiratory patterns, enabling the model to accurately detect and analyze the relevant breathing patterns for pediatric rehabilitation (Table 2).

Table 2. Features extracted by each signal.

Features extraction by each signal
RMS
Peak 1 Height
Peak 2 Freq
Peak 2 Height
Peak 3 Freq
Peak 3 Height
Spectral Power 0.1–0.5
Spectral Power 0.5–1.0
Spectral Power 1.0–2.0
Spectral Power 2.0–5.0

To implement the model, we worked with the Edge Impulse, Inc. software. This software allows recording, processing the signals, generating the classification through a neural network and exporting the model to a microcontroller, allowing classification in real time.

2.6 Model

The features already presented were used to train the model. This data was divided into a training percentage (80%) and a validation percentage (20%) in order to train and test the model. However, a k-fold approach was not employe. In total, 100 records were divided for each respiratory pattern, 80 for training and 20 for validation. The neural network used was a multilayer feed-forward network. The first layer corresponds to the 33 features that feed the network input. The second is a dense layer with 20 neurons. The third is a dense layer with 10 neurons and finally the last layer with four neurons that represent the four labels corresponding to the respiratory patterns. A number of 50 cycles was considered, along with a learning rate of 0.0005 (see Fig. 7).

3 Results

The confusion matrix of the implemented model is reported, specifying the performance for each type of respiratory pattern evaluated. A high sensitivity was obtained for each of the pattern types. In the Buteyko pattern it was 100%, Breath stecking 100%, Pursed Lips 86.4% and for Tidal Volume 100%. (See Fig. 6).

An overall accuracy of 96.25% along with a loss function of 0.18 was obtained for the validation stage of the study after 50 iterations (See Fig. 7). The F1 score values for the Buteyko pattern was 1.0, for the Breath stecking pattern 1.0, for Pursed Lips was 0.93 and for the tidal volume pattern was 0.99 (See Table 3 and Figs. 4, 5).

Table 3. Metrics for Classifier

Metric	Value
Area under ROC Curve	1.00
Weighted average Precision	0.98
Weighted average Recall	0.97
Weighted average F1 score	0.98

4 Discussion

Chronic respiratory diseases in children, such as asthma, bronchiolitis obliterans (BO), and bronchopulmonary dysplasia (BPD), [13], are conditions that can severely affect lung function as patients grow. This gradual decline in lung function may increase the risk of developing chronic obstructive pulmonary disease (COPD) in adulthood [9,10]. As a result, improving respiratory health during the preschool years or even earlier has become essential to address this problem. To ensure that children with chronic respiratory diseases obtain the benefits of pulmonary rehabilitation (PR), it is essential that they perform a combination of aerobic exercise, strength training, and respiratory muscle training. These activities not only improve lung function, but also improve exercise capacity and quality of life for children with these conditions. However, performing these therapies at home has been found to be challenging, suggesting the need for effective and supervised implementation to ensure their effectiveness.

Recent research has confirmed that PR plays a crucial role in the treatment of patients with lung conditions, emphasizing the importance of physical training in their recovery [33]. Using a virtual gaming system to innovate pulmonary rehabilitation: safety, adherence and enjoyment in severe chronic obstructive pulmonary disease [21]. Despite the clear benefits of PR, maintaining physical activity levels in patients has been noted to be a significant challenge, underscoring the need to

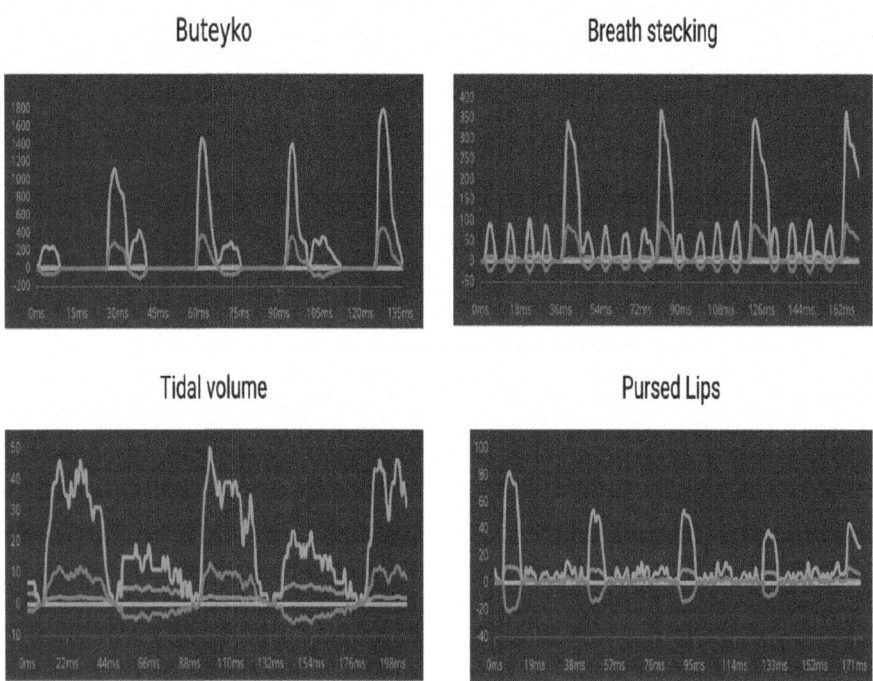

Fig. 4. Real graphic representation of each respiratory pattern studied.

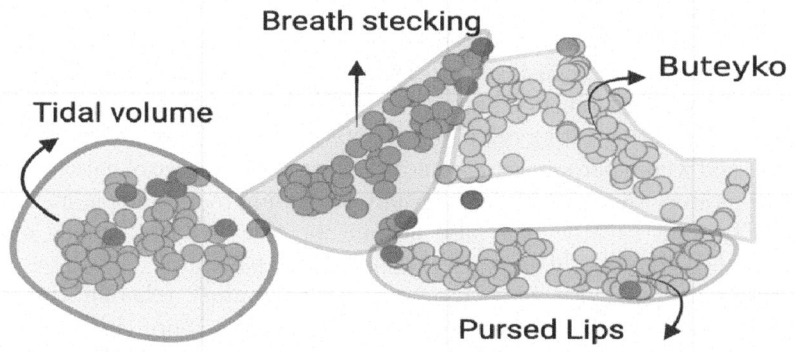

Fig. 5. Cluster of points by pattern. In green, the correctly classified patterns. The incorrect ones are in red. (Color figure online)

	BUTEYCKO	BREATH STACKING	PURSED LIPS	TIDAL VOLUME	UNCERTAIN
BUTEYCKO	100%	0%	0%	0%	0%
BREATH STACKING	0%	100%	0%	0%	0%
PURSED LIPS	0%	0%	86.4%	9.1%	4.5%
TIDAL VOLUME	0%	0%	0%	100%	0%

Fig. 6. Confusion matrix for multi-class classification.

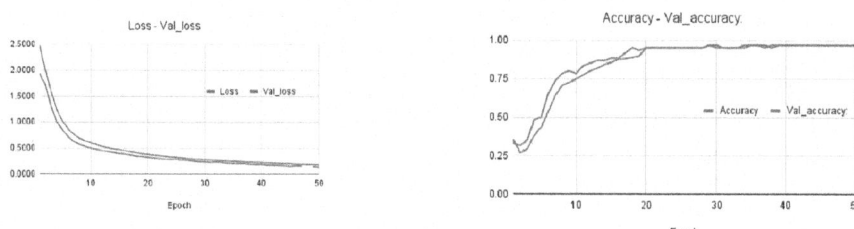

Fig. 7. Training and validation set loss function and accuracy for multi-class classification

implement effective strategies to promote adherence to pulmonary rehabilitation programs.

The games used in rehabilitation often correspond to adaptations of commercial devices or games that are used recreationally and have been personalized by health professionals specialized in the subject for therapeutic purposes. However, they can be designed with a specific focus on the recovery of specific pathologies or well-defined movements. The recognition of respiratory patterns with high therapeutic value using a joystick activated by respiratory flow allows us to expand the horizon of pulmonary rehabilitation proposals. Although the proposal for a joystick activated by respiratory flow has been explored by other research groups, they do not integrate validated active breathing patterns into pulmonary rehabilitation programs, focusing the commands for the avatar's movements on two more basic patterns, such as blowing and suck [25]. On the other hand, Playphysio is an application with a cell phone interface that uses a device similar to a spirometer and whose purpose is to promote long-term adherence of patients with chronic respiratory disease (https://play.physio/our-approach). Using a similar device, ZEPHYRx Breathe Easy also has an app that features, among other things, gamified training (https://www.zephyrx.com/patients).

Other interesting video game proposals for respiratory rehabilitation are a microcontroller that integrates a positive expiratory pressure (PEP) device; These devices are validated equipment and have a recognized effect on respiratory rehabilitation. Through deep and controlled breathing techniques, they can, for example, help eliminate excess mucus in the lungs and improve lung function in people with chronic respiratory diseases, such as cystic fibrosis. Integrating this into a controller that converts the air pressure exerted by a patient using the PEP device during physical therapy exercises into a digital signal to control a variety of games and actions within each game [20], becoming a tool with greater versatility and in which our respiratory pattern recognition model can be integrated.

The recognition of respiratory patterns is not new in the literature and there are several studies on the application of neural networks for the detection of these, which allows, among other things, to monitor patients who may have some alteration in their respiratory pattern, such as, for example, sleep apnea, which can potentially be serious for the health of hospitalized patients [16]. The researchers managed to implement a complex algorithm that has the ability to merge and divide the regions of the different types of breathing, eupnea, bradypnea, tachypnea, apnea and movement, thus allowing a detailed and exhaustive analysis of the data.

Although the use of sensors and microcontrollers is similar, we believe that exploring gameplay with the breathing patterns used in pulmonary rehabilitation allows us to enrich not only the gaming experience, but also the therapeutic potential of video game-based pulmonary rehabilitation.

Although technological advances have made it possible to provide new tools and have favored development in the pediatric bronchopulmonary area, we did not find literature on the recognition of specific respiratory patterns to integrate them as commands in a video game. Therefore, we can conclude that the development of Video game joystick turns out to be a novel, innovative and attractive proposal, which acts as a powerful tool to support children who require pulmonary rehabilitation, promoting adherence to respiratory programs in an entertaining way that will allow an important preventive work with a long-term vision.

We acknowledge the need to expand the sample size used in our study. While the present work achieved high sensitivity in detecting relevant respiratory patterns, the sample of 10 participants limits the generalizability of the results and positions the study more as a proof of concept. Furthermore, we understand the importance of aligning our focus with the pediatric population, given that our primary objective is rehabilitation in children. Therefore, in future research, we plan to extend the trials to a pediatric population, which will allow us to validate the effectiveness of our proposal within the target demographic. These improvements, including a more robust sample size and direct application in pediatric patients, will be essential to strengthen the utility and clinical relevance of our proposal.

On the other hand, in future work, it is recommended to employ k-fold cross-validation. This approach would allow for a more thorough evaluation by reducing the risk of overfitting and providing a better understanding of the model's generalization capabilities across different data subsets. Implementing k-fold validation would enhance the robustness of the results, ensuring that the model's performance is consistently validated across multiple partitions of the dataset.

References

1. Almeida, M.M.F.d., Teodoro, R.d.J., Chiavegato, L.D.: Maneuvers and strategies in respiratory physical therapy: time to revisit the evidence (2020)
2. Ambrosino, N., Vagheggini, G., Mazzoleni, S.: Rehabilitation, weaning and physical therapy strategies in chronic critically ill patients. Eur. Respir. J. **23**(3), 420–427 (2019)
3. Barros-Poblete, M., et al.: Consenso chileno de técnicas de kinesiología respiratoria en pediatría. Neumología pediátrica **13**(4), 137–148 (2018)
4. Bhammar, D.M., Jones, H.N., Lang, J.E.: Inspiratory muscle rehabilitation training in pediatrics: What is the evidence? Can. Respir. J. (2022). https://doi.org/10.1155/2022/5680311
5. Bickton, F.M., Shannon, H.: Barriers and enablers to pulmonary rehabilitation in low- and middle-income countries: a qualitative study of healthcare professionals. Int. J. Chronic Obstruct. Pulmonary Disease **17**, 141–153 (2022). https://doi.org/10.2147/Copd.S348663
6. Bott, J., et al.: Guidelines for the physiotherapy management of the adult, medical, spontaneously breathing patient. Thorax **64**(Suppl 1), i1–i52 (2009)
7. Bowler, S.D., Green, A.: Buteyko breathing techniques in asthma: a blinded randomised controlled trial. Med. J. Aust. **169**(11), 575–578 (2018)
8. Bruton, A., Lewith, G.T.: The Buteyko breathing technique for asthma: a review. Complement. Ther. Med. **13**(1), 41–46 (2005)
9. Deolmi, M., et al.: Early origins of chronic obstructive pulmonary disease: prenatal and early life risk factors. Int. J. Environ. Res. Public Health **20**(3), 2294 (2023)
10. Duan, P., et al.: Impact of early life exposures on COPD in adulthood: a systematic review and meta-analysis. Respirology **26**(12), 1131–1151 (2021)
11. Feng, Z., Wang, J., Xie, Y., Li, J.: Effects of exercise-based pulmonary rehabilitation on adults with asthma: a systematic review and meta-analysis. Respir. Res. **22**(1), 33 (2021). https://doi.org/10.1186/s12931-021-01627-w
12. Gemmell, C.G.: Succinic oxidase factor (a toxic complex produced by Staphylococcus pyogenes) acting on mammalian cell metabolism. University of Glasgow (United Kingdom) (1969)

13. Guillerman, R.P., Brody, A.S.: Contemporary perspectives on pediatric diffuse lung disease. Radiol. Clin. **49**(5), 847–868 (2011)
14. Habib, G.M.M., et al.: Systematic review of clinical effectiveness, components, and delivery of pulmonary rehabilitation in low-resource settings. NPJ Primary Care Respir. Med. **30**(1) (2020)
15. Hassan, I., Lee, J.Y., Ahmed, S.: Application of pressure sensors for real-time monitoring of respiratory patterns. Sensors **22**(5), 1831 (2022). https://doi.org/10.3390/s22051831
16. Hong, J.W., Kim, S.H., Han, G.T.: Detection of multiple respiration patterns based on 1D SNN from continuous human breathing signals and the range classification method for each respiration pattern. Sensors **23**(11), 5275 (2023)
17. Kirkby, S., et al.: Benefits of pulmonary rehabilitation in pediatric asthma. Pediatric Pulmonol. **53**(8), 1014–1017 (2018). https://doi.org/10.1002/ppul.24041
18. Llorente, R.P.A., García, C.B., Martín, J.J.D.: Treatment compliance in children and adults with cystic fibrosis. J. Cyst. Fibros. **7**(5), 359–367 (2008)
19. Mazzoleni, S., et al.: Interactive videogame as rehabilitation tool for patients with chronic respiratory diseases: preliminary results of a feasibility study. Respir. Med. **108**(10), 1516–1524 (2014). https://doi.org/10.1016/j.rmed.2014.07.004
20. Oikonomou, A., Day, D.: Using serious games to motivate children with cystic fibrosis to engage with mucus clearance physiotherapy. In: 2012 Sixth International Conference on Complex, Intelligent, and Software Intensive Systems, pp. 34–39. IEEE (2012)
21. O'donnell, D.E., et al.: Canadian thoracic society recommendations for management of chronic obstructive pulmonary disease–2007 update. Can. Respir. J. J. Can. Thoracic Soc. **14**(Suppl B), 5B (2007)
22. Pleil, J.D., Wallace, M.A.G., Davis, M.D., Matty, C.M.: The physics of human breathing: flow, timing, volume, and pressure parameters for normal, on-demand, and ventilator respiration. J. Breath Res. **15**(4), 042002 (2021)
23. Puppo, H., Torres-Castro, R., Rosales-Fuentes, J.: RehabilitaciÓn respiratoria en niÑos. Revista Médica Clínica Las Condes **28**(1), 131–142 (2017). https://doi.org/10.1016/j.rmclc.2016.12.001
24. Simmich, J., Deacon, A.J., Russell, T.G.: Active video games for rehabilitation in respiratory conditions: systematic review and meta-analysis. JMIR Serious Games **7**(1), e10116 (2019). https://doi.org/10.2196/10116
25. Sobrinho, A.S.F., Scalassara, P.R., Dajer, M.E.: Low-cost joystick for pediatric respiratory exercises. J. Med. Syst. **44**(10), 186 (2020). https://doi.org/10.1007/s10916-020-01655-x
26. Sutanto, Y.S., Makhabah, D.N., Aphridasari, J., Doewes, M., Suradi, Ambrosino, N.: Videogame assisted exercise training in patients with chronic obstructive pulmonary disease: a preliminary study. Pulmonology **25**(5), 275–282 (2019). https://doi.org/10.1016/j.pulmoe.2019.03.007
27. Teleman, B., Vinblad, E., Svedberg, P., Nygren, J.M., Larsson, I.: Exploring barriers to participation in pediatric rehabilitation: Voices of children and young people with disabilities, parents, and professionals. Int. J. Environ. Res. Publi. Health **18**(19) (2021). https://doi.org/10.3390/ijerph181910119
28. Torres Sánchez, I., Megías Salmerón, Y., López López, L., Ortiz Rubio, A., Rodríguez Torres, J., Valenza, M.C.: Videogames in the treatment of obstructive respiratory diseases: a systematic review. Games Health J. **8**(4), 237–249 (2019). https://doi.org/10.1089/g4h.2018.0062
29. Vagedes, J., Co-author, N.: Title of the article. Journal Name **XX**(YY), ZZ–ZZ (2024) DOI if available, available online: URL if available (accessed on [insert date])

30. Vatwani, A.: Pursed lip breathing exercise to reduce shortness of breath. Arch. Phys. Med. Rehabil. **100**(1), 189–190 (2019)
31. Vilozni, D., Bar-Yishay, E., Gur, I., Shapira, Y., Meyer, S., Godfrey, S.: Computerized respiratory muscle training in children with Duchenne muscular dystrophy. Neuromusc. Disord. **4**(3), 249–255 (1994). https://doi.org/10.1016/0960-8966(94)90026-4
32. Vilozni, D., et al.: The role of computer games in measuring spirometry in healthy and "asthmatic" preschool children. Chest **128**(3), 1146–1155 (2005). https://doi.org/10.1378/chest.128.3.1146
33. Wardini, R., et al.: Using a virtual game system to innovate pulmonary rehabilitation: safety, adherence and enjoyment in severe chronic obstructive pulmonary disease. Can. Respir. J. J. Can. Thoracic Soc. **20**(5), 357 (2013)

Recovering Latent Hierarchical Relationships in Image Datasets Through Hyperbolic Embeddings

Ian Roberts[1]($^{\boxtimes}$), Mauricio Araya[1], Ricardo Ñanculef[1], and Mario Mallea[2]

[1] Universidad Técnica Federico Santa María, Valparaíso, Chile
ian.roberts@sansano.usm.cl
[2] Universitat Politècnica de Catalunya, Barcelona, Spain

Abstract. Hyperbolic space has emerged as a promising alternative to Euclidean space for embedding high-dimensional data, including images. In particular, Hyperbolic embeddings have shown to be more effective in discovering hierarchical relationships between data points. However, experiments performed thus far have provided the model with explicit access to this hierarchy during training. This work shows Poincaré embeddings' ability to discover class-subclass relationships in image datasets without direct supervision. In addition, we present applications of this result to content-based image retrieval, demonstrating that Poincaré embeddings can achieve better performance when the relevance of the retrieved elements is measured by taking the class hierarchy into account. Furthermore, we found that Hyperbolic embeddings outperform their Euclidean counterpart for complex class hierarchies. Finally, we discuss the limitations of these results and outline future research directions.

Keywords: Representation Learning · Hyperbolic Geometry · Image Retrieval

1 Introduction

Finding ways of representing data for searching and recovery is one of today's main tasks in modern machine learning and artificial intelligence [2]. Learning aims to group semantically similar images while separating semantically different ones. The simplest version of this task is classifying images according to different classes and creating a measure of distance that can be used as a way to estimate the similarity between different images. This is achieved using embeddings, which are relatively low-dimensional vectors into which we translate unstructured data. This allows us to simplify complex and high-dimensional data into sparse vectors, which we can feasibly analyze or compare.

The generation of these embeddings is a challenging task and is influenced by factors such as the neural network architecture [13], preprocessing techniques

[17], the dimensionality of the embedding space [25], and the mathematical operations (e.g. subtraction) that can be applied on the embeddings in order to support applications (e.g. similarity search) [10]. This last point is of special interest for this work, as significant research has been done lately on the applications of different geometries to the embedding space [15,19,24]. For example, in image classification networks that use Euclidean space, inner layers are used to make the data as linearly separable as possible, leaving the final layer to compute linear boundaries, which essentially separates classes by Euclidean hyperplanes. Therefore, embeddings live in the Euclidean space. Related works have replaced this choice with spherical embeddings by applying a sphere projection operator on the embeddings. Similarly, Hyperbolic embeddings project the output of the last dense layers into a Poincaré ball (see Fig. 1) [10]. In this work, we explore the capabilities of Hyperbolic image embeddings to uncover a latent hierarchical class structure without direct access to these relationships.

Fig. 1. Transformation from Hyperbolic space to Poincaré Disk model. This is achieved by stereographic projection of Hyperbolic space to a unit circle (the Poincaré Disk) [5]

We find that not only Hyperbolic embeddings outperform their Euclidean counterpart in the image retrieval task, but they can also capture more information about the latent hierarchical structure, which can be informally understood as *making better mistakes*.

The article is organized as follows. Section 2 briefly discusses related works. In Sect. 3, we introduce the basic concepts and motivation for using Hyperbolic space to generate hierarchy aware image embeddings. We also present the Hyperbolic layers used in our experiments. We present our methodology in Sect. 4. In Sect. 5 we present the experiments done for training and evaluating the generated embeddings. Finally, in Sect. 6 we show our obtained results.

2 Related Work

Hyperbolic embeddings have been successfully used in natural language processing, driven by the hierarchical structure inherent in this type of data [4,18,19]. This idea can be extended to other tree-like structures such as social networks, human skeletons, and evolutionary relationships between biological entities in phylogenetics [20]. As far as we know, Nickel and Kiela [19] were the first to propose learning an embedding using a Poincaré ball model while considering latent hierarchical structures. They proved that Poincaré embeddings outperformed Euclidean embeddings significantly in data with latent hierarchies, as evaluated on WordNet's noun hierarchy, social networks link prediction, and lexical entailment tasks. This was, however, trained using explicit hypernymy pair relationships, which correspond to the hierarchical relationships present in Wordnet's transitive closure.

Dhingra et al. (2018) expanded on this work by proposing a simpler parametrization technique that maps the output of any (Euclidean) neural net into the Poincaré ball [4], enabling the learning of embeddings for arbitrary objects. They found that the resulting embeddings seemed to encode certain intuitive notions of hierarchy, such as word frequency and phrase constituency. Motivated by Nickel and Kiela's work [19], Ganea et al. (2018) introduced the concept of Hyperbolic Neural Networks [7], which combines the formalism of Möbius gyrovector space with the Riemannian geometry of the Poincaré ball model. This construction provides us with the cornerstone equations presented in Sect. 3. As a result, they derived Hyperbolic versions of important deep learning tools, such as the multinomial logistic regression layer (MLR), feed-forward and recurrent layers. Subsequently, Khrulkov et al. (2019) extended this idea to learning image embeddings, showing that Hyperbolic embeddings offer a superior alternative to Euclidean embeddings in terms of performance for different computer vision tasks, such as few-shot classification and person re-identification [10]. Although not directly related to our work, Anitha et al. (2020) proposed using Hyperbolic neurons in a Hopfield neural network for content-based image retrieval. They found that Hyperbolic Hopfield Neural Networks can guide retrieval and enhance system performance [1].

3 Hyperbolic Geometry

Hyperbolic geometry is a non-Euclidean geometry that studies spaces of constant negative curvature. What is interesting about Hyperbolic space is that it is well-suited to model hierarchical data. Hierarchies can often be represented as taxonomical trees, where the class containing all other elements is the tree's root, and child nodes spread from there over the space, eventually reaching the leaf nodes. Fortunately, Hyperbolic space can naturally embed tree-like structures with arbitrarily low distortion [9,22]. This property holds because the volume of Hyperbolic space expands exponentially, which is analogous to the exponential growth of the number of leaves in a tree as depth increases [19]. In contrast, this expansion is only polynomial in Euclidean space. Consider, for example, a 2D Hyperbolic space with constant curvature $c = -1$. Taking r as the radius, the area of the disk is given as $2\pi(\cosh r - 1)$. Since $\cosh r = \frac{1}{2}(e^r + e^{-r})$, the area grows exponentially with r, while in Euclidean space the area only grows quadratically.

Several isometric models can represent Hyperbolic spaces [21]. This work uses the Poincaré ball model (or Poincaré disk in the 2D case). Thoroughly, let $\mathbb{D}_c^n = \{\mathbf{x} \in \mathbb{R}^n : c\|\mathbf{x}\|^2 < 1, c \geq 0\}$ be the n-dimensional ball of radio $\frac{1}{\sqrt{c}}$. We can define the tangent space $T_x \mathbb{D}_c^n$ of $\mathbb{D}_c^n$ at x as the first order linear approximation of $\mathbb{D}_c^n$ around x. The Poincaré ball model $(\mathbb{D}_c^n, g_c^{\mathbb{D}})$, is defined by the Riemannian metric $g^c = (g_x^c)_{x \in \mathbb{D}_c^n}$ on $\mathbb{D}_c^n$ as a collection of inner-products $g_x^c : T_x \mathbb{D}_c^n \times T_x \mathbb{D}_c^n \to \mathbb{R}$ varying smoothly with x. The Hyperbolic metric tensor g^c is conformal (it defines the same angles) to the Euclidean one, with a conformal factor of $\lambda_x^c = \frac{2}{1-c\|x\|^2}$

3.1 Hyperbolic Neural Networks

Neural networks are machine learning models built as the composition of multiple nonlinear functions known as layers. Each layer outputs a representation or embedding that captures different patterns or attributes of the input data. By training the layer's parameters, neural networks can learn features that uncover the underlying data structure or relate the input data with some desired output. Different layers, including fully connected, convolutional, recurrent, and attention-based, have been proposed to handle different problems.

Below, we present the layers that allow us to work with hyperbolic geometry. The *ToPoincare* layer, which has no trainable weights, allows the transformation of an Euclidean embedding to Hyperbolic embedding. We also introduce the hyperbolic version of the softmax layer, which allows data to be classified through hyperbolic hyperplanes, in this case, with learnable weights. See Fig. 2. In addition, we will define the distance used for image retrieval when the embeddings are hyperbolic.

ToPoincaré Layer. Let $v \in \mathbb{R}^n$ be a data representation extracted from a neural network layer. A *ToPoincaré Layer* is the bijective map $\exp_0^c : T_0 \mathbb{D}_c^n = \mathbb{R}^n \to \mathbb{D}_c^n$, defined as the exponential map [10] evaluated at 0:

$$\exp_0^c(v) = \tanh\left(\sqrt{c}\|v\|\right) \frac{v}{\sqrt{c}\|v\|} \tag{1}$$

Hyperbolic Multiclass Logistic Regression (HMLR) Layer. This layer reformulates Euclidean multiclass logistic regression (MLR) from the perspective of distances to separating hyperplanes, as in [12]. Given K classes, MLR fits a separating hyperplane for each such class, modelling the class posterior probabilities as:

$$p(y = k \mid x) \propto \exp\left((\langle a_k, x \rangle - b_k)\right), \text{where} b_k \in \mathbb{R};\ x, a_k \in \mathbb{R}^n,\ \forall k \in \{1, \ldots, K\}.$$

Thanks to Theorem 5 of Ganea et al. (2018) [7], the Hyperbolic version of MLR can be formulated. Given $p_k \in \mathbb{D}_c^n$ and $a_k \in T_{p_x}\mathbb{D}_c^n \setminus \{0\}$:

$$p(y = k \mid x) \propto \exp\left(\frac{\lambda_{p_k}^c \|a_k\|}{\sqrt{c}} \sinh^{-1}\left(\frac{2\sqrt{c}\langle -p_k \oplus_c x, a_k\rangle}{(1 - c\|-p_k \oplus_c x\|^2)\|a_k\|}\right)\right) \quad \forall x \in \mathbb{D}_c^n, \quad (2)$$

where

$$x \oplus_c y := \frac{(1 + 2c\langle x, y\rangle + c\|y\|^2)x + (1 - c\|x\|^2)y}{1 + 2c\langle x, y\rangle + c^2\|x\|^2\|y\|^2}. \quad (3)$$

We optimize a_k, using the re-parametrization $a_k = (\lambda_0^c/\lambda_{p_k}^c) a_k'$ where $a_k' \in T_0\mathbb{D}_c^n = \mathbb{R}^n$, so that we can optimize a_k' as an Euclidean parameter [7]. On the other hand, as the parameters p_k live in the hyperbolic space, the back-propagated gradient for these parameters is a Riemannian gradient [6]. We then compute the Riemannian gradient $\nabla_{p_k}^R \mathcal{L} = (\lambda_{p_k}^c)^{-2} \nabla_{p_k} \mathcal{L}$ indicating a direction in the tangent space $T_{p_k}\mathbb{D}_c^n$ and update the parameters along the corresponding geodesic in $\mathbb{D}_c^n$ with $\exp_0^c$ [3].

Distance. For $x, y \in \mathbb{D}_c^n$, the induced distance function is defined as:

$$d_c(x, y) = \frac{2}{\sqrt{c}} \tanh^{-1}\left(\sqrt{c}\|(-x) \oplus_c y\|\right). \quad (4)$$

We observe that $\lim_{c \to 0} d_c(x, y) = 2\|x - y\|$, i.e., for $c \to 0$ we recover Euclidean geometry, and for $c = 1$ we recover the geodesic (shortest path) in Hyperbolic space.

4 Methodology

Previous work [1,10] suggests that capacity-limited Hyperbolic classifiers must encode the latent hierarchy of the dataset even if they are trained only with plain ground-truth classes corresponding to the hierarchy leaves. In other words, in an unsupervised setting where the hierarchical relationships between objects are not specified in advance, Hyperbolic embeddings are expected to outperform Euclidean embeddings in discovering hierarchies, especially when the dimensionality of the embedding is limited. This paper focuses on verifying this claim on image data.

To assess this, we conducted experiments in which we trained and evaluated both Euclidean and Hyperbolic convolutional networks, using a pre-trained VGG16 model [23] as the backbone architecture. We employed the CIFAR10 and CIFAR100 datasets, which contain classes that follow an easily verifiable hierarchical structure[1] [8,26]. The architecture details of both models are illustrated in Fig. 2, where the Hyperbolic layers, as defined in Sect. 3, are represented with solid colours. It is important to note that the Hyperbolic convolutional network

---

[1] Hierarchies structures considered for both CIFAR10 and CIFAR100 can be found on Appendix A.

model is hybrid in nature, with only the last layers operating in Hyperbolic space. The second-to-last layer maps the output of the last Euclidean hidden layer onto the Poincaré ball using the ToPoincaré layer [10]. The final layer is a Hyperbolic Multi-Linear Regression (HMLR) [7,10] layer.

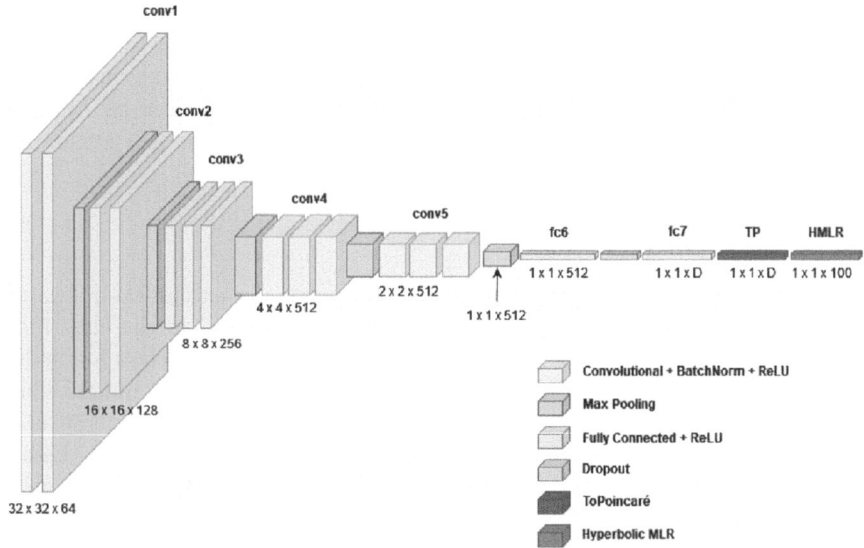

Fig. 2. Hyperbolic VGG16 model. The Euclidean experiment uses the same architecture but replaces the last two layers (ToPoincare and HyperbolicMLR) with its Euclidean counterpart (a Softmax)

To evaluate performance over different dimensionality sizes, we varied the output size of the penultimate layer, from which we get our embeddings, setting different values for D. Note that the Hyperbolic model also requires a hyperparameter of curvature c, which was set to 0.0005. We find this value experimentally by performing a logarithmic search.

All models were trained on a classification task. Once all models were trained, we evaluated the task of content-based image retrieval by considering class hierarchy. To do this, we propose a modified mean average precision metric [14,16], where the relevance function is extended in order to include hierarchy information per level.

4.1 Hierarchy-Aware Computation of mAP@10 (HmAP)

For each query Q, the effectiveness of the retrieved list $d_1, d_2, \ldots, d_L$ is commonly assessed by computing its Average Precision (AP) within the top k results. The

AP@k is determined as follows [14,16]

$$\text{AP@k} = \frac{1}{\min(m_q, k)} \sum_i^k P@i \cdot R(q, d_i). \tag{5}$$

Here, P@i refers to the precision at position i, and R is a relevance function, often binary, which indicates if the result d_i is a relevant item for the query q. The value m_q corresponds to the total number of relevant images for the query q. P@i indicates the proportion of relevant results within the first i positions of the retrieved list that are relevant ($R \neq 0$) for the query. Mean Average Precision (MAP) at k is then defined as the average among the collection of N evaluation queries

$$\text{mAP} = \frac{1}{N} \sum_{i=1}^N \text{AP}_i. \tag{6}$$

To incorporate hierarchy awareness into the metric, we extend the relevance function $R(q, d_i)$ to reflect the relationship between q and d_i in the class hierarchy. Specifically, for the different levels $\ell = 1, 2, \ldots, L$ of the class hierarchy, we define $R(q, d_i) = 1$ if and only if the class corresponding to d_i and the class corresponding to q has a common ancestor at level ℓ. Otherwise, if the classes cannot be linked with a common ancestor, $R(q, d_i) = 0$ for that level. Essentially, our procedure only modifies how mAP handles retrieval errors. An item that may not be initially relevant at a particular hierarchy level can potentially become relevant if we evaluate $R(q, di)$ considering more abstract concepts.

For example, for the CIFAR10 hierarchy presented in Appendix A we measure the hierarchy mAP at different levels. We first consider a **granularity** of $\ell = 3$ since CIFAR10 has 3 levels. At this level, we only consider an image result as relevant if it has the same class as the query (this is traditional mAP). If we move up to a granularity level of $\ell = 2$, we now consider the second level of the hierarchy. This means that for a query with the class "truck", an image with the class "automobile" will be considered relevant because it belongs to a common superclass. An example of a query with class "automobile" can be seen in Fig. 3, where a green box denotes a relevant item and a red box denotes a non-relevant item.

5 Experiments

In order to verify our claim, we start by training both an Euclidean and Hyperbolic convolutional network, with ImageNet pre-trained VGG16 as the backbone, on the CIFAR10 and CIFAR100 datasets. Datasets were split into training, validation and testing subsets. A validation partition was obtained after splitting the training data with an 80:20 split, and for testing, we used the standard CIFAR10/CIFAR100 testing partition. The Training subset was augmented in memory with horizontal and vertical flips at random, and all subsets were normalized according to the dataset mean and standard deviation. No further preprocessing was applied to the images. Both datasets were trained separately.

(a) Results at granularity 3

(b) Results at granularity 2

Fig. 3. Retrieved items at different levels of granularity for a query of class "Automobile". Relevant results are in green. Mistakes are in red. (Color figure online)

To evaluate the effect of dimensionality on the model's ability to capture and represent the latent hierarchies, we varied the size of the embeddings. As we get the embeddings from the penultimate layer, we varied the number of neurons in this layer from 2 to 32 in steps of powers of two.

We trained all the models for 200 epochs using the Adam optimizer [11] with an initial learning rate of 0.001. The learning rate was reduced by a factor of 10 if validation loss didn't improve for 10 epochs. We allowed for an early stopping of training if validation loss didn't improve for 15 epochs. Euclidean models were faster to converge.

We then evaluated the Hierarchy Aware mean Average Precision (HmAP@10) to measure retrieval. To do this, the testing partition was split following a ratio of 90:10, where the 10% was used as queries, and the remaining 90% was used as the gallery where we searched for similar images. Similarity was computed using the Euclidean distance, for Euclidean embeddings, and the Poincaré Ball distance (4) for Hyperbolic embeddings. We measured mAP@10 for all levels of the hierarchy, ranging from granularity 3 (specific concepts) to 1 (broad concepts), in CIFAR10, and granularity 5 (very specific concepts) to 1 (very broad concepts) in CIFAR100. Our experimental results can be found in Sect. 6. Our code is publicly available at GitHub.

6 Results

Results are summarized in Tables 1 and 2 for CIFAR10 and CIFAR100, respectively.

Table 1. HmAP@10 for Euclidean and Hyperbolic models trained on CIFAR10 dataset, for different levels of granularity. H denotes Hyperbolic results, and E denotes Euclidean results.

	Granularity 3		Granularity 2		Granularity 1	
Embedding Size	H	E	H	E	H	E
2	**0.7681**	0.7350	0.8016	**0.8335**	**0.8682**	0.8570
4	**0.8016**	0.7828	**0.8653**	0.8526	**0.8871**	0.8724
8	**0.8018**	0.7759	**0.8707**	0.8559	**0.8878**	0.8768
16	0.7765	**0.7992**	0.8565	**0.8704**	0.8822	**0.8897**
32	**0.8049**	0.8014	**0.8734**	0.8682	**0.8950**	0.8935
64	**0.8023**	0.7976	**0.8689**	0.8620	**0.8864**	0.8794

For CIFAR10, we see that Hyperbolic embeddings outperform Euclidean embeddings in most cases across different levels of hierarchy, although this difference is not conclusive. This can be attributed to the fact that the number of classes in CIFAR10 is not high enough to begin stressing Einlidean space. We repeat the same experiment using the CIFAR100 dataset to test this hypothesis. The results for CIFAR100 can be seen on Table 2.

Interestingly, we found that Hyperbolic embeddings outperform their Euclidean counterpart by a considerable margin in the lower levels of the hierarchy when the number of classes increases, exceeding the results obtained by the Euclidean baseline in the task of image retrieval. These margins decrease as we increase the embedding size, as we no longer stress Euclidean space. We also find that Hyperbolic space is better adapted to find embeddings of a dataset with a latent hierarchical structure.

Table 2. HmAP@10 for Euclidean and Hyperbolic models trained on CIFAR100 dataset, for different levels of granularity. H denotes Hyperbolic results, and E denotes Euclidean results.

	Granularity 5		Granularity 4		Granularity 3		Granularity 2		Granularity 1	
Embedding Size	H	E	H	E	H	E	H	E	H	E
2	**0.1338**	0.0829	**0.3130**	0.2502	0.4934	**0.5005**	0.7508	**0.7515**	0.8235	**0.8344**
4	**0.2804**	0.1810	**0.4766**	0.3800	**0.6548**	0.5791	**0.8315**	0.7804	**0.8898**	0.8596
8	**0.3389**	0.2514	**0.5023**	0.4469	**0.6720**	0.6366	**0.8430**	0.8198	**0.8860**	0.8721
16	**0.3504**	0.3366	**0.5251**	0.5037	**0.6795**	0.6640	**0.8395**	0.8302	0.8784	**0.8850**
32	**0.4155**	0.3860	**0.5759**	0.5472	**0.7194**	0.6923	**0.8677**	0.8506	**0.9006**	0.8921
64	**0.4029**	0.3407	**0.5634**	0.5101	**0.7106**	0.6645	**0.8587**	0.8381	**0.8998**	0.8858

7 Conclusion and Discussion

In this work, we investigated the use of hyperbolic image embeddings to recover latent hierarchical relationships. To do this, we replaced some of the deeper layers in state-of-the-art convolutional neural networks with their Hyperbolic equivalents to create Hyperbolic embeddings. The approach investigated here is thus extensible to other state-of-the-art models that support the replacement of its last layers.

At the same time, our experimental results confirm our hypothesis that Hyperbolic embeddings work better than their Euclidean counterpart when considering the hierarchical structure of data. The hypothesis holds particularly true for datasets with a large number of interrelated classes, where Hyperbolic embeddings effectively exploit the limitations of the Euclidean space to accommodate tree-like structures.

As aggregating multiple image embeddings into class centroids has shown to be highly effective for image retrieval, in future work, we plan to evaluate the use of Hyperbolic centroids to take advantage of the Hyperbolic space to accommodate latent hierarchies.

Acknowledgments. This research was partially funded by National Agency for Research and Development (ANID, Chile), grant numbers FONDEF IT21I0019 and ANID-Basal Project FB0008.

A Hierarchical Structures of CIFAR10 and CIFAR100

(see Figs. 4 and 5).

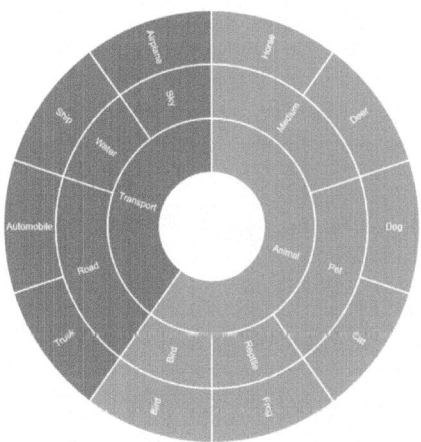

Fig. 4. CIFAR10 Hierarchy

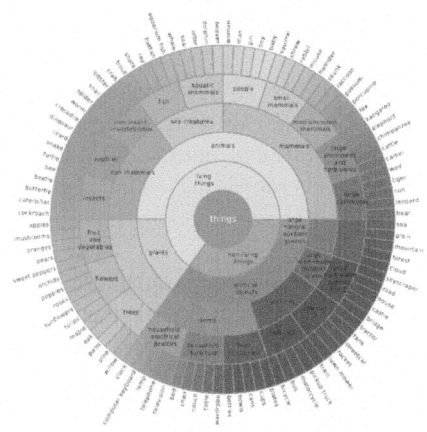

Fig. 5. CIFAR100 Hierarchy [8].

References

1. Anitha, K., Dhanalakshmi, R., Naresh, K., Devi, R.: Hyperbolic hopfield neural networks for image classification in content-based image retrieval. Inter. J. Wavelets, Multiresolution Inform. Process. **19**, 2050059 (2020). https://doi.org/10.1142/S0219691320500599
2. Bengio, Y., Courville, A., Vincent, P.: Representation learning: a review and new perspectives. IEEE Trans. Pattern Anal. Mach. Intell. **35**(8), 1798–1828 (2013). https://doi.org/10.1109/TPAMI.2013.50
3. Bonnabel, S.: Stochastic gradient descent on riemannian manifolds. IEEE Trans. Autom. Control **58**(9), 2217–2229 (2013). https://doi.org/10.1109/TAC.2013.2254619
4. Dhingra, B., Shallue, C.J., Norouzi, M., Dai, A.M., Dahl, G.E.: Embedding text in hyperbolic spaces. CoRR abs/ arXiv: 1806.04313 (2018)
5. Freiherr von Gagern, M.: Creating Hyperbolic Ornaments. Ph.D. thesis, Technische Universität München (2014). https://mediatum.ub.tum.de/1210572
6. Ganea, O., Becigneul, G., Hofmann, T.: Hyperbolic entailment cones for learning hierarchical embeddings. In: Dy, J., Krause, A. (eds.) Proceedings of the 35th International Conference on Machine Learning. Proceedings of Machine Learning Research, 10–15 Jul, vol. 80, pp. 1646–1655. PMLR (2018). https://proceedings.mlr.press/v80/ganea18a.html
7. Ganea, O., Bécigneul, G., Hofmann, T.: Hyperbolic neural networks. CoRR abs/ arXiv: 1805.09112 (2018)
8. Garnot, V.S.F., Landrieu, L.: Metric-guided prototype learning. CoRR abs/ arXiv: 2007.03047 (2020)
9. Gromov, M.: Hyperbolic Groups, pp. 75–263. Springer New York (1987). https://doi.org/10.1007/978-1-4613-9586-7_3
10. Khrulkov, V., Mirvakhabova, L., Ustinova, E., Oseledets, I.V., Lempitsky, V.S.: Hyperbolic image embeddings. CoRR abs/arXiv: 1904.02239 (2019)
11. Kingma, D.P., Ba, J.: Adam: A method for stochastic optimization. arXiv preprint arXiv:1412.6980 (2014)

12. Lebanon, G., Lafferty, J.: Hyperplane margin classifiers on the multinomial manifold. In: Twenty-First International Conference on Machine Learning; - ICML 2004 (2004). https://doi.org/10.1145/1015330.1015333
13. Lecun, Y.: Generalization and network design strategies. In: Connectionism in perspective. Elsevier (1989)
14. Liu, T.Y., et al.: Learning to rank for information retrieval. Foundat. Trends® Inform. Retrieval **3**(3), 225–331 (2009)
15. Meng, Y., et al.: Spherical text embedding. CoRR abs/ arXiv: 1911.01196 (2019)
16. Musgrave, K., Belongie, S., Lim, S.-N.: A metric learning reality check. In: Vedaldi, A., Bischof, H., Brox, T., Frahm, J.-M. (eds.) ECCV 2020. LNCS, vol. 12370, pp. 681–699. Springer, Cham (2020). https://doi.org/10.1007/978-3-030-58595-2_41
17. Nawi, N.M., Atomi, W.H., Rehman, M.: The effect of data pre-processing on optimized training of artificial neural networks. Proc. Technol. **11**, 32–39 (2013). https://doi.org/10.1016/j.protcy.2013.12.159, https://www.sciencedirect.com/science/article/pii/S2212017313003137, 4th International Conference on Electrical Engineering and Informatics, ICEEI 2013
18. Newman, M.: Power laws, pareto distributions and zipf's law. Contemporary Phys. **46**(5), 323–351 (2005).https://doi.org/10.1080/00107510500052444
19. Nickel, M., Kiela, D.: Poincaré embeddings for learning hierarchical representations. CoRR abs/ arXiv: 1705.08039 (2017)
20. Peng, W., Varanka, T., Mostafa, A., Shi, H., Zhao, G.: Hyperbolic deep neural networks: A survey. CoRR abs/ arXiv: 2101.04562 (2021)
21. Ratcliffe, J.G.: Foundations of hyperbolic manifolds. Grad. Texts Math. (2007). https://doi.org/10.1007/978-0-387-47322-2
22. Sarkar, R.: Low distortion delaunay embedding of trees in hyperbolic plane. In: van Kreveld, M., Speckmann, B. (eds.) GD 2011. LNCS, vol. 7034, pp. 355–366. Springer, Heidelberg (2012). https://doi.org/10.1007/978-3-642-25878-7_34
23. Simonyan, K., Zisserman, A.: Very deep convolutional networks for large-scale image recognition. International Conference on Machine Learning Representations - ICLR (2014). https://doi.org/10.48550/arXiv.1409.1556
24. Wilson, R.C., Hancock, E.R., Pekalska, E., Duin, R.P.: Spherical and hyperbolic embeddings of data. IEEE Trans. Pattern Anal. Mach. Intell. **36**(11), 2255–2269 (2014). https://doi.org/10.1109/TPAMI.2014.2316836
25. Yin, Z., Shen, Y.: On the dimensionality of word embedding. CoRR abs/ arXiv: 1812.04224 (2018)
26. Zhu, X., Bain, M.: B-CNN: branch convolutional neural network for hierarchical classification. CoRR abs/ arXiv: 1709.09890 (2017)

SwinDehazing: Haze Removal Using U-Net and Swin Transformer

Percy Maldonado-Quispe and Helio Pedrini(✉)

Institute of Computing, University of Campinas, Campinas, SP 13083-852, Brazil
helio@ic.unicamp.br

Abstract. The paper addresses the challenge of image dehazing, crucial for improving visibility in computer vision applications. Traditional methods such as Dark Channel Prior (DCP) and Color Attenuation Prior (CAP) rely on handcrafted features, but struggle with dense haze and diverse conditions. Deep learning, specifically Convolutional Neural Networks (CNNs), has enhanced dehazing, with models such as DehazeNet and AOD-Net showing significant improvements. The paper introduces a novel SwinDehazing-based U-Net architecture, integrating Swin Transformer and U-Net for robust multi-scale feature extraction and spatial resolution recovery. Key contributions include performance on benchmark datasets, a detailed analysis of the model's components, and enhancements such as replacing GeLU with ReLU and employing shifted window partitioning. The proposed SwinDehazing model demonstrates higher dehazing performance, effectively handling varying haze densities and complex scenes.

Keywords: Image Dehazing · Swin Transformer · Vision Transformer

1 Introduction

Image dehazing, the process of removing haze from images to restore their clarity, is a critical problem in computer vision with numerous applications in photography, autonomous driving, and remote sensing. Hazy images suffer from reduced visibility and contrast, which significantly impairs the performance of vision-based systems.

Traditional image dehazing methods, such as the Dark Channel Prior (DCP) [3] and the Color Attenuation Prior (CAP) [24], rely on handcrafted features and physical models to estimate the transmission map and atmospheric light. While these methods have shown effectiveness in various scenarios, they often struggle with dense haze and diverse environmental conditions.

The advent of deep learning has revolutionized image dehazing by enabling end-to-end learning approaches that can capture complex haze patterns directly from data. Convolutional Neural Networks (CNNs), such as DehazeNet [1] and AOD-Net [4], have significantly improved dehazing performance by learning hierarchical features and directly predicting clear images from hazy inputs.

Transformers, initially designed for natural language processing tasks, have recently demonstrated remarkable success in various vision applications due to their ability to model long-range dependencies through self-attention mechanisms. Vision Transformers (ViTs) [2] have shown that treating images as sequences of patches and applying transformer architectures can achieve state-of-the-art results in image classification tasks. Building on this, Swin Transformers [9] proposed a hierarchical transformer architecture with shifted windows, significantly enhancing computational efficiency and performance on dense prediction tasks such as object detection and segmentation.

In this paper, we propose a novel approach to image dehazing by integrating the advantages of Swin Transformer [8] and a U-Net architecture [16]. The Swin Transformer, with its hierarchical design and efficient self-attention mechanism, is well-suited for capturing both local and global features at multiple scales. By incorporating Swin Transformer into the encoder-decoder structure of U-Net, our method leverages the powerful feature extraction capabilities of transformers and the effective spatial resolution recovery of U-Nets. This combination allows our model to robustly handle varying haze densities and complex scene structures.

Our main contributions are summarized as follows:

(i) We introduce a SwinDehazing-based U-Net architecture for image dehazing, which effectively captures multi-scale features.
(ii) We propose the use of reflection padding to mitigate the occurrence of unwanted artifacts during the learning of key representations.
(iii) We demonstrate the performance of our model compared to some state-of-the-art methods using benchmark dataset experiments.
(iv) We provide an analysis of the model's components, highlighting the benefits of integrating the SwinDehazing block into the dehazing process.

2 Related Work

Image dehazing, the process of restoring clear images from hazy ones, is a well-studied area in computer vision. Numerous traditional and deep learning-based approaches have been proposed to tackle this challenging problem. In this section, we review related works, focusing on classical methods, convolutional neural network-based methods, and recent advancements in using transformers for image dehazing.

2.1 Classical Methods

Traditional image dehazing techniques typically rely on physical models and handcrafted features. One of the most influential works is by He et al. [3], who introduced the Dark Channel Prior (DCP) method. This approach leverages the observation that in most haze-free patches, at least one color channel has some pixels with very low intensities. By estimating the atmospheric light and transmission map, DCP effectively removes haze from images. Another notable method is the Color Attenuation Prior (CAP) proposed by Zhu et al. [24], which models the scene depth of the hazy image using the difference between the brightness and the saturation. These classical methods, while effective in many scenarios, often struggle with dense haze and fail to generalize across diverse scenes.

2.2 CNN-Based Methods

With the advent of deep learning, convolutional neural networks have significantly advanced the state of the art in image dehazing. Cai et al. [1] proposed the DehazeNet, an end-to-end CNN that directly learns the mapping from hazy images to the transmission map. Building on this, Ren et al. [15] introduced the Multi-Scale Convolutional Neural Network (MSCNN), which improves the robustness of dehazing by capturing features at multiple scales. Li et al. [4] further enhanced dehazing performance with the All-in-One Dehazing Network (AOD-Net), which simplifies the dehazing process by learning a direct mapping from the hazy image to the clear image.

Maldonado-Quispe and Pedrini [11] were inspired by deep learning and combines ASM with CNN to estimate physical parameters. These methods have shown remarkable improvements over traditional approaches, but they are often computationally intensive and may not fully capture long-range dependencies in the image. Additionally, there are several studies focusing on unsupervised haze removal, among which notable works include ZID [7], YOLY [6] and XYZ [10].

2.3 Transformer-Based Methods

Recently, transformers have emerged as a powerful alternative to CNNs for various vision tasks, including image dehazing. Unlike CNNs, which rely on local receptive fields, transformers can model long-range dependencies through self-attention mechanisms.

Vision Transformers (ViTs), introduced by Dosovitskiy et al. [2], have demonstrated the potential of transformers in image classification by treating image patches as sequences and applying transformer architectures to them. Building on this, Swin Transformers [9] introduces a hierarchical structure using shifted windows, enabling efficient local and global feature capture with lower computational cost on dense prediction tasks. Another notable work by Wang et al. [18] presents the Pyramidal Vision Transformer (PVT), which integrates a pyramid structure into the transformer design to enhance feature representation across multiple scales, [18] proposes a flexible backbone that completely eliminates convolutions, focusing on a pyramidal structure that improves dense prediction without the need for traditional convolutions.

For image dehazing, transformer-based models have started to show promising results. Zu et al. [25] proposed the ALCEM module, a CNN-based adaptive local context enrichment module that is integrated into the attention mechanisms and feed-forward networks of transformers, thus introducing a locally enhanced attention (LEA) and a locally continuously improved feed-forward network (LCFN). Zhang et al. [22] proposed DedustNet, an end-to-end learning network using Swin Transformer blocks in combination with the wavelet transform to address dust removal in agricultural images. Wei et al. [21] presented an algorithm that simultaneously performs image dehazing and depth estimation, using ResNet-Edge-Encoder and Transformer-Edge-Decoder blocks to improve the extraction of edge information in images. Zhao et al. [23] introduced MSFA-Net, a network that fuses Transformers and CNNs into an encoder-decoder structure for haze removal. Mao et al. [12] presented MIDNet, an efficient network based on local transformers that uses a pyramidal structure and dense connections for multi-scale feature aggregation and a detail enhancement module that preserves crucial information in dehazed images.

Wasi and Shiney [20] introduced a novel Dehazing Transformer (DHFormer), which leverages multi-scale transformer blocks to effectively capture both global and local features in hazy images. The DHFormer architecture consists of a series of transformer layers that operate on different resolutions of the input image, providing a comprehensive understanding of the haze distribution. Song et al. [17] proposed DehazeFormer, which consists of various improvements using Swin Transformer, such as the modified normalization layer, activation function, and spatial information aggregation scheme with a focus on pixel-level attention. Our approach is distinguished by integrating the advantages of the Swin Transformer with the U-Net structure, incorporating improvements in the activation function and the implementation of reflection padding, which optimizes the elimination of artifacts and improves the overall quality of key feature learning.

3 SwinDehazing

Our model integrates Swin Transformer into the U-Net architecture, combining the strengths of both models for effective image dehazing. The U-Net architecture, known for its encoder-decoder structure, is enhanced by replacing its standard convolutional blocks with SwinDehazing blocks. Figure 1 illustrates the main components of the proposed architecture for image dehazing.

3.1 Encoder

The encoder consists of SwinDehazing blocks that hierarchically downsample the input image while capturing multi-scale features. This process involves multiple stages where the input resolution is progressively reduced, and feature channels are increased. Each SwinDehazing block applies shifted window self-attention and feed-forward layers to extract robust features.

The input image $I \subset \mathbb{R}^{H \times W \times C}$ is divided into non-overlapping patches of size $P \times P$. This results in $\frac{H}{P} \times \frac{W}{P}$ patches, each represented as a $P^2 \times C$ vector, where Eq. 1 denotes the patch embedding operation, converting each patch into a fixed-dimensional vector. Following the patch embedding, each patch is linearly embedded into a fixed-dimensional vector, creating a sequence of patch embeddings.

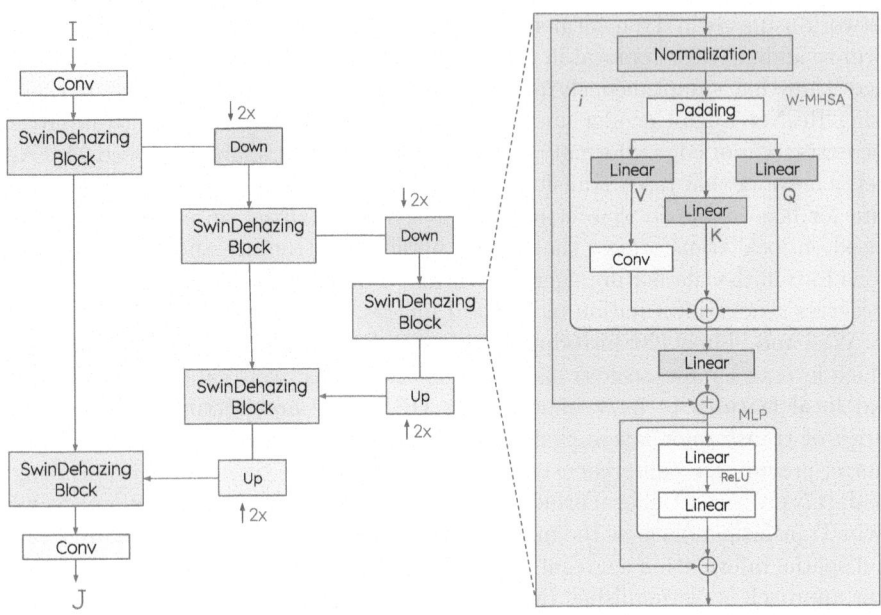

Fig. 1. Graphical overview of the proposed architecture for image dehazing.

$$x_0 = PatchEmbed(I) \tag{1}$$
$$x_1 = Linear(x_0) \tag{2}$$

Here, *Linear* is a fully connected layer that maps the patch vectors to D-dimensional embeddings.

SwinDehazing Block. Each block includes a shifted window self-attention mechanism followed by a MultiLayer Perceptron (MLP) as described in Eq. 3.

$$x_i = SwinBlock(x_{i-1}) \tag{3}$$

Window Multi Head Self-attention: The self-attention mechanism in a shifted window is defined as:

$$Attention(Q, K, V) = Softmax\left(\frac{QK^T}{\sqrt{d_k}}\right) Conv(V) \tag{4}$$

where Q, K and V are the query, key, and value matrices, and d_k is the dimensionality of the key vectors. This block incorporates a shifted window partitioning scheme, where reflection padding helps manage the edges of these windows during MHSA operations. We introduce non-overlapping regions, and reflection padding ensures that when these windows shift, the transitions between windows are smooth and continuous, avoiding the creation of artifacts.

We also recognize that the multiplication in Multi-Head Self-Attention (MHSA) acts as a low-pass filter, a conclusion supported by findings in [13]. Despite the dynamic nature of the spatial information aggregation weights in MHSA, these weights remain positive, effectively functioning as a smoothing mechanism. To complement the spatial information processed by Window Multi-Head Self-Attention (W-MHSA), we apply additional convolution operations to the value V.

MultiLayer Perceptron: This plays an important role in processing and refining features after the self-attention mechanism. After the self-attention layer captures and aggregates information from different parts of the image, the MLP is applied to each patch to further transform the features. The MLP typically consists of two convolution with a GeLU activation function, in between, but we use ReLU activation function because it is a low-level vision task. Additionally, the MLP includes dropout and normalization to prevent overfitting and improve generalization.

The encoder in our image dehazing technique captures multi-scale features using SwinDehazing blocks. The use of shifted window self-attention and hierarchical downsampling ensures that both local and global contexts are effectively represented. This hierarchical feature extraction is crucial for tackling the complexities of haze in images, leading to more accurate and high-quality dehazing results. Throughout the stages, multi-scale feature extraction is achieved by progressively reducing the resolution of the feature maps while increasing the number of channels.

3.2 Bottleneck

At the bottleneck, the most abstracted features are processed through additional SwinDehazing blocks, ensuring a rich representation of the input image's global context.

3.3 Decoder

The decoder mirrors the encoder's structure but in reverse, progressively upsampling the feature maps to reconstruct the high-resolution output image. SwinDehazing blocks are employed here as well, preserving the hierarchical feature information while refining the details.

The decoder in our image dehazing technique plays a role in reconstructing the dehazed image from the multi-scale feature maps refined by the encoder and bottleneck. The decoder is designed to effectively upsample the encoded features and integrate the hierarchical information to produce a high-quality, haze-free image.

The decoder starts by taking the refined feature maps from the bottleneck and progressively upsamples them to reconstruct the original image resolution. This hierarchical upsampling process is essential for gradually restoring the spatial dimensions while maintaining the integrity of the features extracted by the encoder. Each upsampling step is followed by SwinDehazing blocks that help to refine the upsampled features by capturing both local and global context through shifted window self-attention.

A key component of the decoder is the use of skip connections, which link corresponding layers of the encoder and decoder. These connections allow the decoder to access higher resolution features directly from the encoder, preserving important details that might be lost during the downsampling process. The skip connections ensure that the decoder can leverage both the high-level abstract features from the bottleneck and the finer details from the encoder stages, leading to a more accurate reconstruction of the dehazed image.

In the final stages of the decoder, the upsampled feature maps are processed to produce the output image. The last SwinDehazing blocks ensure that the features are fully refined and coherent, preparing them for the final reconstruction step. A convolutional layer is used at the end to transform the final set of features into the desired output image, effectively reconstructing the high-quality, haze-free image.

To recover the desired haze-free image J, we utilize a variation of the Atmospheric Scattering Model (ASM) as proposed by Li et al. [4], which eliminates the need to pre-know the values of the transmission map $t(x)$ and atmospheric light A. Instead, we propose reconstructing the haze-free image by calculating the values of K and b of Eq. 5. This is achieved through our architecture that integrates U-Net with Swin Transformer, leveraging the strengths of both models to effectively estimate these parameters and enhance the dehazing process.

$$J(x) = K(x)I(x) - K(x) + b \qquad (5)$$

3.4 Implementation Details

In this work, we utilized the Adam optimizer with a learning rate of $2e^{-4}$ for the training process. The training dataset comprised input images with dimensions of (256, 256, 3) pixels, and we employed a batch size of 4 for 64 epochs. The original image was divided into patches of size 4, and a window size of 8 was used for processing.

Our approach based on U-Net consists of 5 blocks, where two are responsible for reducing the image's dimensionality, and the last two blocks focus on recovering the haze-free image. For each U-Net layer, we utilized an embedding dimension of [24, 48, 96, 48, 24]. Similarly, we maintained a ratio of 4 for the MLP sections in SwinDehazing, with 8 blocks in each layer (depth).

4 Results

In this section, we present the results of our method, alongside a comparative evaluation with various state-of-the-art haze removal techniques. The methods we compare against include those proposed by He et al. [3], Li et al. [4], Qin et al. [14], and Song et al. [17].

4.1 Datasets

Experiments were conducted on the REalistic Single Image DEhazing (RESIDE) dataset [5], which is renowned for its extensive use in haze removal evaluations. The dataset includes two test subsets: SOTS (Synthetic Object Testing Set) and HSTS (Hybrid Subject Testing Set). The SOTS subset comprises 500 indoor hazy images generated using a physical model with manually tuned parameters. In contrast, the HSTS subset contains 10 synthetic haze images and 10 real-world blurry images captured from various scenes.

Additionally, the RESIDE dataset provides training subsets: the OTS (Outdoor Training Set) includes 69,129 hazy images created from 2,061 ground truth images, while the ITS (Indoor Training Set) comprises 13,990 hazy images derived from 1,399 ground truth images. The distribution of data is detailed in Table 1.

Table 1. Summary of the RESIDE dataset, which is divided into two subsets: Standard and Beta.

Subset	Images	Type
Outdoor Training Set	69,129	Beta
Indoor Training Set	13,990	Standard
Synthetic Objective Testing Set	500	
Hybrid Subjective Testing Set	20	

During the training and validation phases, we utilized the RESIDE dataset, specifically the ITS and OTS subsets, which contain 13,990 and 15,194 (first part) samples respectively. Following the completion of training, we evaluated our model's performance on the indoor and outdoor subsets of the SOTS dataset.

4.2 Metrics

To evaluate the performance of our model and other dehazing methods, we utilized the Structural Similarity Index (SSIM) and Peak Signal-to-Noise Ratio (PSNR) as image quality metrics, which are widely adopted in haze removal research [19]. SSIM measures the structural similarity between the haze-free reference image and the dehazed image produced by our method, reflecting the preservation of image details and structural information. PSNR, on the other hand, quantifies the ratio between the maximum possible signal power and the noise power in the images, providing an indication of the overall image quality.

Higher SSIM and PSNR values indicate better fidelity to the haze-free reference image, signifying higher quality and greater similarity of the dehazed images to the original, clear images. Models that achieve higher scores on these metrics are considered more effective in restoring the visual quality of hazy images.

4.3 Quantitative Comparison

Table 2 presents a quantitative comparison of the results, providing a comprehensive comparison of various image dehazing methods [1,3,4,11,14,17,20,24] evaluated on the SOTS Indoor and HSTS Outdoor subsets of the RESIDE dataset, using two metrics: PSNR and SSIM. Higher values of these metrics indicate better performance in restoring haze-free images, the methods that achieved better results in both metrics are highlighted in **bold**.

The methods are grouped into three categories: Classical, CNN-based, and Transformer-based approaches. Classical methods, represented by DCP and CAP, show lower performance compared to more advanced techniques, with DCP achieving a PSNR of 16.62 and SSIM of 0.809 on indoor images, and 18.76 and 0.694 on outdoor images. CAP achieves slightly higher values for PSNR but lower SSIM on indoor images, while the opposite occurs with outdoor images.

In the CNN-based category, methods such as AOD-Net, DehazeNet, and FFA-Net demonstrate significant improvements over classical methods. For example, FFA-Net achieves a PSNR of 26.50 and SSIM of 0.897 on indoor images, and 28.40 and 0.930 on outdoor images, showcasing its higher performance. LightCNN also shows competitive results. Transformer-based methods, including DHFormer and DehazeFormer(-S), exhibit even higher performance. DehazeFormer(-S) achieves the highest PSNR scores among all methods, with 32.82 on SOTS Indoor, and 30.63 on HSTS Outdoor. Our method, SwinDehazing, also shows excellent performance, with PSNR and SSIM values of 32.25 and 0.961 on indoor images, and 29.32 and 0.948 on outdoor images, indicating its effectiveness in image dehazing tasks.

4.4 Qualitative Comparison

For a qualitative analysis, we examine the images produced by various comparative dehazing methods. Figure 2 showcases four hazy outdoor images from the HSTS subset of the RESIDE dataset. The DCP technique, proposed by He et al. [3], which leverages prior knowledge, demonstrates limitations in effectively removing haze, particularly in regions resembling cloudy skies. This problem is apparent in images 2, 3, and 4, where noticeable haze remnants and dark patches (indicative of over-enhancement) persist, suggesting that the haze could not be fully removed.

The method proposed by Li et al. [4] addresses the challenge of reducing over-enhancement in bright areas, but it introduces the issue of color amplification. In

Table 2. Quantitative comparison of various image dehazing methods on the SOTS Indoor and HSTS Outdoor subsets, evaluated using PSNR and SSIM metrics.

Method	Type	SOTS Indoor		HSTS Outdoor	
		PSNR↑	SSIM↑	PSNR↑	SSIM↑
DCP [3]	Classical	16.62	0.809	18.76	0.694
CAP [24]		16.98	0.714	17.23	0.709
AOD-Net [4]	CNN	20.51	0.816	24.14	0.920
DehazeNet [1]		19.82	0.821	24.75	0.927
FFA-Net [14]		26.50	0.897	28.40	0.930
LightCNN [11]		19.78	0.842	24.18	0.853
DHFormer [20]	Transformer	22.93	0.903	26.83	0.924
DehazeFormer(-B) [17]		**32.82**	0.957	**30.63**	0.941
SwinDehazing (Ours)		32.25	**0.961**	29.32	**0.948**

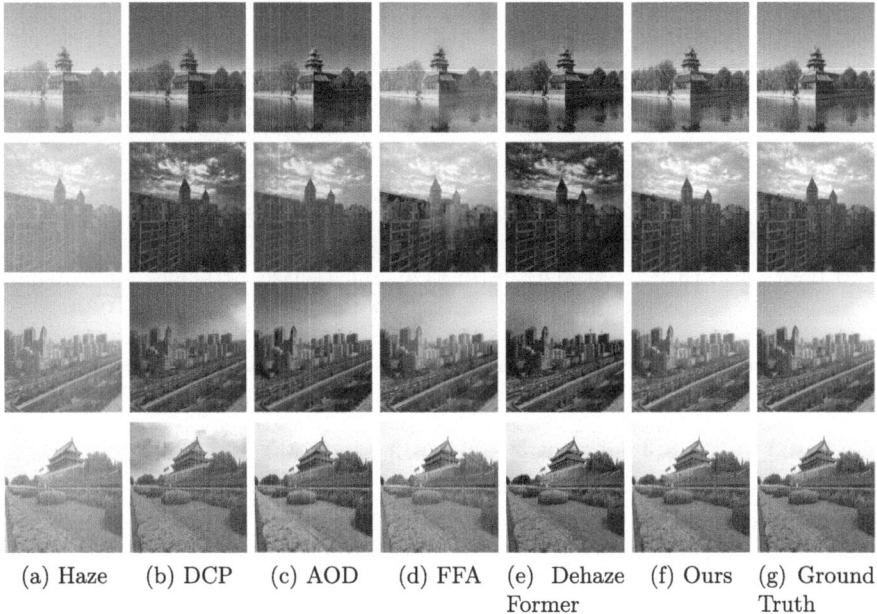

(a) Haze (b) DCP (c) AOD (d) FFA (e) Dehaze Former (f) Ours (g) Ground Truth

Fig. 2. Qualitative comparisons on HSTS Outdoor dataset for different methods.

several cases, this results in a noticeable shift towards more intense color tones, particularly evident in images 1 and 3. Conversely, the technique proposed by Qin et al. [14] shows a more accurate recovery of the original colors, as observed in images 1, 3, and 4. However, it should be noted that image 2 reveals irregular illumination patterns, which are perceptible to the human eye.

In contrast, the images generated by our method still exhibit some limitations in terms of color enhancement, particularly visible in images 1 and 3. Nevertheless, the recovered images demonstrate a notable level of sharpness and clarity with respect to the haze, indicating the effectiveness of our approach in preserving significant image details.

Figure 3 demonstrates the effectiveness of these dehazing techniques on artificially generated indoor images using an atmospheric dispersion model. The displayed images exhibit dense haze, presenting a significant challenge for the proposed techniques to achieve optimal haze removal.

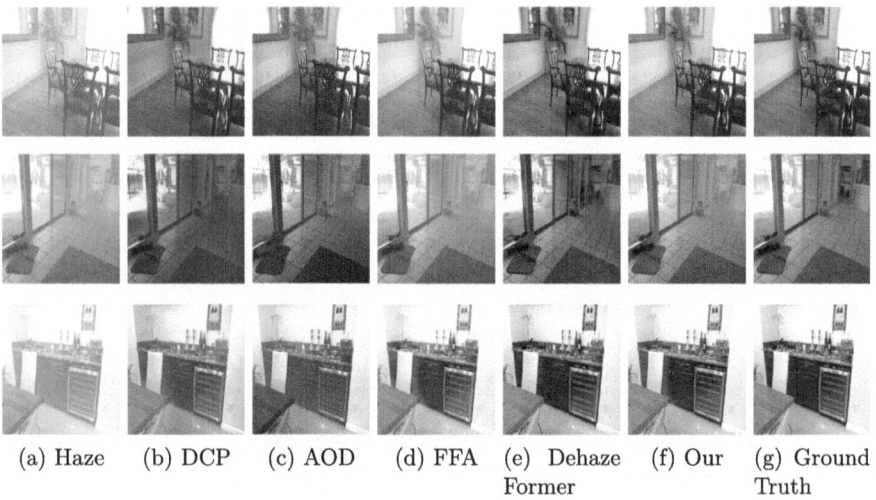

(a) Haze (b) DCP (c) AOD (d) FFA (e) Dehaze Former (f) Our (g) Ground Truth

Fig. 3. Qualitative comparisons on SOTS Indoor dataset for different methods.

Li et al.'s approach [4] focused on dehazing by enhancing colors, although it often results in significant image darkening, especially when dealing with dark-colored scenes, similar to the findings by He et al. [3]. On the other hand, the method proposed by Qin et al. [14] attempted to eliminate haze while preserving overall image information. However, Song et al. [17] achieved efficient haze removal, although at the expense of over-brightening the image in its haze elimination process, leading to reduced color intensity. Nevertheless, our method enhances haze removal while maintaining color intensity. However, it still faces challenges with images under low light conditions, tending to darken such areas during haze removal.

5 Discussion

In our study, the combination of Swin Transformers with the U-Net architecture has proven to be a viable strategy for image haze removal. The ability of

Swin Transformers to capture global and local features, along with the hierarchical and connection structure of U-Net, allows for a restoration of haze-affected regions. In addition, the incorporation of reflection padding has helped mitigate edge effects in convolution, improving reconstruction quality without introducing unwanted artifacts. The use of ReLU activation functions in the MLP has enhanced the nonlinearity of the model, allowing for better fitting and generalization capabilities in complex tasks.

The combination of these techniques has resulted in a significant improvement in the clarity and detail of the recovered images. The experiments demonstrate that our approach not only improves the images visually, but also quantitatively by a minimum percentage. This work highlights the potential of the integration of Swin Transformers and U-Net, in addition to the implemented variations.

6 Conclusions

This paper introduces some enhancements to the Swin Transformer for application in image dehazing, resulting in the proposed SwinDehazing model, which demonstrates higher performance on well-known datasets. Key improvements include the replacement of the commonly used GeLU activation function with ReLU to mitigate adverse effects that, while negligible for high-level vision tasks, are critical for low-level vision tasks.

Additionally, to enhance the capabilities of Multi-Head Self-Attention (MHSA), we employ a shifted window partitioning scheme with reflection padding and incorporate a spatial information aggregation method using convolution. Furthermore, we use the ASM modification equation to more effectively recover the haze-free image, leveraging the strengths of our SwinDehazing architecture.

Acknowledgements. The authors would like to thank the Coordination for the Improvement of Higher Education Personnel (CAPES) and National Council for Scientific and Technological Development (CNPq grant #304836/2022-2) for their financial support.

References

1. Cai, B., Xu, X., Jia, K., Qing, C., Tao, D.: DehazeNet: an End-to-End System for Single Image Haze Removal. IEEE Trans. Image Process. **25**(11), 5187–5198 (2016)
2. Dosovitskiy, A., et al.: An Image is Worth 16 × 16 Words: Transformers for Image Recognition at Scale. In: 9th International Conference on Learning Representations, ICLR 2021, Virtual Event, Austria, 3-7 May 2021. OpenReview.net (2021)
3. He, K., Sun, J., Tang, X.: Single image haze removal using dark channel prior. In: IEEE Conference on Computer Vision and Pattern Recognition, pp. 1956–1963 (2009)

4. Li, B., Peng, X., Wang, Z., Xu, J., Feng, D.: AOD-net: all-in-one dehazing network. In: IEEE International Conference on Computer Vision, pp. 4780–4788 (2017)
5. Li, B., et al.: Benchmarking single-image dehazing and beyond. IEEE Trans. Image Process. **28**(1), 492–505 (2019)
6. Li, B., Gou, Y., Gu, S., Liu, J.Z., Zhou, J.T., Peng, X.: You only look yourself: unsupervised and untrained single image dehazing neural network. Inter. J. Comput. Vis., 1–14 (2021)
7. Li, B., Gou, Y., Liu, J.Z., Zhu, H., Zhou, J.T., Peng, X.: Zero-shot image dehazing. IEEE Trans. Image Process. **29**, 8457–8466 (2020)
8. Liu, Z., et al.: Swin transformer V2: Scaling Up Capacity and Resolution. In: International Conference on Computer Vision and Pattern Recognition (2022)
9. Liu, Z., et al.: Swin transformer: hierarchical vision transformer using shifted windows. In: IEEE/CVF International Conference on Computer Vision (2021)
10. Maldonado-Quispe, P., Pedrini, H.: XYZ unsupervised network: a robust image dehazing approach. In: 19th The International Conference on Computer Vision Theory and Applications, vol. 3, pp. 694–701. Rome, Italy (Feb 2024)
11. Maldonado-Quispe, P., Pedrini, H.: Image dehazing using a simple convolutional autoencoder. In: IEEE Latin American Conference on Computational Intelligence, pp. 1–6. Recife-PE, Brazil (2023)
12. Mao, D., Gao, S., Lü, H., Zhang, C., Zhou, Y.: Multi-scale image dehazing network based on local transformer. J. Comput.-Aided Design Comput. Graph. (0)
13. Park, N., Kim, S.: How do vision transformers work? In: International Conference on Learning Representations, pp. 1–26 (2022)
14. Qin, X., Wang, Z., Bai, Y., Xie, X., Jia, H.: FFA-Net: feature fusion attention network for single image dehazing. In: AAAI Conference on Artificial Intelligence, vol. 34(7), pp. 11908–11915 (2020)
15. Ren, W., Liu, S., Zhang, H., Pan, J., Cao, X., Yang, M.-H.: Single image dehazing via multi-scale convolutional neural networks. In: Leibe, B., Matas, J., Sebe, N., Welling, M. (eds.) ECCV 2016. LNCS, vol. 9906, pp. 154–169. Springer, Cham (2016). https://doi.org/10.1007/978-3-319-46475-6_10
16. Ronneberger, O., Fischer, P., Brox, T.: U-Net: convolutional networks for biomedical image segmentation. In: Navab, N., Hornegger, J., Wells, W.M., Frangi, A.F. (eds.) MICCAI 2015. LNCS, vol. 9351, pp. 234–241. Springer, Cham (2015). https://doi.org/10.1007/978-3-319-24574-4_28
17. Song, Y., He, Z., Qian, H., Du, X.: Vision transformers for single image dehazing. IEEE Trans. Image Process. **32**, 1927–1941 (2023)
18. Wang, W., et al.: Pyramid vision transformer: a versatile backbone for dense prediction without convolutions. In: IEEE/CVF International Conference on Computer Vision, pp. 568–578 (Oct 2021)
19. Wang, Z., Bovik, A.C., Sheikh, H.R., Simoncelli, E.P.: Image quality assessment: from error visibility to structural similarity. IEEE Trans. Image Process. **13**(4), 600–612 (2004)
20. Wasi, A., Shiney, O.J.: DHFormer: A Vision Transformer-Based Attention Module for Image Dehazing. arXiv: 2312.09955 (2023)
21. Wei, X., Ye, X., Mei, X., Wang, J., Ma, H.: Enforcing high frequency enhancement in deep networks for simultaneous depth estimation and dehazing. Appl. Soft Comput. **163**, 111873 (2024)
22. Zhang, S., Tao, Z., Lin, S.: DedustNet: A Frequency-dominated Swin Transformer-based Wavelet Network for Agricultural Dust Removal (2024). https://arxiv.org/abs/2401.04750

23. Zhao, D., Mo, B., Zhu, X.: Multi-scale feature aggregation network for single-image dehazing. IET Image Process. (2024)
24. Zhu, Q., Mai, J., Shao, L.: A fast single image haze removal algorithm using color attenuation prior. IEEE Trans. Image Process. **24**(11), 3522–3533 (2015)
25. Zu, B., Cao, T., Li, Y., Li, J., Ju, F., Wang, H.: SwinT-SRNet: swin transformer with image super-resolution reconstruction network for pollen images classification. Eng. Appl. Artif. Intell. **133**, 108041 (2024)

A Proposal for Explainable Fruit Quality Recognition Using Multimodal Models

Felipe Nuñez[1], Billy Peralta[1(✉)], Orietta Nicolis[1], Luis Caro[2], and Marco Mora[3]

[1] Facultad de Ingeniería, Universidad Andres Bello, Antonio Varas 880, 7500735 Providencia, Región Metropolitana, Chile
f.nezbustamante@uandresbello.edu, {billy.peralta,o.nicolis}@unab.cl
[2] Departamento de Ingeniería Informática, Universidad Católica de Temuco, Rudecindo Ortega 2950, Temuco, Chile
lcaro@uct.cl
[3] Laboratory of Technological Research in Pattern Recognition (LITRP), Universidad Católica del Maule, Avenida San Miguel 3605, Talca, Chile
mmora@ucm.cl

Abstract. The fruit industry in Chile has achieved global recognition for its productivity and leadership in fruit exportation, being the main exporter in the Southern Hemisphere, especially of cherries, grapes, and blueberries. Agricultural automation is a growing trend aimed at reducing laborious work and the consumption of time and personnel. Advances in artificial intelligence are enabling the automation of various processes, such as fruit categorization, though there are still gaps in the precision of classifying fruits in good and bad condition, particularly when considering specialized multimodal models. This work addresses this gap by combining convolutional neural network models and the multimodal CLIP technique, evaluating the effectiveness of convolutional architectures such as ResNet50, Xception, and MobileNet. The experiments show interesting results among different architectures, with ViT-B/16 model standing out for its higher precision in this task.

Keywords: Multimodal classification · artificial vision · CLIP

1 Introduction

The fruit industry in Chile has achieved notable global recognition due to its productivity and leadership in fruit exports. This country is regarded as the main exporter in the Southern Hemisphere, particularly for cherries, grapes, and blueberries. Additionally, in the category of processed products, it is prominent for the export of apples, grapes, and dried plums. It is important to note that Chile not only focuses on one type of fruit but has specialized in the production and export of more than 50 different species [21], demonstrating its solid positioning in the international market and highlighting the quality and diversity of Chilean fruit.

The automation of agricultural processes is an increasing trend in the industry [24] and, as is well known, tasks in this sector are labor-intensive and time-consuming, requiring a significant amount of personnel. Thanks to the advancement of artificial intelligence and its various methods available today, many of the previously mentioned processes are being automated, achieving promising results, as can be seen in the following study on fruit categorization [10] and in another research [16], both of which are subject to ongoing improvement.

Moreover, we can observe the use of artificial intelligence in various areas and how these tools provide support when performing certain tasks. Image recognition has been progressing for some time, and this study proposes the use of two image recognition systems. The first, a more extensively studied system with effective results, is the convolutional neural network [20], which requires pre-training of images so the network can learn and make decisions with high precision. The challenge is that it requires expertise to adapt the network to specific needs and a robust set of images to achieve optimal results.

However, in addition to the convolutional neural network, the CLIP model [22] will be incorporated to analyze fruit quality using both visual and textual information. CLIP will allow for a more comprehensive and contextualized evaluation of fruit quality by combining images and textual descriptions. This complementary approach will provide an additional perspective and enhance the results obtained.

The experiments were performed in several popular fruits such as apples, bananas, and lemons, using different architectures, which will allow for the evaluation of their performance and comparison of the results obtained. In this way, this study aims to contribute to the improvement of fruit quality assessment processes in the Chilean fruit industry, leveraging the potential of artificial intelligence and image recognition systems.

2 Related Works

A comprehensive review of computer vision systems in the agri-food industry, detailing their development, fundamental technologies, and instrumentation is provided by Ma et al. [13]. It highlights the applications of computer vision in quality assessment of agricultural and food products from 2007 to 2013. Additionally, the paper discusses future trends, particularly the integration of computer vision with spectroscopy, to enhance food safety and quality assessments. Ashok and Vinad [1] present a non-destructive method using a Probabilistic Neural Network (PNN) to classify apples based on external quality. It distinguishes healthy from damaged apples with accuracies between 86.52% and 88.33% using extracted features from 65 training and testing images.

Recently, Behera et al. [2] review various machine learning and image processing techniques for identifying, classifying, and grading fruits from 2010 to 2019. The study highlights the automation needs due to high production and limited skilled labor in India, discussing different approaches' achievements and limitations and suggesting directions for future research. Similarly, Dhiman et al.

[6] study various machine learning techniques, such as k-nearest neighbors and neural networks, used to analyze image-based features like shape, size, color, and texture for fruit quality classification, highlighting their effectiveness and time efficiency.

In general, the majority of algorithms used in literature for assessing the fruits' quality are based on convolutional neural networks [7,17]. In particular, Gujuraj et al. [7] presents a fully automated, deep learning-based system for mango grading using CNN and computer vision. It enhances grading accuracy by analyzing ripening stages, shape, texture, color, and defects of mangoes. The system classifies mangoes into three quality grades and has been validated with high accuracy across multiple varieties.

Meenu et al. [14] introduce an efficient real-time automatic citrus fruit grading and sorting machine using adaptive deep learning based on computer vision. The custom CNN model improves postharvest operations like washing and weight-based grading, significantly reducing human labor while optimizing accuracy and performance for sustainable postharvest automation.

Differently, Dhiman et al. [5] use recurrent neural network for a multi-fruit quality assessment system using a recurrent neural network. The proposed model efficiently processes images of nine fruit types to predict their quality based on comprehensive feature extraction, achieving high accuracy and performance metrics over existing systems.

Chakraborty et al. [3] outlines the development of a real-time automatic citrus fruit grading and sorting machine, leveraging a computer vision-based adaptive deep learning model. Employing a custom CNN model for image classification, the machine operates with high efficiency, facilitating postharvest operations such as washing and weight-based grading.

Raja et al. [23] develop a system using CNN and transfer learning to improve fruit categorization, particularly in determining fruit freshness. It introduces a custom CNN architecture and employs a pre-trained VGG model for this purpose. Tested on 70 fruit types, including apples and oranges, the system achieves up to 99.99% accuracy in validation, showcasing its potential for real-time farming applications.

A discussion of the use of convolutional neural networks and transfer learning for automated fruit quality prediction is provided by Mputu et al. [19] The proposed models classify fruits as fresh or rotten with near-perfect accuracy, leveraging deep learning to minimize human effort and error in agriculture. The findings could significantly impact real-time farming applications by reducing costs and labor.

The works reviewed highlight a great modern emphasis on the use of deep learning techniques in fruit recognition, including fruit quality. However, these works do not cover the generation of a specialized output such that it allows giving more details of the characteristics of a fruit in poor condition that surpass a discrete output of few values, typically used in classification. By indicating a characteristic textually, it is possible that the human can have more clues about the real state of the fruit.

3 Theoretical Framework

In this work we apply Xception, ResNet and MobileNet networks. We also use GradCAM to analyze network performance, as well as a CLIP network.

3.1 Xception

Xception is a convolutional neural network with a depth of 71 layers. In this study, Xception was chosen due to its impressive accuracy, which reaches 98.98% in a similar context according to a previous study [18]. Before using Xception in this work, adjustments were made to the model parameters. Additionally, the input images were resized to a specific size of 299 × 299 pixels, as recommended in the official Xception documentation [4]. This choice was based on confidence in its proven performance and ability to handle fruit sorting tasks with high accuracy.

3.2 ResNet50

ResNet50 is another convolutional neural network, but in this case, it is 50 layers deep. Like Xception, ResNet50 can load pre-trained versions of a previously tuned network. The choice of ResNet50 in this study was based on its high performance, which also reaches 96.64% accuracy in previous research [11]. Before use, adjustments were made to the model parameters to adapt it to the fruit classification task. The input images were resized to a size of 224 × 224 pixels, following the recommendations in the official ResNet documentation [8]. ResNet50 was selected due to its demonstrated ability to achieve high accuracy on similar tasks.

3.3 Mobilenet

MobileNet is a neural network with approximately 3.4 million parameters, allowing acceptable performance in several applications. MobileNet is known for its efficiency and ability to execute on devices with limited resources. To ensure best performance, it is recommended to resize input images to a specific size of 224 × 224 pixels, according to the guidelines provided in the MobileNet documentation [9]. The choice of MobileNet is based on its ability to balance performance and computational efficiency, making it suitable for computer vision applications on mobile or resource-constrained devices.

3.4 GradCAM

GradCAM [25] (Gradient-weighted Class Activation Mapping) is a method that is used to analyze areas of an image where a convolutional neural network focus to classify it. This approach typically outputs a heat map to identify areas of the image that are being affected, allowing a result to be determined with greater precision.

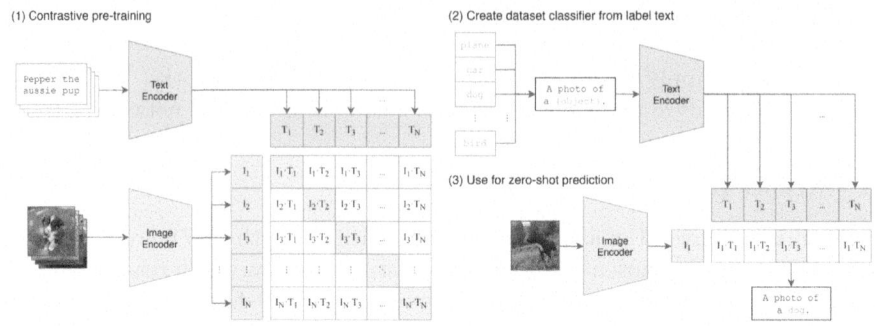

Fig. 1. Diagram of CLIP. (https://openai.com/research/clip)

Grad-CAM operates by taking advantage of gradients generated from the features extracted in the last convolutional layer of the model. Using these gradients, areas of greatest importance in the image are highlighted, creating a heat map that demonstrates how the model is paying attention to different parts of the image. In our context, Grad-CAM is particularly valuable when analyzing spoiled fruits, where specific details are critical to assess deterioration.

3.5 CLIP

CLIP was developed in 2021 by OpenAI [22]. This model is a multimodal neural net- work that stands out for its ability to work simultaneously with text and images. Figure 1 represents three key stages in the operation of CLIP. First, contrastive pre-training involves training CLIP on images and text descriptions so that it learns to relate them, mapping similar representations to related content. Then, a classifier dataset is developed from the label text, using the text descriptions to train a classifier that can label images based on their textual content. This step establishes a strong link between the text and the images. Finally, this connection is used for Zero-Shot predictions, where CLIP can apply its prior understanding to make accurate predictions about images it has not seen before.

4 Experimental Design

4.1 Datasets

For the tests, two different data sets were used, the first [16] and the second [15], each containing different types of fruits, with various amounts of images for each category. The fruits and quantities of images used in each set can be seen in Table 1a and 1b.

Each data set contains more than 1000 images for each kind of fruit, with their respective label as Good Fruit or Bad Fruit. To guarantee robust results,

Table 1. Details of two dataset used in experiments.

Fruit	Quality	Numbers of images
Apple	Good	1000
Apple	Bad	1000
Lime	Good	1000
Lime	Bad	1000
Guava	Good	1000
Guava	Bad	1000
Orange	Good	1000
Orange	Bad	1000
Pomogrante	Good	1000
Pomogrante	Bad	1000
Total		10000

(a) First data set

Fruit	Quality	Numbers of images
Apple	Good	1000
Apple	Bad	1000
Lime	Good	1000
Lime	Bad	1000
Guava	Good	1000
Guava	Bad	1000
Orange	Good	1000
Orange	Bad	1000
Pomogrante	Good	1000
Pomogrante	Bad	1000
Total		10000

(b) Second data set

cross validation was applied in which the images were separated into five folders, maintaining a proportion of 70% for training and 30% for testing. Random internal 20% of training data is used for validation. Each image was labeled according to its condition, whether in good condition or poor condition.

Each data set contains more than 1000 images for each kind of fruit, with their respective label as Good Fruit or Bad Fruit. To guarantee robust results, cross validation was applied in which the images were separated into five folders, maintaining a proportion of 70% for training and 30% for testing. Random internal 20% of training data is used for validation. Each image was labeled according to its condition, whether in good condition or poor condition.

4.2 Setting of Neural Networks

Setting of CNN. This model's input is defined with a specific shape of 256 by 256 pixels. Convolution layers extract patterns from image features using different filter sizes to capture various details. A global Average Pooling layer averages features across the entire image, reducing parameters and preventing overfitting. Features are then flattened into a 1024-vector for use in fully connected layers, which have connections between all neurons in adjacent layers and include a Dropout layer for regularization. The output layer consists of two neurons with softmax activation, representing fruit classes and predicting the probability of each class. The model is trained using the training data over 50 epochs considering the fit on the validation set.

Setting of MobileNet. This network utilizes the pre-trained MobileNet model as a starting point, originally designed for image analysis. Initially, we freeze its weights to retain prior knowledge. Then, we construct a new model to process input images, employing techniques such as data augmentation and normalization. After passing through the base model, we apply Global Average Pooling and Dropout to mitigate overfitting. The output connects to a dense layer with a single unit for binary classification. This model is compiled, trained, and evaluated on our specific data, tailored to our task of analyzing fruit states. Image sizes are set at 224 × 224 pixels. The model is trained using the training data over 30 epochs considering the fit on the validation set.

Setting of Xception Net. This model undergoes preprocessing directly within Keras using configurations from the Xception model. Comprising 71 layers, this network employs separate convolutions for deep channel and spatial convolutions. Notably, the default size for this network is 150 × 150 pixels, though other dimensions may be utilized, the aforementioned size typically yields optimal results. The model is trained using the training data over 30 epochs considering the fit on the validation set.

Setting of ResNet50. The advantage of using this model lies in the multitude of categories it can handle, preloaded with weights from the ImageNet database. The default size for this model is 224 by 224 pixels, requiring a reshape for fruit images. The model is trained using the training data over 12 epochs considering the fit on the validation set. Each epoch involves adjusting model weights and parameters to enhance predictive capability based on calculated loss, aiming for optimized performance in binary classification of images depicting fruits in good and bad condition.

4.3 Setting of CLIP

The experiments conducted with CLIP were performed using an NVIDIA RTX 3060 GPU with 24 GB of RAM allowing analysis of fruit conditions using CLIP's text-image neural network model. The CLIP technique employed two preprocessed models: initially, the "Vit-B/32" model was utilized, followed by the "Vit-B/16" model, involving tokenization of descriptions and images as outlined in the preceding CLIP process section, yielding more precise results regarding fruit condition. The Vit models refers to variants of visual transformers models [12]. We detail each used model as following:

1. **ViT-B/32:** This model is characterized by its focus on computational efficiency by reducing the input image resolution by a factor of 32. This means that images are divided into larger patches during the model's attention phase. This version may be particularly suitable for datasets with higher-resolution images where decreasing quality does not significantly affect the classification task.
2. **ViT-B/16:** This model reduces the input resolution by a factor of 16, allowing for greater attention to fine details in the images. Images are divided into smaller patches, which can be beneficial for datasets with high-resolution images containing crucial details for classification.

Given CLIP's generally accurate outcomes, specific descriptions were necessary to determine fruit condition with greater precision, incorporating textual descriptions about quality for both good and bad fruit classes. We carefully incorporate the class information and some visual feature in the textual description.

We show some examples of textual description of fruits in bad conditions:

- This is a bad banana because it's rotten.
- This is a bad banana because it's moldy.
- This is a bad banana it's black.
- This is a bad banana because it's decomposed.

We also show several examples of textual description of fruits in good conditions:

- This is a good banana because it's fresh.
- This is a good banana because it's sweet.
- This is a banana in good condition, but it is green.
- This is a good banana because it has a nice yellow color.

5 Results

In this section, in the first experiment we evaluate the effectiveness of the different visual classification neural network models. In the second experiment, we qualitatively analyse the visual outputs of classifiers using Grad-CAM [25]. In the third experiment, we mix the visual classifiers models and GradCAM. We use the CLIP model to generate a textual description of state of fruits. However, a valid question is about how to evaluate the accuracy of textual description of fruits?. We propose to measure in an indirect way, we consider the classification performance of textual descriptions.

These textual description are trained with a very simple threshold classifier in order to evaluate indirectly the effectiveness of these textual outputs. Our rationale is that if the textual description are accurate, these could be used to classify the fruits instead of images. In the opposite case, these textual descriptions are not accurate.

5.1 Results of Neural Network Classifiers

In this section, we present the results obtained from the evaluation of different neural network models on both datasets. To assess their performance, we implemented four different neural network architectures, including the standard convolutional neural network and more advanced models such as ResNet50, Xception, and MobileNet, analyzed using the Grad-CAM technique. In the training, we consider a batch size of 32, while we consider a typical cross-entropy loss function using Adam optimizer. The results provide a detailed insight into how each model performed in the task of classifying fruits in good and bad condition. To

Fig. 2. Classification results of Convolutional Neural Network

analyze the results more effectively, both accuracy and precision were calculated for the convolutional neural network model, while accuracy alone was calculated for the other models. These values are indicative of the models' ability to make precise predictions and avoid false positives in fruit classification.

The classification results of convolutional neural network are presented in Fig. 2. These results are based on dataset 1 and reveal high effectiveness in classification, with success rates exceeding 96%.

Table 2 shows the performance comparison among the CNN, ResNet50, Xception, and MobileNet architectures using dataset 1. We also add the number of parameters in order to analyze complexity of architectures. In this comparison, the ResNet50 network stood out with a precision of 98.50%, surpassing the other architectures.

Table 2. Results of classification of tested neural networks for first dataset

Network model	Accuracy	network parameters
ResNet50	0.9850	23.536.641
CNN	0.9626	134.247.246
X-ception	0.9621	20.863.529
MobileNet	0.8920	3.208.001

Table 3. Results of classification of tested neural networks for second dataset

Network model	Accuracy
ResNet50	0.9812
CNN	0.9749
X-ception	0.9232
MobileNet	0.8586

On the other hand, the results obtained by applying the models to a second dataset are presented in Table 3. These results provide a comprehensive view of how the different models perform, reinforcing the consistency of the results obtained previously.

These results indicate that, while the standard convolutional neural network surprisingly demonstrated solid performance. We hypothesise that it is due to the lack of bias in relation to other alternatives. However, ResNet50 achieves the highest precision in classifying fruits in good and bad condition. Therefore, we use this network in the subsequent experiments.

5.2 Qualitative Analysis with GradCAM

After training the neural network, the GradCAM technique is applied to the model to identify the focus areas according to ResNet50. As we can see in Fig. 3, it illustrates where ResNet50 focus in order to perform classification. In the case of the fruit in good condition, the model successfully identifies and recognizes the correct fruit, even if there are other objects in the image. On the other hand, for the fruit in bad condition, Grad-CAM is used to precisely identify the affected areas on the fruit, allowing a detailed visualization of the deteriorated parts.

Integrating Grad-CAM into our project not only broadens our understanding of how the machine learning model makes decisions but also provides a valuable visual perspective for users. This technique highlights how neural networks focus on key areas of the image to achieve their predictions.

5.3 Results of Integration of CLIP and Classification Models

Below, we present the results obtained from evaluating the CLIP model using the dataset related to images of bananas in several states. We choose only one

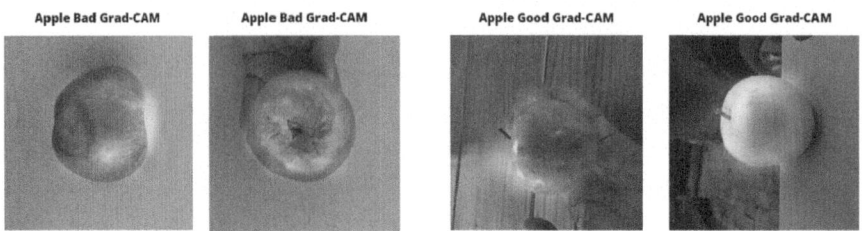

Fig. 3. Grad-CAM Technique.

fruit, banana, to evaluate this proposal, because we need to carefully label each image with text related to state of fruit.

Initially, we conducted a series of tests using the "ViT-B/16" model. We choose this network because of its better results in validation dataset. In this approach, four detailed descriptions of fruits in poor condition and another four of fruits in good condition were provided to the model. Additionally, threee images corresponding to each state were randomly selected. These setups allowed CLIP to more accurately assess the relationship between images and textual descriptions. Figure 4 shows several examples of pairs images and textual description for bananas. Note that we use natural descriptions for each class, which results in multiple descriptions for a particular image. We hope that this redundancy can be helpful in getting better performance from CLIP, which is focused to natural textual descriptions.

These initial configurations were essential for evaluating the model's ability to accurately identify the state of the fruits using visual and textual information. Figure 5 presents a cosine similarity matrix generated to observe how CLIP establishes relationships between a random pair, image and text. We consider a cosine similarity because the textual description and the image are represented by embedded vectors by CLIP in the same multimodal space. In this setting, a value of 0 indicates no relationship and a value of 0.35 is the maximum level of relationship between text and image. Typically, the network is able to obtain a reasonable association between quality and embeddding given by CLIP.

Furthermore, a quantitative analysis was conducted to assess the CLIP model's capability in classifying bananas based on their quality, distinguishing between good and bad conditions. For this evaluation, two labels were established: good condition and bad condition, with the purpose of categorizing 50 random images of bananas under each of these conditions. The analysis focused on measuring two essential metrics, accuracy and precision, which are fundamental for assessing the model's performance.

Using Python's Sklearn library, comparisons were made between the 100 images of bananas and a set of eight different descriptions. These descriptions were used as references to determine whether a banana was in good or bad condition. The images were evenly divided, with 50 images representing bananas in good condition and the other 50 representing bananas in bad condition.

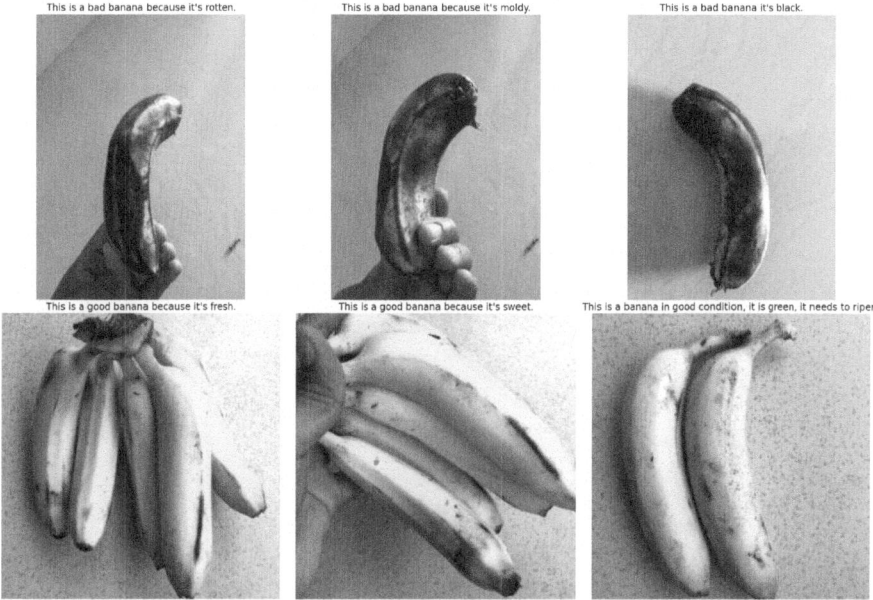

Fig. 4. Samples of pairs image and textual descriptions for bananas.

The results obtained are presented graphically in Table 4. We obtain a good performance considering a simple threshold classifier, despite to use only textual descriptions.

Table 4. Experiment of evaluation of CLIP outputs

Network model	Accuracy	Accuracy class good	Accuracy class bad
CLIP	0.75	0.65	0.80

As an observation, the results could be improved by integrating the CLIP model with one of the other models used in this work, which could lead to more accurate results in the classification of the fruit in question. Note that the classification models has very good perfomance, which suggests to use a multimodal model like CLIP in a second stage.

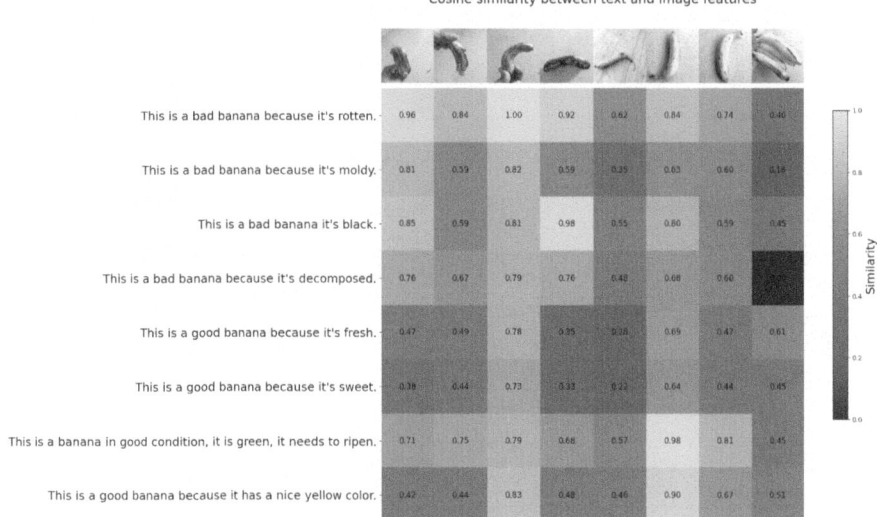

Fig. 5. Cosine Similarity Matrix CLIP model ViT-B/16

6 Conclusions

Throughout this research, various architectures of standard convolutional neural networks and advanced models such as ResNet50, Xception, and MobileNet, along with the innovative CLIP (Contrastive Language-Image Pretraining) technique in its two preprocessed models, have been explored. The main goal of this research was to understand and compare the performance of these different models in the task of classifying the condition of fruits.

Initially, traditional convolutional neural networks were employed, including a basic convolutional network, ResNet50, Xception, and MobileNet. These models demonstrated their inherent ability to extract and learn relevant features from images. However, it was observed that pretrained architectures, such as ResNet50, Xception, and MobileNet, slightly outperformed the basic convolutional network in terms of precision and in capturing subtle details and features in the images. This underscores the importance of transfer learning in solving complex image classification problems.

On the other hand, the incorporation of CLIP in the assessment of banana quality added an exciting dimension to the project. CLIP demonstrated its ability to relate images and descriptions in natural language, bringing a unique perspective to image understanding. This suggests that careful selection of descriptions can have a significant impact on the outcomes achieved. As future work, we plan to conduct tests with a larger number of images could contribute to a more robust and representative evaluation of the model's performance.

Acknowledgments. B. Peralta thanks the support of the National Center for Artificial Intelligence CENIA FB210017, Basal ANID.

References

1. Ashok, V., Vinod, D.: Automatic quality evaluation of fruits using probabilistic neural network approach. 2014 International Conference on Contemporary Computing and Informatics (IC3I), pp. 308–311 (2014). https://api.semanticscholar.org/CorpusID:15933462
2. Behera, S., Rath, A., Mahapatra, A., Sethy, P.: Identification, classification grading of fruits using machine learning computer intelligence: a review. J. Ambient Intell. Humanized Comput. (2020). https://doi.org/10.1007/s12652-020-01865-8
3. Chakraborty, S.K., et al.: Development of an optimally designed real-time automatic citrus fruit grading-sorting machine leveraging computer vision-based adaptive deep learning model. Eng. Appl. Artif. Intell. **120**, 105826 (2023)
4. Chollet, F.: Xception: deep learning with depthwise separable convolutions. In: Proceedings of the IEEE Conference on Computer Vision and Pattern Recognition, pp. 1251–1258 (2017)
5. Dhiman, B., Kumar, Y., Hu, Y.C.: A general purpose multi-fruit system for assessing the quality of fruits with the application of recurrent neural network. Soft. Comput. **25**(14), 9255–9272 (2021)
6. Dhiman, B., Kumar, Y., Kumar, M.: Fruit quality evaluation using machine learning techniques: review, motivation and future perspectives. Multimedia Tools Appli. **81**, 1–23 (2022). https://doi.org/10.1007/s11042-022-12652-2
7. Gururaj, N., Vinod, V., Vijayakumar, K.: Deep grading of mangoes using convolutional neural network and computer vision. Multimedia Tools Appli. **82**(25), 39525–39550 (2023)
8. He, K., Zhang, X., Ren, S., Sun, J.: Deep residual learning for image recognition. In: Proceedings of the IEEE Conference on Computer Vision and Pattern Recognition, pp. 770–778 (2016)
9. Howard, A.G., et al.: Mobilenets: Efficient convolutional neural networks for mobile vision applications. arXiv preprint arXiv:1704.04861 (2017)
10. Häni, N., Roy, P., Isler, V.: Minneapple: a benchmark dataset for apple detection and segmentation. IEEE Robotics Autom. Lett. **5**(2), 852–858 (2020). https://doi.org/10.1109/LRA.2020.2965061
11. Kathepuri, S.: Recognition and classification of fruits using deep learning techniques. Ph.D. thesis, Dublin, National College of Ireland (2020)
12. Liu, Y., et al.: A survey of visual transformers. IEEE Trans. Neural Netw. Learn. Syst. (2023)
13. Ma, J., et al.: Applications of computer vision for assessing quality of agri-food products: a review of recent research advances. Crit. Rev. Food Sci. Nutr. **56**(1), 113–127 (2016)
14. Meenu, M., Kurade, C., Neelapu, B.C., Kalra, S., Ramaswamy, H.S., Yu, Y.: A concise review on food quality assessment using digital image processing. Trends Food Sci. Technol. **118**, 106–124 (2021)
15. Meshram, V., Patil, K.: Fruitnet: Indian fruits dataset with quality (good, bad & mixed quality). Mendeley Data **1** (2021)
16. Meshram, V., Patil, K.: Fruitnet: Indian fruits image dataset with quality for machine learning applications. Data Brief **40**, 107686 (2022)

17. Mohapatra, D., Das, N., Mohanty, K.K., Shresth, J.: Automated visual inspecting system for fruit quality estimation using deep learning. In: Innovation in Electrical Power Engineering, Communication, and Computing Technology: Proceedings of Second IEPCCT 2021, pp. 379–389. Springer (2021)
18. Morshed, M.S., Ahmed, S., Ahmed, T., Islam, M.U., Rahman, A.A.: Fruit quality assessment with densely connected convolutional neural network. In: 2022 12th International Conference on Electrical and Computer Engineering (ICECE), pp. 1–4. IEEE (2022)
19. Mputu, H.S., Mawgood, A.A., Shimada, A., Sayed, M.S.: Real-time tomato quality assessment using hybrid cnn-svm model. IEEE Embedded Syst. Lett., 1–1 (2024). https://doi.org/10.1109/LES.2024.3370634
20. Muresan, H., Oltean, M.: Fruit recognition from images using deep learning. Acta Universitatis Sapientiae, Informatica **10**(1), 26–42 (2018)
21. ODEPA: Frutas frescas y procesadas. https://www.odepa.gob.cl/rubros/frutas-frescas-y-procesadas, Accessed 19 April 2023
22. Radford, A., et al.: Learning transferable visual models from natural language supervision. In: International Conference on Machine Learning, pp. 8748–8763. PMLR (2021)
23. Raja, S.P., et al.: Fruit quality prediction using deep learning strategies for agriculture. Inter. J. Intell. Syst. Appli. Eng. **11**(2s), 301–310 (2023)
24. Sa, I., Ge, Z., Dayoub, F., Upcroft, B., Perez, T., McCool, C.: Deepfruits: a fruit detection system using deep neural networks. Sensors **16**(8), 1222 (2016)
25. Selvaraju, R.R., Cogswell, M., Das, A., Vedantam, R., Parikh, D., Batra, D.: Gradcam: visual explanations from deep networks via gradient-based localization. In: Proceedings of the IEEE International Conference on Computer Vision, pp. 618–626 (2017)

Negative Sampling for Triplet-Based Loss: Improving Representation in Self-supervised Representation Learning

Manuel Alejandro Goyo[1]([✉]) and Mauricio Hidalgo[1,2]

[1] Universidad Técnica Federico Santa María, Valparaíso, Chile
manuel.goyo@sansano.usm.cl, mauricio.hidalgob@usm.cl
[2] FDI - Universidad Finis Terrae, Santiago, Chile
mhidalgo@uft.cl

Abstract. Significant strides have been made in artificial neural networks across various fields, necessitating extensive labeled data for effective training. However, the acquisition of such annotated data is both costly and labor-intensive. To address this challenge, Self-Supervised Representation Learning (SSRL) has emerged as a promising solution. One prominent SSRL method, Contrastive Self-Supervised Learning (CSL), enhances feature representations by discerning similarities and differences among samples in the feature space. Yet, accurately identifying dissimilar samples remains a persistent issue, limiting CSL's effectiveness. In response, an innovative enhancement to CSL is proposed in this paper. Explicit negative sampling strategies using a binary classification algorithm within the feature space are introduced to distinguish between similar and dissimilar features precisely. Additionally, Triplet Loss, originally designed for tasks such as person re-identification and face recognition, is incorporated to further refine feature learning. Experimental evaluations on the CIFAR-10 and SVHN datasets validate the proposed method's superiority in content-based image retrieval (CBIR) and classification tasks. Significant improvements are demonstrated in metrics such as mean average precision (MAP), accuracy, recall, precision, and F1-score compared to existing techniques. This framework contributes to the advancement of SSRL by enabling scalable neural network training on large datasets with minimal annotation, effectively bridging the gap between supervised and unsupervised learning paradigms.

Keywords: Self Representation Learning · Self Supervision · Triplet Loss · Negative Sampling · CBIR · Image Classification · Computer Vision

1 Introduction

In recent years, significant advancements have been demonstrated by artificial neural networks across diverse fields such as economics [40], medicine [36], and industry [30]. Traditional approaches to training these networks have heavily relied on labeled data, which provides specific information about the data's characteristics [21]. However, the acquisition of annotated data is often costly and requires specialized expertise to ensure accurate annotations [20]. To address these challenges, Self-Supervised Representation Learning (SSRL) has increasingly been adopted as a promising alternative. SSRL aims to train robust feature extractors without relying on labeled data. Instead, unsupervised or self-supervised learning techniques are leveraged to learn meaningful representations from the data itself [10]. A prominent method within SSRL is Contrastive Self-Supervised Learning, which has gained traction for its effectiveness in training models without explicit labels [19]. This approach operates by enforcing the model to bring similar samples closer together and push dissimilar samples farther apart in the feature space (see Fig. 1) [22].

Contrastive Self-Supervised Learning relies on a similarity metric, such as Euclidean or cosine distance, to quantify the proximity between samples in the feature space [22]. Notably, methods trained using this approach have achieved state-of-the-art performance in various applications, including computer vision and natural language processing tasks [2,4,13,14,29,44,46]. These methods optimize the feature space representations to enhance the model's ability to discern subtle differences and similarities within the data. By adopting Contrastive Self-Supervised Learning in SSRL, researchers aim to bridge the gap between supervised and unsupervised learning paradigms. This approach not only facilitates scalable training on large datasets at minimal annotation cost but also enhances the versatility and generalization capabilities of neural network models across various domains. In a Contrastive Self-Supervised Learning framework, identifying similar and dissimilar samples is crucial, especially in the absence of labeled data. Commonly, similar samples are selected using augmentations within a batch, preserving semantic information to generate comparable instances [13,14]. However, accurately identifying dissimilar samples presents a significant challenge. Some approaches avoid explicit identification of dissimilar samples by calculating distances to remaining points in a batch [2] or to the closest ones [44], then using the average similarity for the loss function. Yet, concerns arise about whether this average effectively captures truly dissimilar values, potentially compromising the contrastive learning objective by including insufficiently dissimilar elements. Other strategies focus exclusively on reinforcing similarities among data points through parameter updates [4,13]. However, this method may struggle to establish robust decision boundaries, leading to overlaps between different categories and diminished model performance. Other researchers employ explicit negative mining strategies. For example, multiple views per image are considered to distinguish true negatives from false negatives [18], or negative samples are estimated from the distribution over pairs [34]. These methods draw inspiration from metric learning, where identifying hard negatives accelerates error correction during training [32,35], enhancing the model's ability to discern subtle differences and learn discriminative features effectively. Our hypothesis is as follows: Better performance in content-based image retrieval (CBIR) and

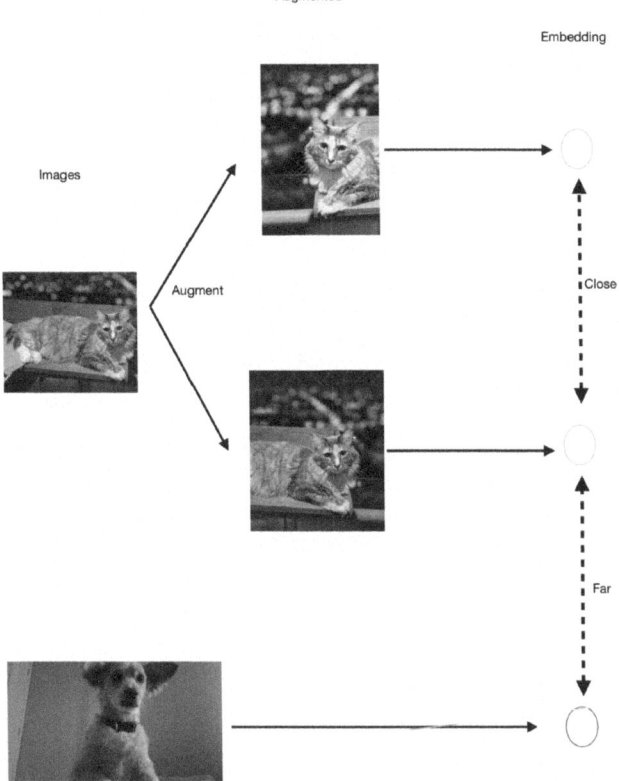

Fig. 1. Illustration of Contrastive Self-Supervised Learning: Similar samples (green) are grouped closer together, while dissimilar samples (red) are pushed farther apart in the feature space. (Color figure online)

classification compared to state-of-the-art methods will be yielded by creating negative samples using a binary classification algorithm to generate subsamples, which are then combined to form the negative.

To test our hypothesis, our approach involves explicitly creating negative samples by applying a binary classification algorithm in the feature space on batch data. This method differentiates data with similar features, even if not semantically similar, from dissimilar data. It then identifies a representative, or centroid, from the similar data and designates it as the negative sample. The centroid effectively encapsulates the information of all the represented vectors, making it a robust negative example. Furthermore, after identifying similar and dissimilar data through this negative sampling, we employ the Triplet Loss function. Originally introduced for applications like person re-identification and face recognition [6,35], Triplet Loss minimizes the similarity between an anchor sample and a negative sample while maximizing the similarity between the anchor and a positive sample, subject to a margin constraint. For clarity, consider the triplet set $(x_i, x_i^+, x_i^-)_{i=1,\cdots,m}$ using one query data and one sample. The triplet loss is defined as:

$$\mathcal{L} = \sum_{i=1}^{m} \max\left(sim\left(x, x_i^-\right) - sim\left(x_i, x_i^+\right), m\right)$$

where sim denotes similarity and m is a margin parameter. The key contributions of our work include:

- Designing a straightforward yet effective similarity-based sampling strategy for generating negative samples.
- Proposing a self-supervised training method based on Triplet Loss for robust representation learning

We show through experiments on two datasets that our approach surpasses existing methods in tasks such as content-based image retrieval (CBIR) and classification, evidenced by superior performance metrics like mean average precision (MAP) and classification metrics (accuracy, recall, precision, F1).

2 Related Work

Understanding the landscape of representation learning is crucial in assessing the advancements and nuances in self-supervised paradigms. In unsupervised representation learning, methods are primarily categorized into generative and discriminative approaches [2,7]. Generative methods aim to create distributions over data and latent embeddings, capturing detailed pixel-level representations. Techniques such as auto-encoding [23,42] and adversarial learning [12] have been fundamental in generating high-fidelity synthetic data. However, these methods are computationally intensive and may not always prioritize effective representation learning, particularly in scenarios where generating high-detail images is unnecessary. In contrast, discriminative methods, particularly contrastive learning [2,4,44], have emerged as leaders in the field of self-supervised learning. These methods focus on learning representations by contrasting instances within the same dataset. By reducing the distance between similar instances (positive pairs) and increasing distances between dissimilar instances (negative pairs), contrastive learning effectively captures semantic relationships in data without relying on labeled examples. Techniques like SimCLR [2], MoCo [3,14], and BarlowTwins [46] exemplify this approach by leveraging large-scale negative sampling, momentum-updated memory banks, and cross-correlation matrices to enhance representation learning. These methodologies not only improve model scalability and generalization but also facilitate robust feature extraction suitable for diverse applications. Furthermore, various auxiliary tasks have been explored to guide representation learning. Techniques such as relative patch prediction [7,8], colorization [26,47], image inpainting [33], geometric transformations [9,11], and image super-resolution [27] aim to augment learning by imposing additional constraints on the learning process. Despite their structured architectures [24], these methods often underperform compared to the superior results achieved by contrastive learning approaches [3,39]. The triplet loss approach, initially devised for tasks like person re-identification and face recognition [6,35], has evolved into a cornerstone of contrastive learning. This methodology focuses on selecting informative triplets—comprising an anchor, a positive example, and a negative example—to enhance the discriminative power of learned representations [16,41,45]. Recent adaptations like truncated triplet loss [44] and Trip-ROMA [28] further refine this approach, demonstrating its flexibility and effectiveness across various learning scenarios. These advancements include strategies for enhancing triplet selection across batches [43], integrating unsupervised triplet loss-based learning into broader frameworks [41], and mitigating the effects of false negatives in triplet loss optimization [34]. In conclusion, contrastive learning and triplet loss methodologies represent significant advancements in representation learning, offering robust solutions for self-supervised learning tasks. These approaches not only enhance the interpretability and efficiency of learned representations but also pave the way for innovative applications in image retrieval, classification, and beyond. As the field continues to evolve, further exploration and refinement of these methodologies promise continued progress in leveraging unlabelled data for sophisticated learning tasks.

3 Algorithm

We introduce an algorithm leveraging a contrastive approach to enhance the similarity between similar data and reduce the similarity between dissimilar data. However, a significant challenge in this approach is identifying negative instances in the absence of labels [14]. To address this, we propose a triplet approach where for each x in the dataset, we construct a triplet (x_a, x_p, x_n). Here, the novel aspect lies in creating the negative x_n, which needs to be sufficiently challenging for effective convergence in the triplet loss model. As common in deep learning literature [2,4,13,14,46], given a batch of elements x from the dataset, x_a and x_p are two instances derived from an augmentation method (see Fig. 2a). To resolve the issue of selecting negative examples, we train an unsupervised binary classification algorithm using the feature space representations z_a and z_p (see Fig. 2a) obtained from an encoder. Specifically, for each element $z_a[i]$, we identify elements in z_p that are classified similarly by our binary classifier, forming a set z'_p (see Fig. 2b). This set z'_p includes elements indistinguishable from our classifier, comprising a mix of semantically similar (positive) elements to $z_a[i]$ and predominantly negative data. Inspired by the concept of hard negative sampling, which corresponds to false-positive samples that are determined by training data [1,37], we compute distances between $z_a[i]$ and all elements in z'_p, retaining only the k closest ones (see Fig. 2c). These elements represent a filtered subset of z'_p, focusing on those most challenging to differentiate. To ensure we do not select the closest positive example, we exclude the closest element and compute the centroid of the remaining points. This centroid effectively encapsulates characteristics from both positive and many negative instances, serving as a robust representation of the filtered dataset. Its ability to challenge discriminative models stems from its composite nature, enhancing the model's ability to differentiate between similar and dissimilar instances.

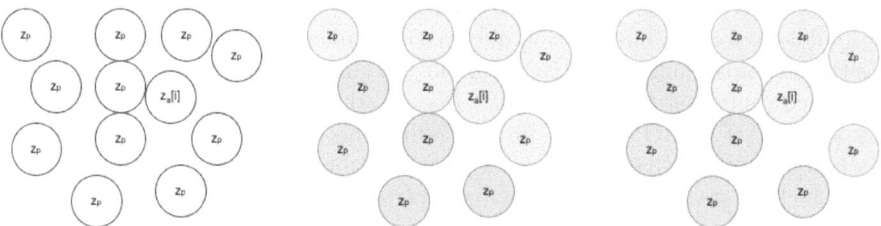

(a) x_a and x_p in feature spaces to train binary classification

(b) binary classification of z_a and z_p with color red and blue

(c) the k-closest ones values close to $z_a[i]$ in green excluding the closest one

Fig. 2. Iterative process to create negative like centroid of green values (Color figure online)

Mathematically, our method is expressed as:

$$\mathcal{L}_1(x_a, x_p, x_n) = \max\left(\text{sim}(z_a, z_n) - \text{sim}(z_a, z_p) + m, 0\right) \quad (1)$$

Here, $z_a = f(\tau_1(x))$ represents the anchor, and $z_p = f(\tau_2(x))$ represents the positive sample. The function f is an encoder neural network, while τ_1 and τ_2 are augmentation transforms drawn from the set T. The function sim() measures the similarity between two vectors, with cosine similarity used by default.

The i-th element of z_n is computed as:

$$z_n[i] = \text{Centroid}(\text{k-nearest}_{-1}(z'_p))$$

Here, the vector z'_p is obtained from the elements of z_p that have the same binary label as $z_a[i]$, provided by a clustering algorithm (GaussianMixture) trained for binary classification using $[z_a, z_p]$. The term k-nearest$_{-1}(z'_p)$ refers to the k-nearest elements of z_p to $z_a[i]$ excluding the closest one. The function Centroid then calculates the centroid of these $(k-1)$-nearest elements.

Finally, the total loss function is defined as:

$$\mathcal{L}_{\text{oss}} = \mathbb{E}_x\left(\mathcal{L}_1(x_a, x_p, x_n)\right) \quad (2)$$

This proposed solution called *NesTrip* addresses the limitations of existing methods by introducing a more effective way of constructing negative representations, thereby aiming to enhance overall performance in representation learning, particularly in Content-Based Image Retrieval and Classification tasks.

4 Main Results

We are going to present the methodology to train our algorithm and our results:

4.1 Training

Data Augmentation: One type of augmentation involves spatial/geometric transformation of data, such as cropping and resizing (with horizontal flipping), rotation [11], and cutout [5]. The other type of augmentation involves appearance transformation, such as color distortion (including color dropping, brightness, contrast, saturation, hue) [17,38], Gaussian blur, and Sobel filtering.

Algorithm: Our algorithm is based on [2]. The final algorithm can be found in Algorithm 1.

Datasets: We use two different datasets to validate the results. The CIFAR-10 dataset comprises 60,000 32 × 32 color images categorized into 10 classes, each containing 6,000 images. It is divided into 50,000 training images and 10,000 test images [25], and The SVHN (Street View House Numbers) dataset is a real-world image dataset specifically designed for developing machine learning and object recognition algorithms with minimal data preprocessing and formatting requirements. It consists of images containing digits, with 10 classes representing each digit from 0 to 9. The dataset is split into 73,257 digits for training, and 26,032 digits for testing [31].

Metrics:

- Mean Average Precision (MAP) is a crucial metric in image retrieval tasks, providing a comprehensive measure of a system's effectiveness across multiple queries. It assesses the average precision at each relevant image's position in the ranked list and computes the mean of these values. Relevant images are defined based on query relevance, and precision is calculated by dividing the number of relevant images retrieved up to a certain position by the total number of retrieved images up to that position. To calculate MAP@K, a variant of MAP where only the top K retrieved items are considered, you can use the following formula:

$$\text{MAP@K} = \frac{1}{|Q|} \sum_{q=1}^{|Q|} \frac{\sum_{k=1}^{K} \text{Precision@k}_q \times \text{Relevance}(k)}{\min(K, |R_q|)}$$

  where: $|Q|$ is the total number of queries, Precision@k_q is the precision at position k for query q, Relevance(k) is a binary indicator function that is 1 if the item at position k is relevant and 0 otherwise, $|R_q|$ is the number of relevant items for query q, and K is the cutoff rank.
- Accuracy, Recall, Precision, and F1-score are fundamental metrics for evaluating classification tasks. Accuracy measures the proportion of correctly classified instances among all instances, providing an overall assessment

of the model's performance. Recall quantifies the proportion of true positive instances correctly identified by the model among all actual positive instances. Precision measures the proportion of true positive instances among all instances predicted as positive, offering insights into the model's precision in positive predictions. F1-score, the harmonic mean of precision and recall, balances the trade-off between precision and recall, providing a single metric that reflects both measures' performance. These metrics collectively offer a comprehensive understanding of the classification model's effectiveness in correctly identifying instances belonging to different classes.

Evaluation: The evaluation was carried out using two different methods. Firstly, the CBIR method was employed, where the last output layer of the encoder was used to generate a feature vector for each image. Subsequently, the closest images in the training set were retrieved for each image in the test set, aiming to measure the results of the k nearest neighbors using the Mean Average Precision at K (MAP@K) metric. Secondly, a linear evaluation was conducted. In this approach, only a linear layer was added to the encoder, and then the model was retrained to perform classification using the available labels while keeping the encoder weights frozen.

Other protocols: Our encoder is based in Deep Residual Network (Resnet50) [15]. The batch size is 64. The k-nearest neighbor used is 15. The maximum epoch is 200, we use stochastic gradient descent with a learning rate of 0.6 and a cosine learning rate decay schedule.

4.2 Baselines

We will compare our method with four notable state-of-the-art self-supervised techniques.

- **SimCLR** [2]: is a simple framework for contrastive learning of visual representations. It involves applying two different data augmentations, $\tau \sim \mathcal{T}$ and $\tau' \sim \mathcal{T}$, to each data sample, producing two correlated views. A base encoder and a projection head are trained to maximize agreement using a contrastive loss. After training, the projection head is removed, and the encoder is used to generate representations, denoted as h, for downstream tasks. A key aspect of SimCLR is the use of a learnable nonlinear transformation between the representation and the contrastive loss, which significantly improves the learned representations.
- **SimSiam** [4]: is designed to maximize the similarity between two augmentations of the same image while preventing collapsing solutions. It uses two augmented views of an image processed by the same encoder network (comprising a backbone and a projection MLP). One view goes through a prediction MLP, while a stop-gradient operation is applied to the other. The goal is to maximize similarity between the two outputs. SimSiam does not use negative pairs or a momentum encoder. The authors show the existence of collapsing solutions and highlight the importance of the stop-gradient operation in

avoiding these, suggesting a different optimization problem than traditional contrastive learning.
- **BYOL** [13]: is a method for self-supervised image representation learning that uses two neural networks, termed the online and target networks, which learn from each other. The online network is trained to predict the target network's representation of an augmented view of the same image. Simultaneously, the target network is updated using a slow-moving average of the online network. This slow-moving average encourages the online network to encode increasing amounts of information and reduces the likelihood of collapsed solutions.
- **BarlowTwins** [46]: introduces an objective function that avoids collapse by evaluating the cross-correlation matrix between the outputs of two identical networks fed with distorted versions of the same sample. The objective is to make this matrix as close to the identity matrix as possible, ensuring that the embedding vectors of the distorted versions are similar while minimizing redundancy among their components.

Table 1. MAP Results

Model with CIFAR-10	1000	100	10	1
SimCLR [2]	0.687	0.7732	0.8221	0.881
SimSiam [4]	0.691	0.8054	0.8475	0.904
BYOL [13]	0.6917	0.7832	0.8377	0.905
BarlowTwins [46]	0.4323	0.5753	0.6689	0.791
NesTrip	0.7362	0.8272	0.8601	0.905
Model with SVHN	1000	100	10	1
SimCLR [2]	0.2593	0.3899	0.508	0.657
SimSiam [4]	0.5188	0.7177	0.812	0.874
BYOL [13]	0.3217	0.4636	0.584	0.715
BarlowTwins [46]	0.3758	0.5671	0.69	0.78
NesTrip	0.5703	0.6961	0.7579	0.822

5 Analysis

To analyze the performance of NesTrip across different experiments in content retrieval and classification, we examined two sets of tables: one presenting results in terms of MAP (Mean Average Precision) (Table 1) and another showcasing linear evaluation metrics such as accuracy, recall, precision, and F1-score (Table 2).

Table 2. Linear Evaluation Results

Model with CIFAR-10	Accuracy	Recall	Precision	F1
SimCLR [2]	0.9014	0.9014	0.9014	0.9016
SimSiam [4]	0.8587	0.8587	0.8692	0.8607
BYOL [13]	0.9028	0.9028	0.9027	0.9028
BarlowTwins [46]	0.8328	0.8328	0.8331	0.8328
NesTrip	0.9284	0.9284	0.9286	0.9283
Model with SVHN	Accuracy	Recall	Precision	F1
SimCLR [2]	0.8130	0.8130	0.8138	0.8127
SimSiam [4]	0.2233	0.2233	0.4027	0.1127
BYOL [13]	0.8090	0.8090	0.8102	0.8089
BarlowTwins [46]	0.8456	0.8456	0.8468	0.8457
NesTrip	0.8563	0.8563	0.8577	0.8565

5.1 MAPS Results

On the CIFAR-10 dataset, NesTrip demonstrated outstanding performance, surpassing the compared models in all scenarios MAP@1000, MAP@100, and MAP@10. This indicates that NesTrip achieves superior average precision when retrieving relevant images in the top-ranking positions, which is crucial for tasks where the initial retrievals are most important. Although SimSiam and BYOL showed close performances in MAP@1, NesTrip led in the most critical positions. On the SVHN dataset, NesTrip also demonstrated robust performance, particularly in MAP@1000, where it outperformed all compared models, indicating its strength in maintaining precision when retrieving a large number of images. However, in MAP@100, MAP@10, and MAP@1, SimSiam presented slightly higher results, though NesTrip remained competitive. This consistent performance across different retrieved set sizes showcases NesTrip's effectiveness in image retrieval tasks.

5.2 Linear Evaluation Results

In linear evaluation, NesTrip excelled across all metrics (accuracy, recall, precision, and F1-score) on both CIFAR-10 and SVHN, demonstrating its superiority in supervised classification tasks. On CIFAR-10, NesTrip achieved an accuracy of 92.84%, surpassing all compared models, highlighting its effectiveness in image classification. This strong performance was also reflected in the SVHN dataset, where NesTrip led in all metrics, with an accuracy of 85.63%, slightly higher than that of BarlowTwins.

NesTrip's superior performance in linear evaluation, coupled with its competitive results in MAP metrics, underscores its capability to be effective in both image retrieval and classification tasks, reaffirming its position as a robust and versatile model in computer vision tasks.

Algorithm 1. Triplet Loss Training Algorithm

1: **Input:** Unlabeled dataset **X**
2: **Output:** Trained model
3: Initialize encoder network f
4: Define hyperparameters: learning rate α, margin m
5: **while** Training not converged **do**
6:     **for** each batch **x** in **X do**
7:         Apply data transformations τ_1 and τ_2 to create $\tau_1(x)$ and $\tau_2(x)$
8:         Compute embeddings $\mathbf{z_a} = f(\tau_1(x))$ and $\mathbf{z_p} = f(\tau_2(x))$
9:         Train a binary classification GaussianMixture model on $\mathbf{z_a}$ and $\mathbf{z_p}$
10:         **for** each element $z_a[i]$ in $\mathbf{z_a}$ **do**
11:             Select $\mathbf{z'}_p$, the elements of $\mathbf{z_p}$ with the same binary label as $z_a[i]$
12:             Compute distances between z_{p_i} and each element in $\mathbf{z'}_p$, and select the k nearest neighbors
13:             Exclude $z_a[i]$ itself from the neighbors, and compute the centroid $\mathbf{z_n}[i]$ of the remaining neighbors
14:         **end for**
15:         Calculate $\mathcal{L}_1$ using Eq. 1 with $\mathbf{z_a}$, $\mathbf{z_p}$, and $\mathbf{z_n}$
16:         Calculate total loss $\mathcal{L}_{\text{oss}}$ using a final loss function (e.g., including $\mathcal{L}_1$ and other regularization terms)
17:         Update model parameters using backpropagation
18:     **end for**
19: **end while**
20: **Return:** Trained model

6 Conclusions

The advancement of artificial neural networks has revolutionized various fields including economics, medicine, and industry, driven primarily by their ability to learn from labeled data. However, the reliance on annotated data for model training poses significant challenges due to its high cost and expert involvement. To mitigate these challenges, Self-Supervised Representation Learning (SSRL) has emerged as a viable alternative, aiming to train robust feature extractors without the need for labeled data. Contrastive Self-Supervised Learning (CSL), a prominent method within SSRL, has garnered considerable attention for its effectiveness in training models without explicit labels. By enforcing the model to bring similar samples closer together and push dissimilar samples apart in the feature space, CSL optimizes feature representations to enhance the model's capability in discerning subtle differences within the data. Our approach addresses key challenges in CSL by introducing explicit negative sampling strategies using a binary classification algorithm in the feature space. This method allows us to distinguish between data with similar and dissimilar features, effectively enhancing the discriminative power of the model. By leveraging Triplet Loss, originally proposed for tasks like person re-identification and face recognition, we further refine our approach to maximize the similarity between an anchor and positive samples while minimizing it with negative samples, subject to a specified margin constraint. Through comprehensive experiments on CIFAR-10 and SVHN datasets, we demonstrated that our approach outperforms existing methods in both content-based image retrieval (CBIR) and classification tasks. Our method achieved superior performance metrics such as mean average precision (MAP), accuracy, recall, precision, and F1-score, underscoring its effectiveness in practical applications. Our study contributes to advancing SSRL by proposing a robust framework that combines explicit negative sampling and Triplet Loss to improve representation learning. By bridging the gap between supervised and unsupervised learning paradigms, our approach not only enhances model generalization but also offers scalable solutions for training neural networks on large-scale datasets at minimal annotation cost. Future research directions may explore further optimizations and extensions of our method across diverse domains and datasets.

Acknowledgments. This work was supported in part by the Agencia Nacional de Investigación y Desarrollo (doctoral scholarship 21221059).

References

1. Bucher, M., Herbin, S., Jurie, F.: Hard negative mining for metric learning based zero-shot classification. In: Hua, G., Jégou, H. (eds.) ECCV 2016. LNCS, vol. 9915, pp. 524–531. Springer, Cham (2016). https://doi.org/10.1007/978-3-319-49409-8_45
2. Chen, T., Kornblith, S., Norouzi, M., Hinton, G.: A simple framework for contrastive learning of visual representations. In: International Conference on Machine Learning, pp. 1597–1607. PMLR (2020)
3. Chen, X., Fan, H., Girshick, R., He, K.: Improved baselines with momentum contrastive learning. arXiv preprint arXiv:2003.04297 (2020)
4. Chen, X., He, K.: Exploring simple siamese representation learning. In: Proceedings of the IEEE/CVF Conference on Computer Vision and Pattern Recognition, pp. 15750–15758 (2021)
5. DeVries, T., Taylor, G.W.: Improved regularization of convolutional neural networks with cutout. arXiv preprint arXiv:1708.04552 (2017)
6. Ding, S., Lin, L., Wang, G., Chao, H.: Deep feature learning with relative distance comparison for person re-identification. Pattern Recogn. **48**(10), 2993–3003 (2015)
7. Doersch, C., Gupta, A., Efros, A.A.: Unsupervised visual representation learning by context prediction. In: Proceedings of the IEEE International Conference on Computer Vision, pp. 1422–1430 (2015)
8. Doersch, C., Zisserman, A.: Multi-task self-supervised visual learning. In: Proceedings of the IEEE International Conference on Computer Vision, pp. 2051–2060 (2017)

9. Dosovitskiy, A., Springenberg, J.T., Riedmiller, M., Brox, T.: Discriminative unsupervised feature learning with convolutional neural networks. Adv. Neural Inform. Process. Syst. **27** (2014)
10. Ericsson, L., Gouk, H., Loy, C.C., Hospedales, T.M.: Self-supervised representation learning: introduction, advances, and challenges. IEEE Signal Process. Mag. **39**(3), 42–62 (2022)
11. Gidaris, S., Singh, P., Komodakis, N.: Unsupervised representation learning by predicting image rotations. arXiv preprint arXiv:1803.07728 (2018)
12. Goodfellow, I., et al.: Generative adversarial nets. Adv. Neural Inform. Process. Syst. **27** (2014)
13. Grill, J.B., et al.: Bootstrap your own latent-a new approach to self-supervised learning. Adv. Neural. Inf. Process. Syst. **33**, 21271–21284 (2020)
14. He, K., Fan, H., Wu, Y., Xie, S., Girshick, R.: Momentum contrast for unsupervised visual representation learning. In: Proceedings of the IEEE/CVF Conference on Computer Vision and Pattern Recognition, pp. 9729–9738 (2020)
15. He, K., Zhang, X., Ren, S., Sun, J.: Deep residual learning for image recognition. In: Proceedings of the IEEE Conference on Computer Vision and Pattern Recognition, pp. 770–778 (2016)
16. Hermans, A., Beyer, L., Leibe, B.: In defense of the triplet loss for person re-identification. arXiv preprint arXiv:1703.07737 (2017)
17. Howard, A.G.: Some improvements on deep convolutional neural network based image classification. arXiv preprint arXiv:1312.5402 (2013)
18. Huynh, T., Kornblith, S., Walter, M.R., Maire, M., Khademi, M.: Boosting contrastive self-supervised learning with false negative cancellation. In: Proceedings of the IEEE/CVF Winter Conference on Applications of Computer Vision, pp. 2785–2795 (2022)
19. Jaiswal, A., Babu, A.R., Zadeh, M.Z., Banerjee, D., Makedon, F.: A survey on contrastive self-supervised learning. Technologies **9**(1), 2 (2020)
20. Jing, L., Tian, Y.: Self-supervised visual feature learning with deep neural networks: A survey. IEEE Trans. Pattern Anal. Mach. Intell. **43**(11), 4037–4058 (2020)
21. Karamanolakis, G., Mukherjee, S., Zheng, G., Hassan, A.: Self-training with weak supervision. In: Proceedings of the 2021 Conference of the North American Chapter of the Association for Computational Linguistics: Human Language Technologies, pp. 845–863 (2021)
22. Kaya, M., Bilge, H.Ş: Deep metric learning: a survey. Symmetry **11**(9), 1066 (2019)
23. Kingma, D.P., Welling, M.: Auto-encoding variational bayes. arXiv preprint arXiv:1312.6114 (2013)
24. Kolesnikov, A., Zhai, X., Beyer, L.: Revisiting self-supervised visual representation learning. In: Proceedings of the IEEE/CVF Conference on Computer Vision and Pattern Recognition, pp. 1920–1929 (2019)
25. Krizhevsky, A.: Learning multiple layers of features from tiny images. Tech. rep. (2009)
26. Larsson, G., Maire, M., Shakhnarovich, G.: Learning representations for automatic colorization. In: Leibe, B., Matas, J., Sebe, N., Welling, M. (eds.) ECCV 2016 LNCS, vol. 9908, pp. 577–593. Springer, Cham (2016). https://doi.org/10.1007/978-3-319-46493-0_35
27. Ledig, C., et al.: Photo-realistic single image super-resolution using a generative adversarial network. In: Proceedings of the IEEE Conference on Computer Vision and Pattern Recognition, pp. 4681–4690 (2017)
28. Li, W., et al.: Trip-roma: Self-supervised learning with triplets and random mappings. Trans. Mach. Learn. Res. (2022)

29. Li, W., et al.: Trip-ROMA: Self-supervised learning with triplets and random mappings. Trans. Mach. Learn. Res. (2023). https://openreview.net/forum?id=MR4glug5GU
30. Li, Z., Liu, F., Yang, W., Peng, S., Zhou, J.: A survey of convolutional neural networks: analysis, applications, and prospects. IEEE Trans. Neural Netw. Learn. Syst. **33**(12), 6999–7019 (2021)
31. Netzer, Y., Wang, T., Coates, A., Bissacco, A., Wu, B., Ng, A.Y.: Reading digits in natural images with unsupervised feature learning. In: NIPS Workshop on Deep Learning and Unsupervised Feature Learning 2011 (2011)
32. Oh Song, H., Xiang, Y., Jegelka, S., Savarese, S.: Deep metric learning via lifted structured feature embedding. In: Proceedings of the IEEE Conference on Computer Vision and Pattern Recognition, pp. 4004–4012 (2016)
33. Pathak, D., Krahenbuhl, P., Donahue, J., Darrell, T., Efros, A.A.: Context encoders: feature learning by inpainting. In: Proceedings of the IEEE Conference on Computer Vision and Pattern Recognition, pp. 2536–2544 (2016)
34. Robinson, J., Chuang, C.Y., Sra, S., Jegelka, S.: Contrastive learning with hard negative samples. arXiv preprint arXiv:2010.04592 (2020)
35. Schroff, F., Kalenichenko, D., Philbin, J.: Facenet: a unified embedding for face recognition and clustering. In: Proceedings of the IEEE Conference on Computer Vision And Pattern Recognition, pp. 815–823 (2015)
36. Shahid, N., Rappon, T., Berta, W.: Applications of artificial neural networks in health care organizational decision-making: A scoping review. PLoS ONE **14**(2), e0212356 (2019)
37. Simo-Serra, E., Trulls, E., Ferraz, L., Kokkinos, I., Fua, P., Moreno-Noguer, F.: Discriminative learning of deep convolutional feature point descriptors. In: Proceedings of the IEEE International Conference on Computer Vision, pp. 118–126 (2015)
38. Szegedy, C., et al.: Going deeper with convolutions. In: Proceedings of the IEEE Conference on Computer Vision and Pattern Recognition, pp. 1–9 (2015)
39. Tian, Y., Sun, C., Poole, B., Krishnan, D., Schmid, C., Isola, P.: What makes for good views for contrastive learning? Adv. Neural. Inf. Process. Syst. **33**, 6827–6839 (2020)
40. Tkáč, M., Verner, R.: Artificial neural networks in business: two decades of research. Appl. Soft Comput. **38**, 788–804 (2016)
41. Turpault, N., Serizel, R., Vincent, E.: Semi-supervised triplet loss based learning of ambient audio embeddings. In: ICASSP 2019-2019 IEEE International Conference on Acoustics, Speech and Signal Processing (ICASSP), pp. 760–764. IEEE (2019)
42. Vincent, P., Larochelle, H., Bengio, Y., Manzagol, P.A.: Extracting and composing robust features with denoising autoencoders. In: Proceedings of the 25th International Conference on Machine Learning, pp. 1096–1103 (2008)
43. Wang, G., Wang, G., Zhang, X., Lai, J., Yu, Z., Lin, L.: Weakly supervised person re-id: differentiable graphical learning and a new benchmark. IEEE Trans. Neural Netw. Learn. Syst. **32**(5), 2142–2156 (2020)
44. Wang, G., Wang, K., Wang, G., Torr, P.H., Lin, L.: Solving inefficiency of self-supervised representation learning. In: Proceedings of the IEEE/CVF International Conference on Computer Vision, pp. 9505–9515 (2021)
45. Wang, X., Zhang, H., Huang, W., Scott, M.R.: Cross-batch memory for embedding learning. In: Proceedings of the IEEE/CVF Conference on Computer Vision and Pattern Recognition, pp. 6388–6397 (2020)

46. Zbontar, J., Jing, L., Misra, I., LeCun, Y., Deny, S.: Barlow twins: self-supervised learning via redundancy reduction. In: International Conference on Machine Learning, pp. 12310–12320. PMLR (2021)
47. Zhang, R., Isola, P., Efros, A.A.: Colorful image colorization. In: Leibe, B., Matas, J., Sebe, N., Welling, M. (eds.) ECCV 2016. LNCS, vol. 9907, pp. 649–666. Springer, Cham (2016). https://doi.org/10.1007/978-3-319-46487-9_40

Seed-Based Superpixel Re-Segmentation for Improving Object Delineation

Lucca S. P. Lacerda[1], Felipe C. Belém[1,2],
Zenilton Kleber Gonçalves do Patrocínio Júnior[1], Alexandre X. Falcão[2],
and Silvio J. F. Guimarães[1]

[1] Pontifical Catholic University of Minas Gerais, Belo Horizonte, Brazil
silvio.jamil@gmail.com
[2] University of Campinas, Campinas, Brazil

Abstract. Superpixels are an effective image segmentation strategy, whose results apply and assist in classification tasks. However, by aiming for maximum performance with a minimum quantity of regions, the object delineation may be compromised, demanding re-segmentation. In this paper, we propose and evaluate three re-segmentation strategies that rely on a novel and accurate superpixel framework, named SICLE, and require minimal intervention from the user (*i.e.*, up to three clicks). Our qualitative and quantitative results show significant improvement over the previous superpixel segmentation for separating the object of interest from the background.

Keywords: Seed-based approach · Superpixel re-segmentation · Graph

1 Introduction

One may define superpixel as a disjoint collection of connected picture elements (*i.e.*, pixels) that share a common property, such as color or texture. Often, numerous superpixels are generated such that the object of interest may be accurately built from its comprising parts. Furthermore, from such grouping, it is possible to extract high-level information while reducing the workload magnitude. Consequently, several works recur to superpixel segmentation methods for assisting to their particular tasks [3].

The authors thank the Pontifícia Universidade Católica de Minas Gerais – PUC-Minas, Coordenação de Aperfeiçoamento de Pessoal de Nível Superior – CAPES – (Grant PROAP 88887.842889/2023-00 – PUC/MG, Grant STIC-AMSUD 88887.878869/2023-00, Grant PDPG 88887.708960/2022-00 – PUC/MG - Informática, and Finance Code 001), the Conselho Nacional de Desenvolvimento Científico e Tecnológico – CNPq (Grants 407242/2021-0, 306573/2022-9, 442950/2023-3 and 304711/2023-3), Fundação de Apoio à Pesquisa do Estado de Minas Gerais – FAPEMIG (Grant APQ-01079-23, Grant APQ-05058-23 and PCE-00417-24) and Fundação de Apoio à Pesquisa do Estado de São Paulo – FAPESP (Grant 2023/14427-8).

© The Author(s), under exclusive license to Springer Nature Switzerland AG 2025
R. J. Barrientos and S. A. Velastin (Eds.): CIARP 2024, LNCS 15368, pp. 148–161, 2025.
https://doi.org/10.1007/978-3-031-76607-7_11

(a) original (b) two regions (c) three regions (d) selection (e) new regions

Fig. 1. Example of re-segmentation: (a) original image; (b) image with two superpixels; (c) image with three superpixels; (d) identification of the selected superpixel to be re-segmented (in blue); and (e) result of the re-segmentation by our strategy. (Color figure online)

One example of such methods is the *Superpixels through Iterative CLEarcutting* (SICLE) [5], which uses a three-step pipeline strategy for generating superpixels: (i) seed oversampling; (ii) superpixel generation; and (iii) seed removal. In (i), a set of initial points (*i.e.*, seeds) is built by selecting a significantly higher quantity than the desired number of superpixels. Then, in step (ii), superpixels are generated through the computation of optimum-path forests rooted at the seeds by the *Image Foresting Transform* (IFT) [11]. Finally, in step (iii), the seeds are ranked based on a mathematical criterion (*i.e.*, relevance), and a portion of those lowest ranked are removed. Then, steps (ii) and (iii) are iteratively performed until the desired number of superpixels is reached. This strategy not only achieves state-of-the-art delineation in segmentation (according to [3]) but also permits generating several segmentations of different superpixel resolutions, named scales, in a single execution.

Another IFT-based superpixel method is the *Recursive Iterative Spanning Forest* (RISF) [13], which generates a hierarchical superpixel segmentation by executing several instances of the *Iterative Spanning Forest* (ISF) [19] framework. ISF follows a similar pipeline as SICLE, defined by three independent steps: (i) seed sampling; (ii) superpixel generation; and (iii) seed recomputation. However, differently from SICLE, step (i) selects some seeds approximately to the number of desired superpixels, and step (iii) minimizes superpixel intra-dissimilarity by reallocating seeds to a novel position. Consequently, the algorithm stops when achieving a strict number of iterations.

Superpixel properties should be attained within the fewest regions as possible [14], especially the ability to build the object accurately from its inner superpixels. However, as seen in Fig. 1, such quantity can severely compromise delineation. Thanks to user selection, it was possible to provide a nice delineation of the brown horse, as illustrated in Fig. 1(e).

It is important to emphasize that, although the minimal quantity of superpixels often results in several leaks, these errors occur whenever color distinction is again minimal, independently of the amount required. Thus, a new segmentation is performed on top of the previous to correct such a result, usually erroneous segmentation. This procedure is named **re-segmentation** and is widely used in literature [10,12,17].

Similarly to superpixels, correcting a previous segmentation should be achieved with minimal user effort. Thus, instead of reevaluating all regions, which would require a new segmentation from scratch and would not guarantee the desired correction, the user may indicate the "*problematic*" superpixel that needs correction, resulting in a more effective approach. Thus the correct borders remained unchanged from such isolation, while the wrong ones (in the desired region) are susceptible to improvement. Yet, the correction procedure must also demand minimal user intervention (*e.g.*, few clicks), even when performed in a part of the image.

By isolating the problematic superpixel, and considering the user clicks, one could recur to one of three strategies within the former: (i) use a simple, but effective, seed-based approach; (ii) apply the primer method; or (iii) use a hierarchical approach aiming for the optimal grouping. As the segmentation is an ill-posed problem since the context is relevant the human (expert) in some cases must be included in the loop with minimal effort. In this work, we study these three approaches while requiring minimal user intervention, limited to one intervention in approaches (ii) and (iii); and three user interferences in approach (i).

Before applying the re-segmentation, two main steps are needed: (a) superpixel segmentation; and (b) selection of the problematic superpixel. For the superpixel segmentation, we can apply any method for generating superpixels. Here, for instance, we have used SICLE thanks to its high delineation quality. However, if there is any problem in the object delineation, the user may select the problematic one for its improvement through re-segmentation. In this work, we have studied three different seed-based methods, one for each strategy: IFT, guided by user markers, SICLE, user-marker-free, and RISF, hierarchical and user-marker-free.

Quantitative and qualitative results show that the proposed approach achieved better results than generating the desired amount directly from SICLE. Thus, the main contributions of this work are two-fold: (i) a proposal of seed-based superpixel re-segmentation methods for improving object delineation by incorporating the human-in-the-loop for identifying the problematic superpixel; and (ii) an assessment involving non-hierarchical and hierarchical methods for helping the re-segmentation.

This work is organized as follows. Section 2 introduces the mathematical concepts required in this work, whose details are presented in Sect. 3. Experimental setup and results are shown in Sect. 4 and we conclude this work in Sect. 5.

2 Theoretical Background

2.1 Graph Notions

An *image* I can be represented as a pair $I = (\mathcal{P}, \mathbf{F})$ in which $\mathcal{P} \subseteq \mathbb{Z}^2$ denotes the set of *picture elements* (*i.e.*, pixels), and $\mathbf{F}$ maps every $p \in \mathcal{P}$ to a feature vector $\mathbf{F}(p) \in \mathbb{R}^m$, in which I is a colored image (*e.g.*, RGB or CIELAB colorspaces) when the number of spectral planes m is 3. One may create a *digraph* $\mathsf{G} =$

$(\mathcal{V}, \mathcal{E})$ from I, such that $\mathcal{V} \subseteq \mathcal{P}$ and $\mathcal{E}$ are the *vertex* and *(directed) edge* sets, respectively. Moreover, $\mathcal{E} \subset \mathcal{N}^2$ defines the *adjacency* relation between node pairs $(v_i, v_j) \in \mathcal{E}$, like the 4- or 8-adjacency.

A *path* $\pi_{s \leadsto t} = \langle s = v_1, v_2, \ldots, v_n = t \rangle$ is a finite sequence of adjacent nodes, in which $(v_i, v_{i+1}) \in \mathcal{E}$ for $1 \leq i < n$, and is *trivial* when $n = 1$. We may omit the path's *origin* by writing π_t, whenever it is irrelevant for the context. A path $\pi_p = \pi_q \cdot \langle q, p \rangle$ is an extension of π_p through the *concatenation* of π_q by the arc $(q, p) \in \mathcal{E}$.

2.2 Image Foresting Transform

The *Image Foresting Transform* (IFT) [11] is a framework for the development of image processing operators based on connectivity and has been used to reduce several tasks as optimum-path forest computations over the image graph. In this work, we consider the IFT version restricted to a *seed set* $\mathcal{S} \subset \mathcal{V}$, that generates superpixels through seed-competition for the non-seed vertices $\mathcal{V} \setminus \mathcal{S}$.

For a given arc $(s,t) \in \mathcal{E}$, it is possible to assign a non-negative *arc-cost* value $\mathbf{w}_*(s,t) \in \mathbb{R}^+$ through an *arc-cost function* $\mathbf{w}_*$. A common approach is to compute the ℓ_2-norm between the nodes' features—i.e., $\|\mathbf{F}(s) - \mathbf{F}(t)\|_2$ for $s, t \in \mathcal{V}$. Consider Π_G the set of all possible paths in G. Then, a *connectivity function* $\mathbf{f}_*$ maps every path in Π_G to a *path-cost value* $\mathbf{f}_*(\pi_t) \in \mathbb{R}^+$. One of the most effective connectivity functions for object delineation is the $\mathbf{f}_{\max}$ function:

$$\mathbf{f}_{\max}(\langle x \rangle) = \begin{cases} 0 & \text{if } x \in \mathcal{S}, \\ +\infty & \text{otherwise} \end{cases} \quad (1)$$

$$\mathbf{f}_{\max}(\pi_y \cdot \langle y, x \rangle) = \max\{\mathbf{f}_{\max}(\pi_y), \mathbf{w}_*(y, x)\}$$

Another example of connectivity function is the $\mathbf{f}_\Sigma$ function, as presented below:

$$\begin{aligned} \mathbf{f}_\Sigma(\langle x \rangle) &= \mathbf{f}_{\max}(\langle x \rangle) \\ \mathbf{f}_\Sigma(\pi_{s \leadsto y} \cdot \langle y, x \rangle) &= \mathbf{f}_\Sigma(\pi_{s \leadsto y}) + \iota(\mathbf{w}_*(s,x))^\beta + \|y - x\|_2 \end{aligned} \quad (2)$$

in which $\iota \in \mathbb{R}^+$ controls the superpixel irregularity, and $\beta \in \mathbb{R}^+$ controls the superpixel's boundary adherence. A path π_t^* is said to be *optimum* if, for any other path $\tau_t \in \Pi_\mathsf{G}$, $\mathbf{f}_*(\pi_t^*) \leq \mathbf{f}_*(\tau_t)$.

Let $\mathbf{C}$ be a *cost map* in which assigns, to every path $\pi_x \in \Pi_\mathsf{G}$, its respective path-cost value $\mathbf{f}_*(\pi_x)$. The IFT algorithm minimizes $\mathbf{C}(x) = \min_{\forall \pi_x \in \Pi_\mathsf{G}}\{\mathbf{f}_*(\pi_x)\}$ whenever $\mathbf{f}_*$ satisfies certain conditions [8]. First, the IFT assigns path-costs to all trivial paths accordingly and, then, it computes optimum paths in a non decreasing order, from the seeds to the remaining nodes in the graph. Therefore, independently if $\mathbf{f}_*$ suffices the desired properties in [8], the IFT always generates a spanning forest and, consequently, each superpixel is a tree rooted in a unique seed $s \in \mathcal{S}$. During the segmentation process, a *predecessor map* $\mathbf{P}$ is generated and defined. Such map assigns any node $x \in \mathcal{V}$ to its *predecessor* y in the optimum path $\pi_y^* \cdot \langle y, x \rangle$, or to a distinctive marker $nil \notin \mathcal{V}$—in such case, y is said to be a *root* of **V**. In this work, every seed is a root of **V**. One may see that **V** is a representation of an *optimum-path forest*, and it allows to recursively obtain the optimum-path root **R**(x) of x and its root's label **L**(**R**(x)).

2.3 Recursive Iterative Spanning Forest (RISF)

RISF is an ISF-based method, as defined in [19], an ISF-based method can operate over either pixels or voxels. It is necessary to understand that an ISF-based method primarily includes the following elements: an initial sampling strategy, an adjacency relation, a path-cost function for the differential IFT algorithm [9] computation, and a seed recomputation procedure [13].

In the ISF framework, superpixel segmentation is formulated within the more general Image Foresting Transform (IFT) framework together with an iterative scheme similar to SLIC [1], as first presented in [2]. RISF is a general formalization that extends ISF to Region Adjacency Graphs (RAGs). This formalization is called Recursive Iterative Spanning Forest (RISF) since it calls ISF multiple times with a reducing number of regions (scales), creating a new RAG from the resulting regions for each subsequent execution, up to achieving the desired number of regions. The obtained regions (superpixels/supervoxels) per scale define a hierarchical segmentation [13].

2.4 Superpixels Through Iterative Clearcutting (SICLE)

Since SICLE generates the over-segmentation used in this study, a brief explanation about it is needed to understand the partial segmentation used in the process. SICLE operates by seed oversampling and repeating connectivity-based superpixel delineation (*i.e.*, IFT executions) and, whenever is provided, object-based seed removal until it reaches the desired number of superpixels [5]. The authors in [5] present two optimized variants: SICLE-IRREG and SICLE-COMP, which provide irregularly—and compact—shaped superpixels with top delineation performance.

SICLE is a generalization of two state-of-the-art methods, *Dynamic and Iterative Spanning Forest* (DISF) [6] and *Object-based DISF* (ODISF) [7]. In contrast to classical methods, which provide fast but often fair delineation performance, SICLE can generate a multiscale segmentation on the fly with accurate border delineation, similar to hierarchical approaches. However, differently from the latter, SICLE can correct previous incorrect estimations by reestablishing competition throughout iterations. If the user possesses an estimation of the object's location, one may improve significantly the performance, as reported by the authors [5].

3 Seed-Based Superpixel Re-Segmentation

Our re-segmentation strategies are detailed in this section. For applying the re-segmentation, as can be seen in Fig. 2, the user must select the desired superpixel

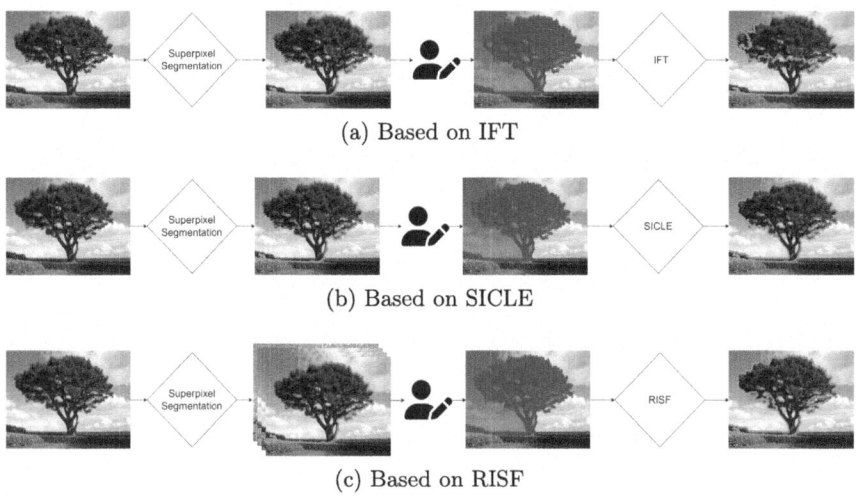

Fig. 2. Proposals for superpixel re-segmentation based on seed-based approaches.

from a set of superpixels, that can be generated by any method, but here, for instance, we have used SICLE thanks to its high delineation quality. In this work, we have studied three different seed-based methods for re-segmenting a superpixel, one for each strategy: (i) IFT guided by user markers [16]; (ii) SICLE that is user-marker-free; and (iii) RISF that is hierarchical and user-marker-free.

Although any state-of-the-art approach in superpixel segmentation may be used, the need for few but accurate superpixels is constant and can result in errors. Thus, the first step consists of generating a superpixel segmentation and maybe correcting the delineation of a selected superpixel, if one desires. To obtain such information, we developed a graphical interface that assists the user in clicking the superpixel of interest and, if necessary, two new seeds for re-segmentation. From such an intervention, we can extract (and isolate) the problematic superpixel as a mask for the next step.

From the generated mask, we assume the need for correction is internal to the superpixel rather than external since such inference resides in a proper superpixel selection. One may note that, based on such an idea, exists a portion of the superpixel borders overlapping the object boundaries, and, by defining it as a mask, such borders remain unaltered, guaranteeing such delineation. On the other hand, the non-delineated object boundaries (thus, incorrect superpixel borders) are either external or internal to the selected superpixel. For the former, it is possible to see that delineation accuracy is not harmed, since any partition does not modify the correct borders. However, the latter case allows the improvement of such incorrect superpixel borders by performing a new, but local, segmentation.

Thus, we selected three different strategies. One involves executing a new IFT execution within the boundaries of the selected superpixel and starting from the selected seeds. This strategy is illustrated in Fig. 2(a). The motivation for such an approach relies on the efficacy of the IFT framework for delineation [11] and, furthermore, being a **localized competition**, one could argue in favor of a simpler and faster approach for approximating the desired delineation, especially due to the user intervention in selecting the seeds.

Another strategy consists of applying SICLE within the superpixel of interest to achieve a better delineation of the object, at the expense of being slightly slower than a single IFT execution. As illustrated in Fig. 2(b), such an approach reduces even more the user effort to a single click since it does not need user-selected seeds. Furthermore, from the reported performance in delineation for different images and objects, a more localized competition, as one could argue, would strongly favor an already effective and efficient method.

Finally, the third approach recurs to RISF for a hierarchical segmentation of the superpixel of interest. The main ideas are illustrated in Fig. 2(c). More formally, by considering the first scale from the initial superpixel segmentation, and respecting the defined mask, RISF aims for the ideal superpixel grouping, through region-merging, for approximating the object borders. Similarly to SICLE, this strategy offers a better delineation but requires more computational time than IFT. On the other hand, and differently from both previous re-segmentation strategies, RISF guarantees that the object (and its borders) can be accurately delineated if a proper selection is found.

Independently of the approach chosen, the final localized segmentation generates more than one region, which demands relabeling for coherence. It is important to emphasize that only IFT and SICLE achieve $N+1$ superpixels, while RISF cannot assure precisely.

4 Experimental Results

The main goal of this work is to propose strategies for improving the object delineation considering the selection of a *problematic* superpixel. Thus, the user must indicate which superpixel does not achieve good delineation according to them.[1] To illustrate the quality of the results, we provide a quantitative analysis by considering a dataset containing one object. For that, *Parasites* [4] dataset (77 images) was used for evaluation, due to the challenging task of segmenting helminth eggs in a liquid solvent, implying in less perceptible borders (*i.e.*, low gradient).

Regarding the qualitative analysis, we present several examples considering both the *Parasites* dataset and some natural images (extracted, in special, from BSDS500 and Weizmann datasets).

4.1 Quantitative Analysis

The experimental protocol can be described as follows. Firstly, we apply a superpixel segmentation method forcing us to obtain only a few

---

[1] Code available https://github.com/IMScience-PPGINF-PucMinas/I-SiCLE.

Table 1. Average results considering images whose initial SICLE segmentation obtained a DICE coefficient in between [0.5, 0.9]. The best result of each metric is depicted in bold.

Metric	Baseline	IFT	RISF	SICLE
DICE ↑	0.793 ± 0.151	0.787 ± 0.125	0.741 ± 0.223	**0.837 ± 0.146**
BR ↑	0.796 ± 0.126	0.746 ± 0.149	0.774 ± 0.139	**0.800 ± 0.123**
UE ↓	0.117 ± 0.096	0.153 ± 0.110	0.115 ± 0.114	**0.078 ± 0.066**
ASA ↑	0.941 ± 0.048	0.923 ± 0.055	0.943 ± 0.057	**0.961 ± 0.033**

regions/superpixels(for instance, two). Without loss of generalization, when the superpixel segmentation provides few segments this method could be considered as a segmentation method. Here, we have applied SICLE thanks to its delineation quality (see [3] for more details). From this segmentation, the user must select the superpixel to be re-segmented and indicate the markers, if needed, since each marker will be in separated new regions.

As previously stated, we evaluate three distinct strategies (IFT, SICLE, and RISF) and consider a single SICLE execution (demanding only 3 superpixels) as the baseline. The motivation lies in estimating the improvement achieved by the pipeline in contrast to directly generating the necessary superpixel quantity. To provide quantitative measurement, several measures are adopted. *Boundary Recall* (BR) [18] calculates the ratio of superpixel borders overlapping the object boundaries while *Under-segmentation Error* (UE) [15] computes the error from multiple object overlap from the generated superpixels. Inversely proportional to the latter, the *Achievable Segmentation Accuracy* (ASA) [18] estimates the maximum accuracy possible of segmenting the objects, if one knows how to select them properly. Finally, we also consider the DICE coefficient which, in this work, measures the similarity between the object of the ground-truth with the superpixel that majorly overlaps it. Aside from UE, for which a lower value is better, we aim for high values for BR, DICE, and ASA.

However, instead of considering the complete Parasites dataset, we chose a subset for evaluation using DICE for filtering. After running a superpixel segmentation approach (for instance, SICLE) to obtain only 2 superpixels, we chose the images (and their results) whose DICE lay in between 50% and 90% (*i.e.*, 30 images). We argue that, for those that achieved less than 50%, re-segmentation would not be enough, since it performed worse than a random selection. On the other hand, for those that presented more than 90% of DICE, we say that it is within the bounds of subjectivity when segmenting the object (*i.e.*, shape approximation when annotating) and, therefore, can be considered an accurate result, not requiring re-segmentation.

Table 1 shows the results obtained in the aforementioned subset of images. For DICE, it is interesting to notice that RISF, an improved IFT-based method, achieved a worse performance than IFT, implying the severe impact of incorrect region merging. Although one may argue that DICE and ASA are alike, we recall that, in this work, we measure considering the superpixel (thus, one) which majorly overlaps the object. Therefore, when analyzing ASA, the improvement of RISF over IFT may indicate that the former, on average, generates several superpixels that combined result in an accurate segmentation, but not a single one that may be classified as the object itself. In conjunction with that, the results of BR and UE show that RISF achieves a better delineation than IFT, but is still worse than the baseline.

Re-segmenting with SICLE surpasses all approaches according to every measure with a lower standard deviation. We argue that SICLE manages to build a single superpixel that comprises most of the object, seeing from the top results achieved in DICE. Moreover, if one accurately selects the object's superpixels, SICLE's re-segmentation achieves better ASA performance than the other methods. Therefore, the user effort is minimized to one click for selecting a major part of the object, and may best approximate the borders by selecting the remaining ones, if desired. Finally, by not being hierarchical as RISF, SICLE corrects the superpixel borders in the baseline, as indicated by the increased BR and reduced UE.

4.2 Qualitative Analysis

The main concept of this work is re-segmenting superpixels to provide a better object delineation. Figure 3 shows that the user must select the superpixel to be divided according to his interest. Here, the user selected the blue superpixel (for each image) as the problematic one which must be re-segmented to improve the border delineation. For example, in the three first rows of Fig. 3 (tree, elk, horses, and egret), the selected superpixels contain the object to be delineated, however, the last row (flower), the selected superpixel contains a small part of the flower which must be separated from the background. Moreover, from these results, we can argue that the SICLE produces, visually, the best results for all images, except for the elk in which IFT is better.

Figure 4 illustrates some results on the Parasite dataset, and the border difference between columns (b) and (c) is due to the SICLE's ability to correct its superpixel borders. One can see that the baseline cannot separate the object of interest from the background (impurity included). When re-segmenting with IFT, the result does not present a stable performance since it tends to sustain the baseline's errors (as indicated in Table 1). Also, RISF cannot properly merge regions to improve the segmentation, even considering SICLE's first scale. Conversely, re-segmentation using SICLE corrects all the exemplified segmentations, best approximating the object borders requiring, as mentioned, a single click, which was insufficient for the other approaches.

We tested an iterative re-segmentation approach using SICLE and examples are presented in Figs. 5 and 6. Simply put, the user generates the initial segmentation requiring only two superpixels and, at each iteration (until satisfaction), the *problematic* superpixel is selected for improvement. In those examples, the SICLE approach cannot generate the expected result, as indicated by (b). However, an accurate object segmentation would be generated if applied once

Seed-Based Superpixel Re-Segmentation for Improving Object Delineation 157

(a) original   (b) selection   (c) baseline   (d) IFT   (e) SICLE   (f) RISF

Fig. 3. Examples of re-segmentation in natural images. The original images are illustrated in (a). The selected superpixel (in blue) is showed in (b). The baseline with several superpixels is showed in (c). For the images in (d)-(f), the selected superpixel in (b) is re-segmented by using IFT, SICLE and RISF, respectively. (Color figure online)

more. Thus, the user required only two interventions, one-click for each iteration, to obtain the desired result. For these examples, a single execution of SICLE-IRREG was performed for (b) and two iterations of SICLE-COMP for (b) to (c) and (c) to (d).

Despite the good results achieved by our re-segmentation proposal, it may fail in some cases as illustrated in Fig. 7, in which instead of separating the impurities from the egg, the re-segmentation process divided the inner part of its outer part. We can argue that this may occur when a gradient exists inside the desired object.

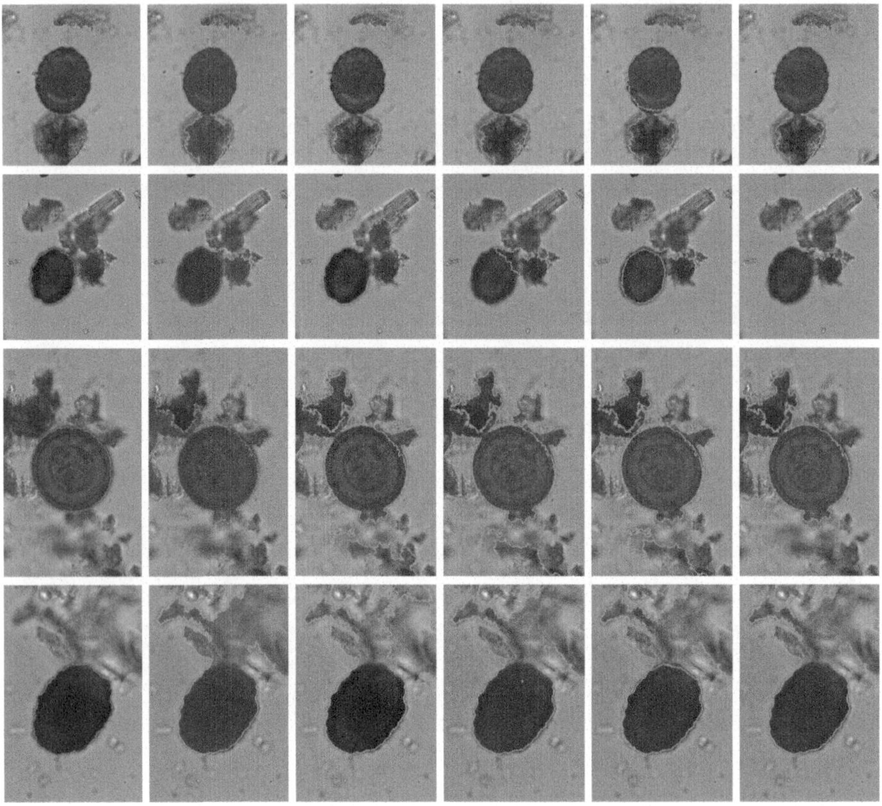

(a) original (b) selection (c) baseline (d) IFT (e) SICLE (f) RISF

Fig. 4. Examples of re-segmentation in the parasites dataset. The original images and the object border (in red) are illustrated in (a). The selected superpixel (in blue) is shown in (b). The baseline with three superpixels is shown in (c). For the images in (d)-(f), the selected superpixel in (b) is re-segmented by using IFT, SICLE, and RISF, respectively, and the results are superimposed with the desired object. (Color figure online)

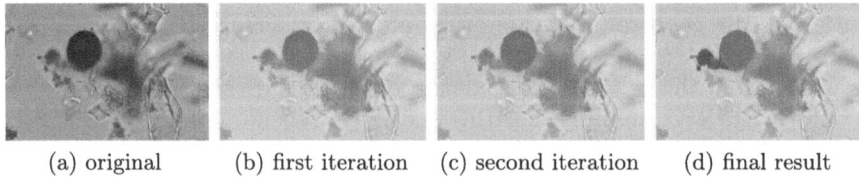

(a) original (b) first iteration (c) second iteration (d) final result

Fig. 5. Example of a sequential superpixel selection. First, the apply a superpixel segmentation illustrated in (b). From (b), we have selected the superpixel in green to be re-segmented and shown in (c). From (c), we have selected the superpixel in blue to be re-segmented which is illustrated in (d). (Color figure online)

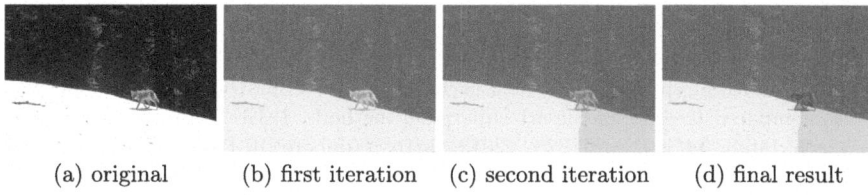

(a) original   (b) first iteration   (c) second iteration   (d) final result

Fig. 6. Example of a sequential superpixel selection. First, the apply a superpixel segmentation illustrated in (b). From (b), we have selected the superpixel in yellow to be re-segmented and shown in (c). From (c), we have selected the superpixel in blue to be re-segmented which is illustrated in (d). (Color figure online)

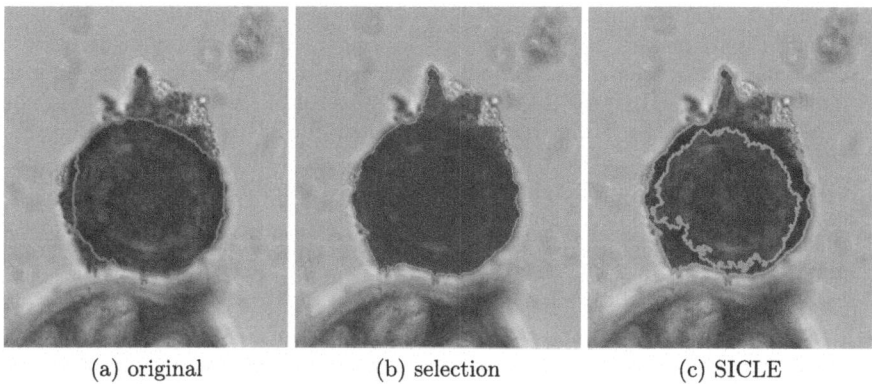

(a) original   (b) selection   (c) SICLE

Fig. 7. Example of re-segmentation failure. Instead of separating the impurities from the egg, the re-segmentation process divided the inner part of its outer part, illustrated in (c).

5 Conclusion and Future Work

In this paper, we proposed three re-segmentation methods designed to improve the delineation of object borders after a superpixel-based segmentation. Each method runs through the initial segmentation, and here, we have used SICLE. Our proposed methods, which involve human interaction at the first stage, demonstrate significant improvements in accurately defining object borders.

The experimental results indicate that these re-segmentation techniques effectively refine the superpixel boundaries, leading to more precise and contextually relevant segmentation. Future work will focus on further refining these methods and exploring automated approaches to reduce the reliance on human interaction, to improve their practicality and scalability for large-scale image segmentation tasks. Moreover, we intend to study how to replace the user with a salient object detection for identifying the superpixel to be re-segmented.

References

1. Achanta, R., Shaji, A., Smith, K., Lucchi, A., Fua, P., Süsstrunk, S.: Slic superpixels compared to state-of-the-art superpixel methods. IEEE Trans. Pattern Anal. Mach. Intell. **34**(11), 2274–2282 (2012). https://doi.org/10.1109/TPAMI.2012.120
2. Alexandre, E.B., Chowdhury, A.S., Falcao, A.X., Miranda, P.A.V.: IFT-SLIC: a general framework for superpixel generation based on simple linear iterative clustering and image foresting transform. In: 2015 28th SIBGRAPI Conference on Graphics, Patterns and Images, pp. 337–344 (2015). https://doi.org/10.1109/SIBGRAPI.2015.20
3. Barcelos, I.B., Belém, F.D.C., João, L.D.M., Patrocínio, Z.K.G.D., Falcão, A.X., Guimarães, S.J.F.: A comprehensive review and new taxonomy on superpixel segmentation. ACM Comput. Surv. **56**(8) (2024). https://doi.org/10.1145/3652509
4. Belém, F., Guimarães, S.J.F., Falcão, A.X.: Superpixel segmentation by object-based iterative spanning forest. In: Vera-Rodriguez, R., Fierrez, J., Morales, A. (eds.) Progress in Pattern Recognition, Image Analysis, Computer Vision, and Applications, pp. 334–341. Springer International Publishing, Cham (2019)
5. Belém, F., Perret, B., Cousty, J., Guimarães, S.J.F., Falcão, A.X.: Efficient multiscale object-based superpixel framework. CoRR abs/2204.03533 (2022). https://doi.org/10.48550/arXiv.2204.03533
6. Belém, F.C., Guimarães, S.J.F., Falcão, A.X.: Superpixel segmentation using dynamic and iterative spanning forest. IEEE Signal Process. Lett. **27**, 1440–1444 (2020). https://doi.org/10.1109/LSP.2020.3015433
7. Belém, F.C., Perret, B., Cousty, J., Guimarães, S.J.F., Falcão, A.X.: Towards a simple and efficient object-based superpixel delineation framework. In: 2021 34th SIBGRAPI Conference on Graphics, Patterns and Images (SIBGRAPI), pp. 346–353 (2021). https://doi.org/10.1109/SIBGRAPI54419.2021.00054
8. Ciesielski, K.C., Falcão, A.X., Miranda, P.A.V.: Path-value functions for which Dijkstra's algorithm returns optimal mapping. J. Math. Imaging Vis. **60**(7), 1025–1036 (2018). https://doi.org/10.1007/s10851-018-0793-1
9. Condori, M.A.T., Cappabianco, F.A.M., Falcao, A.X., Miranda, P.A.V.D.: Extending the differential image foresting transform to root-based path-cost functions with application to superpixel segmentation. In: 2017 30th SIBGRAPI Conference on Graphics, Patterns and Images (SIBGRAPI), pp. 7–14 (2017). https://doi.org/10.1109/SIBGRAPI.2017.8
10. Cui, Q., Pan, H., Zhang, K., Li, X., Sun, H.: Multiscale and multisubgraph-based segmentation method for ocean remote sensing images. IEEE Trans. Geosci. Remote Sens. **61**, 1–20 (2023). https://doi.org/10.1109/TGRS.2023.3247697
11. Falcão, A., Stolfi, J., Lotufo, R.: The image foresting transform: theory, algorithms, and applications. TPAMI **26**(1), 19–29 (2004)
12. Korting, T.S., Dutra, L.V., Fonseca, L.M.G.: A resegmentation approach for detecting rectangular objects in high-resolution imagery. IEEE Geosci. Remote Sens. Lett. **8**(4), 621–625 (2011). https://doi.org/10.1109/LGRS.2010.2098389
13. Lemes Galvão, F., Falcão, A.X., Shankar Chowdhury, A.: RISF: recursive iterative spanning forest for superpixel segmentation. In: 2018 31st SIBGRAPI Conference on Graphics, Patterns and Images (SIBGRAPI), pp. 408–415 (2018). https://doi.org/10.1109/SIBGRAPI.2018.00059
14. Liu, M.Y., Tuzel, O., Ramalingam, S., Chellappa, R.: Entropy rate superpixel segmentation. In: CVPR 2011, pp. 2097–2104 (2011). https://doi.org/10.1109/CVPR.2011.5995323

15. Neubert, P., Chemnitz, P.P.: Superpixel benchmark and comparison (2012). https://api.semanticscholar.org/CorpusID:4190710
16. Rauber, P.E., Falcão, A.X., Spina, T.V., de Rezende, P.J.: Interactive segmentation by image foresting transform on superpixel graphs. In: 2013 XXVI Conference on Graphics, Patterns and Images, pp. 131–138 (2013). https://doi.org/10.1109/SIBGRAPI.2013.27
17. Song, X., Zhou, L., Li, Z., Chen, J., Yan, B., Zeng, L.: Interactive image segmentation based on hierarchical superpixels initialization and region merging. In: 2014 7th International Congress on Image and Signal Processing, pp. 410–414 (2014). https://doi.org/10.1109/CISP.2014.7003815
18. Stutz, D., Hermans, A., Leibe, B.: Superpixels: an evaluation of the state-of-the-art. Comput. Vis. Image Underst. **166**, 1–27 (2018)
19. Vargas-Muñoz, J.E., Chowdhury, A.S., Alexandre, E.B., Galvão, F.L., Vechiatto Miranda, P.A., Falcão, A.X.: An iterative spanning forest framework for superpixel segmentation. IEEE Trans. Image Process. **28**(7), 3477–3489 (2019). https://doi.org/10.1109/TIP.2019.2897941

Towards Interactive Video Segmentation by Dynamic and Iterative Spanning Forest

Danielle Vieira[1], Isabela Borlido Barcelos[1],
Zenilton K. G. Patrocínio Jr[1], Alexandre Falcão[2],
and Silvio Jamil F. Guimarães[1](✉)

[1] Laboratory of Image and Multimedia Data Science (ImScience), Pontifical Catholic University of Minas Gerais, 31980-110, Belo Horizonte, Minas Gerais, Brazil
{ddvieira,isabela.borlido}@sga.pucminas.br, {zenilton,sjamil}@pucminas.br
[2] Laboratory of Image Data Science (LIDS), University of Campinas, 13083-852, Campinas, São Paulo, Brazil
afalcao@ic.unicamp.br

Abstract. Interactive video segmentation aims to segment objects from videos using user information of object location. It allows for segmenting different objects from the same scene and has many applications, such as video editing and scene understanding. While in automatic video segmentation, the major challenges are temporal coherency and occlusion, interactive segmentation models must also handle unseen objects. This work proposes an interactive video segmentation strategy based on seed competition and user-drawn scribbles. Our proposal starts with a seed oversampling strategy and iteratively computes the optimum path forest for the seed set, maintaining the most relevant trees. Our Interactive Video Segmentation by Dynamic and Iterative Spanning Forest (iVSD-ISF) extends the Interactive Dynamic and Iterative Spanning Forest for videos, avoiding object leakage by dynamically creating trees at critical image positions. The proposed method is highly competitive with the state-of-the-art achieving the second highest score, in terms of IoU, considering all studied methods, and the best IoU among the ones without optical flow computation for SegTrackv2.

Keywords: Video segmentation · Graph-based · Interactive

The authors thank the Pontifícia Universidade Católica de Minas Gerais – PUC-Minas, Coordenação de Aperfeiçoamento de Pessoal de Nível Superior – CAPES – (Grant PROAP 88887.842889/2023-00 – PUC/MG, Grant STIC-AMSUD 88887.878869/2023-00, Grant PDPG 88887.708960/2022-00 – PUC/MG - Informática, and Finance Code 001), the Conselho Nacional de Desenvolvimento Científico e Tecnológico – CNPq (Grants 407242/2021-0, 306573/2022-9, 442950/2023-3 and 304711/2023-3), Fundação de Apoio à Pesquisa do Estado de Minas Gerais – FAPEMIG (Grant APQ-01079-23, Grant APQ-05058-23 and PCE-00417-24) and Fundação de Apoio à Pesquisa do Estado de São Paulo – FAPESP (Grant 2023/14427-8).

© The Author(s), under exclusive license to Springer Nature Switzerland AG 2025
R. J. Barrientos and S. A. Velastin (Eds.): CIARP 2024, LNCS 15368, pp. 162–177, 2025.
https://doi.org/10.1007/978-3-031-76607-7_12

1 Introduction

Video segmentation consists of partitioning a video into significant regions and its applications cover high- and low-level tasks, such as video summarization; object detection in video; and tracking of persons. One can find in the literature several algorithms of video segmentation, which mostly are extensions of image segmentation techniques. The main problem of these approaches is the lack of temporal coherence [12,46], which can be solved by considering 26-adjacency elements, so-called voxel, instead of 8-adjacency, and a set of spatially contiguous voxels, that have a similar appearance (intensity, color, texture, etc.), is defined as supervoxel. Thus a voxel has three coordinates (x, y, t), in which time t can be seen as the third dimension. As video segmentation is an ill-posed problem since it depends on the context, the inclusion of the human-in-the-loop is needed to well inform the desired object (or region) to be segmented.

Video segmentation methods can provide the object location without additional information [15,48] based on restricted assumptions. This reduces the user effort but brings drawbacks to these methods. Conversely, interactive (or semi-supervised) segmentation benefits from the user information—as a labeled frame or internal and external line markers (*i.e.*, scribbles)—to better delineate the desired object. Several interactive video segmentation approaches [2,3,21,24,37,42] require one or more frame masks to segment the object and rely on motion information to identify the object. Also, some approaches [2,3,45] propagate the initial labels (provided by the user) using an initial superpixel segmentation, which restricts the method to the superpixel segmentation error, and may cause error propagation.

Despite the good results obtained by the literature, we are seeking for strategies in which the object delineation is reached. In this context, The Image Foresting Transform (IFT), a graph seed-based approach can be seen as the state-of-the-art for this task. The IFT is a framework for designing image operators based on connectivity that efficiently computes an optimum-path forest using a path-cost function in a graph from an initial seed set. The IFT has different applications, such as multiscale shape skeletonization [18], data clustering [11,35], boundary tracking [14], and pixel classification [32]. More recently, several works explored IFT for interactive image segmentation [4,9], superpixel segmentation [6–8,20,40], and supervoxel segmentation [26,41]. In [9], the dynamic estimation of trees within path-cost functions is explored for interactive segmentation. Other works further explore such dynamic estimation for superpixel segmentation [6–8]. In [40], a superpixel segmentation is performed using an IFT-based algorithm, which begins with seed sampling and iteratively moves the seeds to the centroids of the superpixels. In [8], seed oversampling is performed with further iterative seed removal to achieve the desired number of superpixels to avoid the seeds' placement bias. The pipeline in [8] proved to be effective for superpixel segmentation [5] and inspired other works [4,6,7,26]. For instance, the proposal in [4], so-called iDISF, explores the pipeline from [8] for interactive image segmentation and introduces new dynamic path-cost functions and seed removal strategies. While the new functions handle the strong but meaningless local gradients, the seed removal strategies handle the critical region at the objects' boundaries. The work in [26] extends [8] for supervoxel segmentation in videos. With a random seed oversampling strategy and a root-based path cost function, the strategy in [26] can preserve important images' boundaries. Inspired by [26], the authors in [41] propose a streaming strategy for supervoxel segmentation. Instead of segmenting the whole video at once, it divides it into blocks of the same size. The strategy in [41] also requires a unique intersection frame to propagate the last block's supervoxel labels to the current one. Despite existing strategies for segmentation in videos based on IFT, as far as we know, no other IFT-based algorithm is able to segment objects from videos. This work provides a strategy for interactive video segmentation using user-drawn markers restricted to the first video frame.

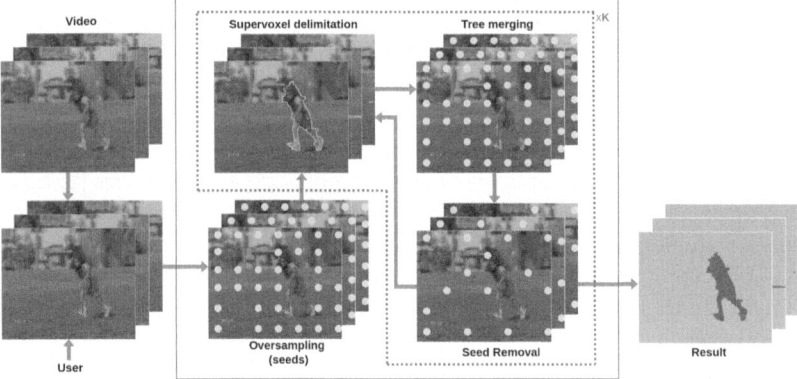

Fig. 1. Methodology of our method iVSDISF with a four-step pipeline: (i) seed oversampling; (ii) supervoxel delimitation; (iii) tree merging; and (iv) seed removal. Where K is the desired number of iterations.

Since iDISF was not conceived for interactive video segmentation purposes, in this work we explore its decoupled structure to propose an IFT-based approach for interactive video segmentation, called *Interactive Video Segmentation by Dynamic and Iterative Spanning Forest* (iVSDISF). Our approach extends iDISF [4] for videos with a four-step pipeline: (i) seed oversampling; (ii) forest computation; (iii) tree merging; and (iv) seed removal, as illustrated in Fig. 1. Similar to iDISF, our approach requires user-drawn markers, which indicate the object and background locations, and perform an initial grid sampling for a high number of seeds. Then, it computes a forest rooted at the seed set and modifies the original IFT algorithm to create seeds that prevent trees from conquering nodes with high local gradients. To ensure a decreasing number of seeds through iterations and avoid the trees' size being a bias in the removal step, the same number of created trees during seed conquering are merged. Finally, we remove irrelevant seeds based on size and color similarity. At the end of the process, trees with the same object label compose an object.

Thus, the main contributions of this work are two-fold: (i) proposal of interactive graph-based video segmentation, called *Interactive Dynamic and Iterative Spanning Forest for Video Segmentation* (iVSDISF) without optical flow computation; and (ii) thanks to the merging step, the labels are well propagated guaranteeing the temporal coherence. The proposed method is highly competitive with the state-of-the-art achieving the second highest score, in terms of *Intersection Over Union* (IoU), considering all studied methods, and the best IoU among the ones without optical flow computation for SegTrackv2 [29] dataset.

The concepts to understand those IFT-based methods, in special, iDISF are described in Sect. 2. Related works are presented in Sect. 3. The iVSDISF is described in Sect. 4. In Sect. 5, we demonstrate that iVSDISF can considerably increase segmentation accuracy with minimum user effort and it is highly competitive to the state-of-the-art. Finally, Sect. 6 provides a conclusion and future work.

2 Theoretical Background

This Section presents the fundamental concepts required to understand our proposal. We first introduce some graph notions and video modeling in graphs. Then, we present the *Image Foresting Transform* (IFT) [19] framework and the *Interactive Dynamic and Iterative Spanning Forest* (iDISF).

2.1 Graph Notation

A *video* V can be represented as a pair $\mathsf{V} = (\mathcal{V}, \mathbf{I})$ in which $\mathcal{V} \subseteq \mathbb{N}^3$ denotes the set of *volume elements* (*i.e.*, voxels), and $\mathbf{I}$ maps every $v \in \mathcal{V}$ to a feature vector $\mathbf{I}(v) \in \mathbb{R}^m$, in which V is a colored video (*e.g.*, RGB or CIELAB colorspaces) when $m = 3$. One may create a *digraph* $\mathsf{G} = (\mathcal{N}, \mathcal{E})$ from V, such as $\mathcal{N}$ and $\mathcal{E}$ are the *vertex* and *edges* sets, respectively, $\mathcal{E} \subset \mathcal{N}^2$ defines the *adjacency* relation between node pairs $(v_i, v_j) \in \mathcal{E}$, and $\mathcal{E}$ contains oriented arcs. Let two vertices $v_i, v_j \in \mathcal{N}$ with spatial positions $\mathcal{V}(v_i) = \langle x_i, y_i, z_i \rangle$ and $\mathcal{V}(v_j) = \langle x_j, y_j, z_j \rangle$, respectively. There is an edge $(v_i, v_j) \in \mathcal{E}$ if $z_i = z_j$ or $z_i = z_j - 1$ (*i.e.*, they are in the same video frame or v_j is in the frame right after v_i).

A *path* $\pi_{s \rightsquigarrow t} = \langle s = v_1, v_2, \ldots, v_n = t \rangle$ is a finite sequence of adjacent nodes, in which $(v_i, v_{i+1}) \in \mathcal{E}$ for $1 \leq i < n$. A path is trivial when $n = 1$. We may omit the path *origin* voxel by writing π_t. A path $\pi_p = \pi_q \cdot \langle p, q \rangle$ is an extension of π_p through the arc $(p, q) \in \mathcal{E}$. Let a path π_t in a set of paths Π_G, a connectivity function $f : \Pi_G \rightarrow \mathcal{C}$ calculates a cost-path, in which the costs $f(\pi)$ for all $\pi_t \in \Pi_G$ are an ordered cost-set $\mathcal{C}$. A path π_t is optimum when $f(\pi_t) \leq f(\pi'_t)$ for any $\pi'_t \in \Pi_G$ irrespective its initial node. A predecessor map $\mathcal{P}$ assigns the predecessor s of each pixel t in a path π_p, in which $(s, t) \in \mathcal{E}$. If t is a root (*i.e.*, the initial node), $\mathcal{P}(t) = nil$. For a path π_t, we denote $R(t)$ as the root of t.

2.2 Image Foresting Transform

Given a path-cost function f and a graph G, the IFT algorithm generates an *optimum-path forest* $\mathcal{P}$, by assigning one *optimum-path* π_q^* to each node $q \in \mathcal{N}$.

For a given seed set $\mathcal{S} \subset \mathcal{N}$, the IFT starts from all nodes being trivial paths with costs defined by Eq. 1.

$$f(\langle q \rangle) = \begin{cases} 0, & \text{if } q \in \mathcal{S} \\ +\infty & \text{otherwise.} \end{cases} \qquad (1)$$

One may see each trivial path as a growing tree $\mathcal{T}_s$ rooted at s. After setting the initial costs (Eq. 1), the trees grow, such that each seed conquers its most strongly connected nodes. When a seed s tries to conquer a node q, adjacent to p, through the optimum path π_p, the algorithm evaluates if $f(\pi_p \cdot \langle p, q \rangle) < f(\pi_q)$ and then substitutes π_q by $\pi_p \cdot \langle p, q \rangle$ when it is true. Several path-cost functions were designed for IFT-based segmentation [17]. For instance, Eq. 2 presents a path-cost function based on the maximum criteria, where $\mu_{\mathcal{T}_{R(p)}} = \frac{1}{|\mathcal{T}_{R(p)}|} \sum_{t \in \mathcal{T}_{R(p)}} \mathbf{I}(t)$ is the mean color vector of a tree $\mathcal{T}_{R(p)}$, rooted at $R(p) \in \mathcal{S}$.

$$f_1(\pi_p \cdot \langle p, q \rangle) = \max\{f_1(\pi_p), \|\mu_{\mathcal{T}_{R(p)}} - \mathbf{I}(q)\|_2\}, \qquad (2)$$

2.3 Interactive Dynamic and Iterative Spanning Forest

The *Interactive Dynamic and Iterative Spanning Forest* (iDISF) is an interactive IFT-based strategy to segment objects in images. First, the user indicates object and background regions by manually drawing scribbles. The object and background pixels marked by the user compose the object $\mathbf{S}_o$ and background $\mathbf{S}_b$ seeds sets, respectively. Then, it oversamples seeds with grid sampling, creating a higher number N_0 of seeds than the desired number $N_f \ll N_0$ of superpixels. In grid sampling, seeds are equally spaced by distance $d = \sqrt{|\mathcal{N}|/N_0}$. Afterward, they are moved to the position with the lowest gradient value among the 8 neighbors. To maintain a minimum distance between these seeds and those drawn by the user, a seed p is created only if $\|p - q\|_2 > \frac{d}{2}$ for all $q \in \{\mathbf{S}_o \cup \mathbf{S}_b\}$. Thus, a seed set $\mathcal{S}$ is composed of N_G seeds sampled by grid strategy along with $\mathbf{S}_o$ and $\mathbf{S}_b$. In iDISF, trees rooted at $\mathcal{S}$ conquer pixels based on a path-cost function. Afterward, roots of irrelevant trees are removed from $\mathcal{S}$. In iDISF, two seed removal strategies were proposed. In this work, we adopt the strategy of removal by class, where trees' relevance is defined based on whether they are positioned at the objects' borders, which is the most critical region for object delineation.

A tree adjacency $\mathcal{B}_1$ may be defined as $\mathcal{B}_1 = \{(\mathcal{T}_s, \mathcal{T}_t) \mid \exists (x,y) \in \mathcal{E}, x \in \mathcal{T}_s, y \in \mathcal{T}_t \text{ and } s \neq t\}$. Let a label function L, in which $L(s) = 1$ if $s \in \mathbf{S}_o$ or 0 otherwise. Based on L, we define a tree-adjacency relation between object and background superpixels (*i.e.*, the object border) $\mathcal{B}_2 = \{(\mathcal{T}_s, \mathcal{T}_t) \mid \exists (x,y) \in \mathcal{E}, x \in \mathcal{T}_s, y \in \mathcal{T}_t \text{ and } L(s) \neq L(t)\}$. Let a class function M that assigns 1 to seeds whose tree is adjacent to trees with the same label and 2 otherwise – *i.e.*, seeds with class 2 are in the object border, while seeds with class 1 are not. The *relevance by class* (Eq. 3) assigns different relevance values for each seed $s \in \mathcal{S}$ with respect to M.

$$V(s) = \begin{cases} \frac{|\mathcal{T}_s|}{|\mathcal{N}|} \min_{\forall (\mathcal{T}_s, \mathcal{T}_t) \in \mathcal{B}_1} \{\|\mu_{\mathcal{T}_s} - \mu_{\mathcal{T}_t}\|_2\}, & \text{if } M(s) = 1 \\ \min_{\forall (\mathcal{T}_s, \mathcal{T}_t) \in \mathcal{B}_2} \{\|\mu_{\mathcal{T}_s} - \mu_{\mathcal{T}_t}\|_2\}, & \text{otherwise} \end{cases} \qquad (3)$$

Let $\mathbb{S}_x \subset \mathcal{S}$ the set of seeds s whose $M(s) = x$, and let $\mu(\mathbb{S}_x)$ and $\sigma(\mathbb{S}_x)$ be its relevance mean and its relevance standard deviation values, respectively. Then, remove from $\mathcal{S}$ seeds s sampled by grid strategy with $V_x(s) < \mu(\mathbb{S}_x) + \sigma(\mathbb{S}_x)$. Additionally, seeds s with $M(s) = 1$ are maintained if any neighboring seeds have been discarded, irrespective of $V(s)$. Note that any $s \in \{\mathbf{S_o} \cup \mathbf{S_b}\}$ is not applicable for being removed. In iDISF, the seed set $\mathcal{S}$ reduces by iteratively computing the optimum-path forest followed by a seed removal step. iDISF stops when the number of desired iterations is reached or $\mathcal{S} = \mathbf{S}_o \cup \mathcal{S}_b$.

3 Related Works

Video segmentation methods can be categorized based on the amount of data (or user effort) required to provide an object mask. Unsupervised methods provide automatic responses without requiring prior information for segmentation. In [13], a visual tracking approach segments video based on multiple regions of the target computed by a Gaussian mixture model in a joint feature-spatial space. In [34], the authors present a tractable approximation of spectral clustering, which relies on an initial segmentation using appearance and optical flow features from self-supervised approaches. In [15,48], deep models leverage motion and appearance information. Specifically, [48] employs two deep networks with pre-trained weights to exploit temporal motion in a top-down manner. In contrast, [15] uses a two-stream network to fuse motion and appearance within a unified framework. The model in [15] is weakly supervised, requiring human annotations only for training. Unsupervised approaches benefit form do not require user effort or additional data, performing segmentation based on restricted assumptions about the application scenario.

Interactive segmentation methods, on the other hand, require either entirely manually annotated frames (masks) [2,3,21,24,37,42] or user-drawn scribbles [45]. For instance, a decision forest classifier establishes correlations between region labels in [2,3]. These correlations may be between superpixels of nonsuccessive video frames [3] or based on a cross-correlation among pixel labels and patch clustering. The authors in [21,45] unify the tracking and segmentation problem using Hough-voting classification and spectral matching, respectively. In [21], the classifier tracks objects and provides information to guide the segmentation process, whereas in [45], the object segmentation of the first frame guides the segmentation process, which provides information to the tracking mechanism. While methods with initial superpixels highly rely on the initial oversegmentation [2,3,45], methods with a unified approach for segmentation and tracking [21,45] may lead to error propagation from one task to another. The video segmentation problem can also be formulated as a Markov Random Field problem [24,37], incorporating appearance and motion cues. In [42], the stationary distributions of the random walks iteratively exploit superpixels of two consecutive frames, in which one frame was labeled by the previous iteration or the initial mask.

In this work, we model the video segmentation problem as an optimum path forest computation problem in the video graph restricted to an initial seed set.

Instead of requiring a full frame mask, our proposal needs just a few strokes for object and background regions. Similar to other IFT-based approaches, our method performs an efficient optimum path forest computation.

4 Interactive Dynamic Iterative Spanning Forest for Video Segmentation

In this work, we propose an IFT-based approach for interactive video segmentation, called *Interactive Video Segmentation by Dynamic and Iterative Spanning Forest* (iVSDISF). Our approach extends iDISF [4] for videos with a four-step pipeline: (i) seed oversampling; (ii) forest computation; (iii) tree merging; and (iv) seed removal. Similar to iDISF, our approach requires user-drawn markers, which indicate the object and background locations, and perform an initial grid sampling for a high number of seeds. Then, it computes a forest rooted at the seed set and modifies the original IFT algorithm to create seeds that prevent trees from conquering nodes with high local gradients. To ensure a decreasing number of seeds through iterations and avoid the trees' size being a bias in the removal step, the same number of created trees during seed conquering are merged. Finally, we remove irrelevant seeds based on size and color similarity. At the end of the process, trees with the same object label compose an object.

4.1 Forest Computation

In this work, we perform a seed oversampling to increase the chance of retaining relevant seeds. Oversampling seeds with subsequent removal of the least relevant ones proved to be an effective strategy to obtain better-positioned seeds for segmentation [4,6–8,26]. Therefore, our strategy adopts the same strategies for seed oversampling and removal by class from iDISF [4].

After oversampling, the seeds compete for the remaining nodes based on a path-cost function (Eq. 2). Therefore, the trees dynamically conquer nodes along the smooth gradient considering the mean tree color. However, since the object can move along the time axis, a tree labeled as an object may grow through the background nodes with similar color, causing segmentation leakage. iVSDISF reduces the leakage restricting the nodes' conquering. Since the algorithm minimizes the internal color differences at each tree, one may argue that, if $f(\pi_{p \leadsto x} \cdot \langle x, y \rangle)$ is too high compared to $f(\pi_q) \forall q \in \mathcal{T}_p$ (*e.g.*, higher than a threshold $\gamma(\mathcal{T}_p)$), the node q may significantly change $\mu_{\mathcal{T}_p}$ and, therefore, may cause a leakage.

Let $\tau(s)$ the standard deviation of the cost $f(\pi_q)$ for $q \in \mathcal{T}_s$ and $\bar{\tau} = \sum_{s \in \mathcal{S}} \frac{\tau(s)}{|\mathcal{S}|}$. Let $R(x) = p$ and $f(\pi_x \cdot \langle x, y \rangle) = \min_{s \in \mathcal{S}} \{f(\pi_{s \leadsto y})\}$. iVSDISF creates paths based on minimum cost restricted to a maximum threshold γ. Therefore, when a tree rooted at $p \in \mathcal{S}$ tries to extend a path π_x through an arc $\langle x, y \rangle$ with $f(\pi_x \cdot \langle x, y \rangle) > \gamma(\mathcal{T}_p)$, y became a new seed in $\mathcal{S} \setminus \{\mathbf{S}_o \cup \mathbf{S}_b\}$, in which $\gamma(\mathcal{T}_p) = \max\{\tau(p) + \bar{\tau}, 0.01\}$. It is important to note that a node y only became a seed if $f(\pi_{p \leadsto x} \cdot \langle x, y \rangle)$ is minimum for all trees and is greater than $\gamma(\mathcal{T}_p)$.

4.2 Trees Merging

During the forest computation, iVSDISF creates seeds to increase the trees' color homogeneity, preventing them from spreading into object and background regions, thereby reducing the leakage. On the other hand, the number of new trees can significantly increase after each iteration. Since the object is the critical region to segment, iVSDISF only creates background seeds in oversampling and seed competition. Therefore, a higher number of seeds tends to result in smaller trees biased by the high number of background ones. Conversely, the new trees may be significantly smaller than the others, thereby being considered irrelevant during the removal step. In both cases, the seed set does not improve through iterations. We prevent both from occurring by including a merging step after the seeds have conquered the remaining nodes. Let $\|\mu_{\mathcal{T}_p} - \mu_{\mathcal{T}_q}\|_2$ be the tree dissimilarity and $\mathcal{B}_1$ be the tree neighboring relation. The trees are merged in pairs according to the smaller dissimilarity with their most similar neighbor until they reach the number of trees from the beginning of the iteration. The similarity is updated after each merging. Since only background seeds are created during seed competition, only background trees from the oversampling step or created during seed competition are merged. The remaining seeds, from user markers, remain unchanged, as these provide highly reliable information.. Afterward, a seed removal by class [4] removes the less relevant trees, improving object segmentation.

5 Experimental Analysis

This work proposes an interactive video segmentation strategy based on seed competition. For segmenting a video using our method, it is necessary to have at least one frame with object and background markers. Hence, for iVSDISF, the first frame and every 10 frames have object or background markers (if there are no objects in the frame). We also set $N_0 = 20000$, 5 iterations, and a maximum of 200 new seeds per iteration.

For our experiments, we select the SegTrackv1 [37] and SegTrackv2 [29] datasets (see Fig. 2 for some samples). SegTrackv1 comprises 6 videos with one object each over 266 frames. Its videos contain the following challenges: color overlap between object and background, inter-frame motion, and complex object deformation.

Conversely, SegTrackv2 extends SegTrackv1 to 14 videos with 24 objects in total over 1066 frames. SegTrackv2 contains the same 6 videos from SegTrackv1, some with 2 labeled objects, and 8 other videos. In addition to the challenges in SegTrackv1, the videos in SegTrackv2 present motion blur, appearance change, slow motion, occlusion, multiple adjacent/interacting objects, and more videos with complex deformation.

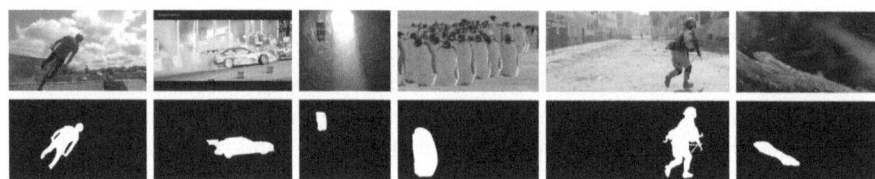

Fig. 2. Example of images from SegTrackv2 dataset.

Table 1. Comparison of different video segmentation methods. The table shows whether each method uses optical flow, whether it is supervised, and their respective mean Intersection over Union (mIoU) scores. The best and second best results are in bold and underlined, respectively.

Method	Optical flow	Supervision	mIoU	Method	Optical flow	Supervision	mIoU
[49]	✓	sup.	62.0	[50]	✓	unsup.	72.1
[33]	✓	unsup.	47.8	[48]	✓	unsup.	74.2
[36]		unsup.	57.3	[34]	✓	unsup.	**74.9**
[27]		unsup.	57.2	[15]		wekly-sup.	61.4
[23]		unsup.	70.1	[16]	✓	inter.	67.2
[30]		unsup.	57.0	[28]	✓	inter.	54.3
[47]	✓	unsup.	58.6	**Ours**		inter.	74.6

5.1 Quantitative Analysis

For a quantitative analysis, we employed IoU (Intersection over Union) and mean pixel error per frame as measures. Let A and B be the sets of object pixels in the segmentation and its ground-truth, respectively. The IoU of A is the intersection between A and B divided by their union. A higher IoU indicates a more precise segmentation, in which the prediction and the ground-truth coincide more. Mean pixel error per frame measures the mean of miss-labeled pixels in the segmentation.

Tables 1, 2, and 3 present a quantitative comparison with several methods. The information about the performance of the other algorithms was extracted from [23,34,42,44]. Therefore, our markers differ from the other interactive methods in Tables 1 and 2. Table 1 reports the mean IoU for the dataset, while Table 2 reports the mean IoU per object. As can be observed in Tables 1 and 2 our method obtains the second best IoU score on average per object and video, showing that it is highly competitive using only a few frames with markers, with the other compared methods that use either an optical flow computation or are supervised, or both. Many optical flow strategies rely on the assumption of the object's movement, thereby failing with static objects or moving backgrounds. Also, they are usually computationally expensive and may employ a supervised strategy, requiring high-quality labeled data for training. Our approach, iVSD-ISF, only requires partially labeled data and does not depend on optical flow.

Table 2. IoU scores on SegTrackv2 dataset. The best and second best results are boldfaced and underlined, respectively. The average scores with ∗ are computed without failed videos, represented by −. Results for [15,16,23] are reported per video and the remaining ones per object.

Video (#frames)	[45]	[1]	[43]	[10]	[21]	[25]	[42]	[31]	[38]	[22]	[44]	[16]	[15]	[23]	Ours
bird of paradise (98)	93.0	82.3	44.3	46.5	5.1	85.3	86.7	89.7	87.1	86.8	90.1	81.4	69.9	**93.7**	91.7
birdfall (30)	**78.7**	9.4	32.5	36.4	56.0	61.6	64.6	65.3	52.9	57.4	46.1	56.5	38.0	64.9	46.6
bmx-person (36)	88.9	44.5	27.9	36.0	2.0	78.8	89.6	67.1	87.9	39.2	**92.2**	75.4	59.1	84.7	83.7
bmx-bike (36)	5.70	0.00	6.04	3.86	–	8.93	11.96	3.2	4.0	32.5	40.1				**48.3**
cheetah-deer (29)	66.1	17.7	33.1	38.7	46.1	56.4	56.6	5.4	25.9	18.8	**66.6**	51.8	59.6	51.8	54.9
cheetah-monkey (29)	35.3	0.7	14.0	19.7	**47.4**	30.0	32.1	9.2	37.2	24.4	46.7				34.0
drift-1 (74)	67.3	42.9	43.5	57.2	62.6	80.8	82.5	68.5	77.9	55.2	**93.4**	74.1	87.6	82.9	84.7
drift-2 (74)	63.7	11.1	11.6	13.8	21.8	37.8	63.2	32.7	27.4	27.2	50.9				**80.5**
frog (279)	56.3	63.4	45.2	38.8	14.5	42.8	48.3	76.1	78.4	67.1	81.2	71.3	57.0	**83.2**	81.3
girl (21)	84.6	62.4	52.4	62.0	53.6	83.0	82.6	86.5	84.2	31.9	**91.6**	86.0	66.7	84.6	88.6
hummingbird-1 (29)	58.3	14.0	28.8	25.1	11.8	48.9	52.9	53.2	67.2	13.7	**76.2**	62.4	65.2	46.4	56.0
hummingbird-2 (29)	50.7	36.8	45.9	44.2	–	41.6	44.7	28.7	**68.5**	25.2	67.5				65.2
monkey (31)	86.0	76.1	61.7	58.7	73.1	86.1	86.4	87.5	87.8	61.9	**92.5**	82.3	80.5	73.9	83.2
monkeydog-monkey (71)	**82.2**	5.0	22.1	25.7	61.0	74.2	76.4	40.5	47.1	68.3	43.2	52.5	32.8	38.1	78.3
monkeydog-dog (71)	21.1	9.0	10.2	3.8	18.9	17.1	23.1	17.1	21.0	18.8	**87.4**				19.9
parachute (51)	**94.4**	92.5	69.9	59.3	85.6	91.1	92.3	93.7	93.3	69.1	94.2	85.9	51.6	93.7	91.1
penguin-1 (42)	94.2	80.2	20.8	40.1	54.5	88.1	87.8	81.6	80.4	72.0	**97.0**	13.9	71.3	24.0	92.6
penguin-2 (42)	91.8	73.1	20.8	37.9	67.0	87.4	86.8	82.0	83.5	80.7	**92.8**				**92.8**
penguin-3 (42)	91.9	46.3	10.3	31.2	7.6	84.8	83.9	78.5	83.9	75.2	**95.1**				93.5
penguin-4 (42)	90.3	51.6	13.0	30.2	54.3	79.2	83.8	76.4	86.2	80.6	**92.0**				90.6
penguin-5 (42)	76.3	53.7	18.9	10.7	29.6	74.2	77.3	47.8	82.3	62.7	**86.3**				83.4
penguin-6 (42)	88.7	70.1	32.3	35.0	2.1	86.7	85.9	84.3	87.3	75.5	**95.2**				89.2
soldier (32)	81.1	71.9	43.0	54.2	70.7	60.8	63.9	55.3	86.8	66.5	**90.8**	69.2	69.8	80.0	78.8
worm (243)	79.3	72.4	27.4	44.3	36.8	67.9	78.7	65.4	**83.2**	34.7	65.0	78.2	50.6	80.0	80.9
Average per object	71.9	45.3	30.7	35.6	40.1*	64.7	68.4	58.2	67.6	51.9	**78.1**	–	–	–	74.6
Average per video	72.3	48.8	37.0	40.4	41.0*	64.2	68.2	61.2	69.3	50.8	**76.9**	67.2	61.4	70.1	74.0

As can also be seen in Table 2, iVSDISF presents either the best or second in half (12 of 24) of the video objects in terms of IoU score, the IoU values are very close to the other compared methods. In the BMX-bike video with 36 frames in which the object is a bicycle and therefore has more small and detailed areas, our method was the one that obtained an IoU close to 50, we can argue that as our method is an IFT-based method, the object delineation is more effective, as stated in [8,40]. In larger videos such as frog and worm, we obtain the second highest IoU score, since there are not many changes in the position of the object thus the trees between one frame and the next one are very close in terms of the position (x, y, t) and the color. Thanks to these issues, our method is able to perform good segmentation without many leaks.

Table 3 compares the performance of different video segmentation methods in terms of mean pixel error, highlighting the best results obtained for each

Table 3. Mean pixel error per frame on SegTrackv1 dataset. The best and second best results are boldfaced and underlined, respectively.

Video (#frames)	[3]	[45]	[10]	[37]	[2]	[24]	[21]	[39]	[43]	[13]	[25]	[42]	Ours
Birdfall (30)	508	**163**	481	252	468	189	466	243	1204	454	215	<u>194</u>	460
Cheetah (29)	855	806	2825	1142	1501	1170	1431	**391**	2765	1216	740	<u>735</u>	777
Girl (21)	<u>1200</u>	1904	7790	1304	1705	2283	6338	1935	10505	1755	1527	1621	**995**
Monkeydog (71)	412	342	5361	563	736	**333**	809	497	2466	683	428	387	<u>335</u>
Parachute (51)	296	275	3105	235	404	<u>228</u>	1028	**187**	2369	502	331	284	331
Penguin (42)	1736	571	11669	1705	19310	**443**	6239	903	9078	6627	883	901	<u>539</u>
Mean Error	834.5	<u>676.8</u>	5205.2	866.8	4020.7	774.3	2718.5	692.7	4731.2	1872.8	687.3	687.0	**572.7**

video evaluated on the SegTrackv1 dataset. As one can see in Table 3, iVSDISF achieved the lowest mean pixel error per frame.

5.2 Qualitative Analysis

In terms of qualitative analysis, we illustrate, in Fig. 3, some results obtained by iVSDISF. It is possible to observe that our method can follow the object in each video, outlining it according to its shape without many leaks. Especially in Bird of Paradise, where the bird changes its shape a lot when turning around, it is possible to follow it, as well as its details such as its small feet. In the video Frog, which has 279 frames, it shows that our method can follow the object without losing any region, using 1 frame with markers every 10 frames. As in shorter videos such as Girl, it was possible to segment the object that has several regions of different colors.

5.3 Limitations

We have identified that our proposal may not handle occlusions in low contrast regions as shown in Fig. 4, where the bush in front of the monkey is part of the object region, instead of being part of the background, since the object is just the monkey. Another limitation of our method is videos in which the object moves very quickly between frames. If there are no images with markers that capture this new rapid movement, segmentation is impaired, as seen in the hummingbird video in Fig. 4. Also, since our approach is interactive, the results depend on the quality of the markers and their frequency per frame.

Towards Interactive Video Segmentation   173

(a) frog (b) girl (c) bird of paradise

Fig. 3. iVSDISF segmentations with 20000 initial seeds and 5 iterations for frames (a) 1,70,126,153,200, and 250; (b) 1,8,12,14,17, and 21; and (c) 3,11,70,75,80, and 98, from top to bottom.

Fig. 4. iVSDISF segmentation with 20000 initial seeds and 5 iterations for frames 1,7,10,14,18, and 21 of monkey video (top row) and frames 2,7,8,10,11, and 15 of hummingbird video (bottom row). Our proposal may not handle occlusion, especially with low-contrast regions (top row) or objects at high speed (bottom row).

6 Conclusion

In this paper, we propose a new interactive video segmentation strategy based on user-drawn seed and scribble competition, called Interactive Video Segmentation by Dynamic and Iterative Spanning Forest (iVSDISF), an extension of Interactive Dynamic and Iterative Spanning Forest (iDISF), an interactive IFT-based strategy to segment objects in images. Our method allows object segmentation in video with only a few frames with user-drawn markers and with few leaks in the object region. Furthermore, according to the results presented, our algorithm is competitive with the other compared methods that also perform object segmentation in video. For future works, we intend to study the impact of the number of images with user-made markers, as well as adapt the second relevance-based seed removal strategy of iDISF for video segmentation of objects. We also intend to study new strategies to improve segmentation when there is occlusion and videos with objects that move very quickly between frames.

References

1. Avinash Ramakanth, S., Venkatesh Babu, R.: SeamSeg: video object segmentation using patch seams. In: CVPR, pp. 376–383 (2014)
2. Badrinarayanan, V., Budvytis, I., Cipolla, R.: Semi-supervised video segmentation using tree structured graphical models. IEEE Trans. Pattern Anal. Mach. Intell. **35**(11), 2751–2764 (2013)
3. Badrinarayanan, V., Budvytis, I., Cipolla, R.: Mixture of trees probabilistic graphical model for video segmentation. IJCV **110**, 14–29 (2014)
4. Borlido Barcelos, I., Belém, F., Miranda, P., Falcão, A.X., do Patrocínio, Z.K.G., Guimarães, S.J.F.: Towards interactive image segmentation by dynamic and iterative spanning forest. In: Lindblad, J., Malmberg, F., Sladoje, N. (eds.) DGMM 2021. LNCS, vol. 12708, pp. 351–364. Springer, Cham (2021). https://doi.org/10.1007/978-3-030-76657-3_25
5. Barcelos, I.B., Belém, F.D.C., João, L.D.M., Patrocínio, Z.K.G.D., Falcão, A.X., Guimarães, S.J.F.: A comprehensive review and new taxonomy on superpixel segmentation. ACM Comput. Surv. **56**(8) (2024). https://doi.org/10.1145/3652509
6. Belém, F., Guimarães, S.J.F., Falcão, A.X.: Superpixel segmentation by object-based iterative spanning forest. In: Vera-Rodriguez, R., Fierrez, J., Morales, A. (eds.) CIARP 2018. LNCS, vol. 11401, pp. 334–341. Springer, Cham (2019). https://doi.org/10.1007/978-3-030-13469-3_39
7. Belém, F.C., et al.: Novel arc-cost functions and seed relevance estimations for compact and accurate superpixels. JMIV **65**(5), 770–786 (2023)
8. Belém, F.C., Guimarães, S.J.F., Falcão, A.X.: Superpixel segmentation using dynamic and iterative spanning forest. IEEE Sig. Process. Lett. **27**, 1440–1444 (2020)
9. Bragantini, J., Martins, S.B., Castelo-Fernandez, C., Falcão, A.X.: Graph-based image segmentation using dynamic trees. In: Vera-Rodriguez, R., Fierrez, J., Morales, A. (eds.) CIARP 2018. LNCS, vol. 11401, pp. 470–478. Springer, Cham (2019). https://doi.org/10.1007/978-3-030-13469-3_55
10. Cai, Z., Wen, L., Lei, Z., Vasconcelos, N., Li, S.Z.: Robust deformable and occluded object tracking with dynamic graph. TIP **23**(12), 5497–5509 (2014)

11. Cappabianco, F.A., X, A.X.F., Yasuda, C.L., Udupa, J.K.: Brain tissue MR-image segmentation via optimum-path forest clustering. Comput. Vision Image Understand. **116**(10), 1047–1059 (2012)
12. Chen, J., Paris, S., Durand, F.: Real-time edge-aware image processing with the bilateral grid. In: ACM SIGGRAPH 2007 Papers. SIGGRAPH '07. ACM, New York, NY, USA (2007)
13. Chockalingam, P., Pradeep, N., Birchfield, S.: Adaptive fragments-based tracking of non-rigid objects using level sets. In: ICCV, pp. 1530–1537. IEEE (2009)
14. Condori, M.A.T., Mansilla, L.A.C., Miranda, P.A.V.: Bandeirantes: a graph-based approach for curve tracing and boundary tracking. In: Angulo, J., Velasco-Forero, S., Meyer, F. (eds.) ISMM 2017. LNCS, vol. 10225, pp. 95–106. Springer, Cham (2017). https://doi.org/10.1007/978-3-319-57240-6_8
15. Dutt Jain, S., Xiong, B., Grauman, K.: FusionSeg: learning to combine motion and appearance for fully automatic segmentation of generic objects in videos. In: CVPR, pp. 3664–3673 (2017)
16. Faktor, A., Irani, M.: Video object segmentation by non-local consensus voting. In: Proceedings of the British Machine Vision Conference. British Machine Vision Association (2014)
17. Falcão, A., Bragantini, J.: The role of optimum connectivity in image segmentation: can the algorithm learn object information during the process? In: Couprie, M., Cousty, J., Kenmochi, Y., Mustafa, N. (eds.) DGCI 2019. LNCS, vol. 11414, pp. 180–194. Springer, Cham (2019). https://doi.org/10.1007/978-3-030-14085-4_15
18. Falcão, A.X., Feng, C., Kustra, J., Telea, A.: Multiscale 2D medial axes and 3d surface skeletons by the image foresting transform. In: Skeletonization: Theory, Methods and Applications, p. 43 (2017)
19. Falcão, A.X., Stolfi, J., de Alencar Lotufo, R.: The image foresting transform: theory, algorithms, and applications. IEEE Trans. Pattern Anal. Mach. Intell. **26**(1), 19–29 (2004)
20. Galvão, F.L., Guimarães, S.J.F., Falcão, A.X.: Image segmentation using dense and sparse hierarchies of superpixels. Pattern Recogn. 107532 (2020). https://doi.org/10.1016/j.patcog.2020.107532
21. Godec, M., Roth, P.M., Bischof, H.: Hough-based tracking of non-rigid objects. Comput. Vis. Image Underst. **117**(10), 1245–1256 (2013)
22. Grundmann, M., Kwatra, V., Han, M., Essa, I.: Efficient hierarchical graph-based video segmentation. In: CVPR, pp. 2141–2148. IEEE (2010)
23. Hu, Y.-T., Huang, J.-B., Schwing, A.G.: Unsupervised video object segmentation using motion saliency-guided spatio-temporal propagation. In: Ferrari, V., Hebert, M., Sminchisescu, C., Weiss, Y. (eds.) ECCV 2018. LNCS, vol. 11205, pp. 813–830. Springer, Cham (2018). https://doi.org/10.1007/978-3-030-01246-5_48
24. Jain, S.D., Grauman, K.: Supervoxel-consistent foreground propagation in video. In: Fleet, D., Pajdla, T., Schiele, B., Tuytelaars, T. (eds.) ECCV 2014. LNCS, vol. 8692, pp. 656–671. Springer, Cham (2014). https://doi.org/10.1007/978-3-319-10593-2_43
25. Jang, W.D., Kim, C.S.: Semi-supervised video object segmentation using multiple random walkers. In: British Machine Vision Conference (2016)
26. Jerônimo, C., et al.: Graph-based supervoxel computation from iterative spanning forest. In: Lindblad, J., Malmberg, F., Sladoje, N. (eds.) DGMM 2021. LNCS, vol. 12708, pp. 404–415. Springer, Cham (2021). https://doi.org/10.1007/978-3-030-76657-3_29
27. Jun Koh, Y., Kim, C.S.: Primary object segmentation in videos based on region augmentation and reduction. In: CVPR, pp. 3442–3450 (2017)

28. Keuper, M., Andres, B., Brox, T.: Motion trajectory segmentation via minimum cost multicuts. In: ICCV, pp. 3271–3279 (2015). https://doi.org/10.1109/ICCV.2015.374
29. Li, F., Kim, T., Humayun, A., Tsai, D., Rehg, J.M.: Video segmentation by tracking many figure-ground segments. In: ICCV (2013)
30. Liu, R., Wu, Z., Yu, S., Lin, S.: The emergence of objectness: learning zero-shot segmentation from videos. In: Advances in Neural Information Processing Systems, vol. 34, pp. 13137–13152 (2021)
31. Märki, N., Perazzi, F., Wang, O., Sorkine-Hornung, A.: Bilateral space video segmentation. In: CVPR, pp. 743–751 (2016)
32. Papa, J.P., Falcão, A.X., Suzuki, C.T.: Supervised pattern classification based on optimum-path forest. Int. J. Imaging Syst. Technol. **19**(2), 120–131 (2009)
33. Papazoglou, A., Ferrari, V.: Fast object segmentation in unconstrained video. In: ICCV, pp. 1777–1784 (2013)
34. Ponimatkin, G., Samet, N., Xiao, Y., Du, Y., Marlet, R., Lepetit, V.: A simple and powerful global optimization for unsupervised video object segmentation. In: WACV, pp. 5892–5903 (2023)
35. Rocha, L.M., Cappabianco, F.A., Falcão, A.X.: Data clustering as an optimum-path forest problem with applications in image analysis. Int. J. Imaging Syst. Technol. **19**(2), 50–68 (2009)
36. Tokmakov, P., Alahari, K., Schmid, C.: Learning video object segmentation with visual memory. In: ICCV, pp. 4481–4490 (2017)
37. Tsai, D., Flagg, M., Nakazawa, A., Rehg, J.M.: Motion coherent tracking using multi-label MRF optimization. IJCV **100**, 190–202 (2012)
38. Tsai, Y.H., Yang, M.H., Black, M.J.: Video segmentation via object flow. In: CVPR, pp. 3899–3908 (2016)
39. Varas, D., Marques, F.: Region-based particle filter for video object segmentation. In: CVPR, pp. 3470–3477 (2014)
40. Vargas-Muñoz, J.E., Chowdhury, A.S., Alexandre, E.B., Galvão, F.L., Miranda, P.A.V., Falcão, A.X.: An iterative spanning forest framework for superpixel segmentation. TIP **28**(7), 3477–3489 (2019)
41. Vieira, D., Barcelos, I.B., Belém, F., Patrocínio, Z.K.G., Falcão, A.X., Guimarães, S.J.F.: Streaming graph-based supervoxel computation based on dynamic iterative spanning forest. In: Vasconcelos, V., Domingues, I., Paredes, S. (eds.) CIARP 2023. LNCS, vol. 14470, pp. 90–104. Springer, Cham (2024). https://doi.org/10.1007/978-3-031-49249-5_7
42. Wang, H., Liu, W., Xing, W.: Video object segmentation via random walks on two-frame graphs comprising superpixels. J. Vis. Commun. Image Represent. **80**, 103293 (2021)
43. Wang, S., Lu, H., Yang, F., Yang, M.H.: Superpixel tracking. In: ICCV, pp. 1323–1330. IEEE (2011)
44. Wang, W., Shen, J., Porikli, F., Yang, R.: Semi-supervised video object segmentation with super-trajectories. IEEE Trans. Pattern Anal. Mach. Intell. **41**(4), 985–998 (2018)
45. Wen, L., Du, D., Lei, Z., Li, S.Z., Yang, M.H.: Jots: Joint online tracking and segmentation. In: CVPR, pp. 2226–2234 (2015)
46. Winnemoller, H., Olsen, S.C., Gooch, B.: Real-time video abstraction. ACM Trans. Graph **25**, 2006 (2006)
47. Yang, C., Lamdouar, H., Lu, E., Zisserman, A., Xie, W.: Self-supervised video object segmentation by motion grouping. In: ICCV, pp. 7177–7188 (2021)

48. Yang, Y., Lai, B., Soatto, S.: Dystab: unsupervised object segmentation via dynamic-static bootstrapping. In: CVPR, pp. 2825–2835. IEEE Computer Society, Los Alamitos, CA, USA (2021). https://doi.org/10.1109/CVPR46437.2021.00285
49. Yang, Y., Loquercio, A., Scaramuzza, D., Soatto, S.: Unsupervised moving object detection via contextual information separation. In: CVPR, pp. 879–888 (2019)
50. Ye, V., Li, Z., Tucker, R., Kanazawa, A., Snavely, N.: Deformable sprites for unsupervised video decomposition. In: CVPR, pp. 2647–2656 (2022). https://doi.org/10.1109/CVPR52688.2022.00268

Data-Driven Evolutionary Algorithms for Optimizing Pumping Stations in Water Distribution Networks: Classifier-Guided Search Space Reduction

Thalía Faúndez-Lizama[1], Nicolás Gajardo-Sepúlveda[1]($\boxtimes$),
Jimmy H. Gutiérrez-Bahamondes[2], Daniel Mora-Melia[3],
and César A. Astudillo[2]

[1] Master's in Operations Management, Faculty of Engineering, Universidad de Talca, 3340000 Curicó, Chile
nigajardo18@alumnos.utalca.cl
[2] Department of Computer Science, Faculty of Engineering, Universidad de Talca, 3340000 Curicó, Chile
jgutierrezb@utalca.cl
[3] Department of Engineering and Construction Management, Faculty of Engineering, Universidad de Talca, 3340000 Curicó, Chile

Abstract. In the context of increasing water scarcity and the need to enhance water distribution network efficiency, this study focuses on optimizing the design of pumping stations using data-driven evolutionary algorithms guided by classification models. These facilities are critical components in water networks due to their high energy consumption and role in maintaining necessary water pressure. The proposed approach integrates a machine learning classifier model within the evolutionary process to pre-label solutions as feasible or infeasible, thereby reducing search space and computational costs associated with hydraulic simulations. Classification models include support vector classifier, artificial neural networks, and random forest. These models are trained on data generated from initial evolutionary optimization, enabling the classifier to predict the viability of new solutions. Results show that all tested classification models reduced the number of objective function evaluations and consequently, required hydraulic simulations. In some cases, the optimal solution defined by the original method was improved, demonstrating dual benefits by reducing both cost and computational effort. This work highlights that data-driven evolutionary optimization can effectively tackle highly complex problems, offering potential future applications in other areas of civil engineering and water resource management. Implementing this methodology promises significant time and resource savings in large-scale optimization processes.

Keywords: Data-Driven Evolutionary Optimization · Classification · Machine Learning · Pumping Station · Search Space Reduction

1 Introduction

In recent decades, water scarcity has increased due to population growth, climate change, and inefficient resource management. In this context, water distribution networks (WDNs) stand out as vital assets for society, driving research into their design, operation, and rehabilitation. WDNs require significant energy consumption, especially in direct injection networks where pumping stations (PS) are large energy consumers. Optimizing these stations is crucial for improving the overall efficiency of the network (Wang et al., 2021). Studies classify the optimization of PS into three main approaches: design, operation, and combined approaches to optimize both design and operation (Müller et al., 2021).

Efficient methods are crucial, especially for highly nonlinear problems. Evolutionary algorithms (EAs), particularly genetic algorithms (GAs), have proven effective (Mora-Melià et al., 2017), used to minimize water loss, optimize hydraulic system performance, and reduce operational and maintenance costs. They have also been integrated into hydraulic simulators to improve WDN designs (Sangroula et al., 2022).

A notable study by Gutiérrez-Bahamondes et al. (2021) simultaneously optimizes the design and operation of PS in WDNs, minimizing the sum of investment and operation costs through a nonlinear programming model solved with a pseudo-genetic algorithm. However, computational complexity arises from the need to hydraulically simulate each analysis period to verify constraints, particularly in large-scale networks.

To address these challenges, methodologies have been proposed that accelerate the evaluation of feasible solutions, reducing the search space, for example, in pipe sizing (Diao et al., 2022). Another approach involves surrogate models that can be classification or regression models (Yaochu Jin, 2021), employing artificial neural networks to replace hydraulic simulations (Sarbu, 2021).

An emerging methodology, Data-driven evolutionary optimization (DDEO), integrates EAs with data-trained algorithms to predict the performance of parameter configurations (Jin et al., 2019). Currently, the relationship of DDEO with WDNs includes applications such as predicting water quality parameters and leak detection (Hu et al., 2021). Although this methodology has been applied to WDNs, its application in optimizing PS design is still evolving.

The main contributions of this work are listed below:

- Development of a new Data-driven evolutionary algorithm (DDEA) that integrates machine learning classifiers to reduce the search space and computational costs associated with hydraulic simulations.
- Application of the proposed methodology to a real case study of a water distribution network, demonstrating significant reductions in objective function evaluations and computational effort.
- Evidence that the proposed approach can improve optimal solutions defined by traditional methods, highlighting its potential for broader applications in civil engineering and water resource management.
- Presentation of a scalable and efficient optimization framework that can generate substantial time and resource savings in large-scale optimization problems.

The remainder of this document is structured as follows: Sect. 2 details the materials and methods used, including the proposed DDEA and integrated machine learning classifiers. Section 3 presents the results of the case study applied to a specific water distribution network, comparing the performance of different classifiers. The implications of the findings are discussed, and the potential of the proposed methodology to reduce computational effort is highlighted. Finally, Sect. 4 concludes the document, summarizing the main contributions and suggesting directions for future research.

2 Materials and Methods

The study by Gutiérrez-Bahamondes et al. (2021) is an example of optimizing complex systems (WDNs and PSs) using evolutionary algorithms, which can be affected by the problem's dimension at a real scale. This author conducted 100 experiments, where an optimal value of 40,259 euros was obtained in 100,000 evaluations of the objective function.

This section presents the DDEA methodology visualized in Fig. 1, which aims to expedite the evolutionary process for problems like the one described above. Additionally, an adaptation of the methodology for subsequent application is shown, following the guidelines of computational implementation for the development of experimentation.

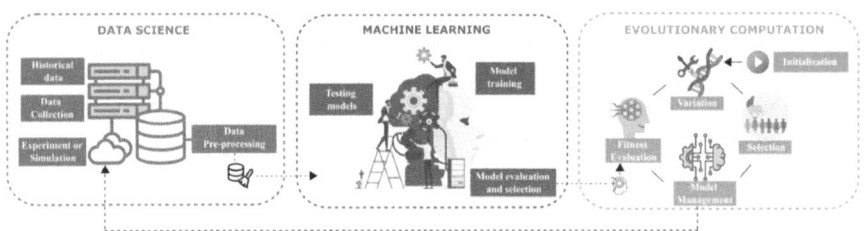

Fig. 1. Data Driven Methodology (Yaochu Jin, 2021).

This study proposes integrating a machine learning classifier within an evolutionary optimization process to predict the feasibility of the solutions under study, evaluating by hydraulic simulation only those labeled as feasible. Considering this, Fig. 2 presents the diagram of the adapted methodology proposed for this context. First, to train the classifier model, a synthetic database is constructed from solutions evaluated by a traditional genetic algorithm during a limited evolution process. These data and their classification (feasibility outcome) are used to train the machine learning model. Second, a second evolutionary optimization is performed, using the last population from the previous stage as the initial population, and with the classifier implemented, only the solutions labeled as feasible by the trained model are evaluated. It is important to note that a solution is infeasible when the pumps of the PS are not capable of providing the necessary pressure for the required flow.

Three DDEAs are developed, as shown in Fig. 2, trained with different databases to compare scenarios in the results section. A set of classifiers from the literature is taken and subjected to training and testing, which identifies the best representatives as the Support vector classifier (SVC), Artificial neural networks (ANN), and Random forest (RF).

Figure 2 presents a diagram of the proposed methodology. First, to train the classifier model, a synthetic database is constructed from solutions evaluated by a traditional genetic algorithm during a limited evolution process. These data and their classification (feasibility outcome) are used to train the machine learning model. Second, a second evolutionary optimization is performed, using the initial population as the last population obtained in the previous stage and evaluating only the solutions classified as feasible by the trained model. The algorithms used and compared in this study are the following: Support vector classifier (SVC), Artificial neural networks (ANN), and Random forest (RF).

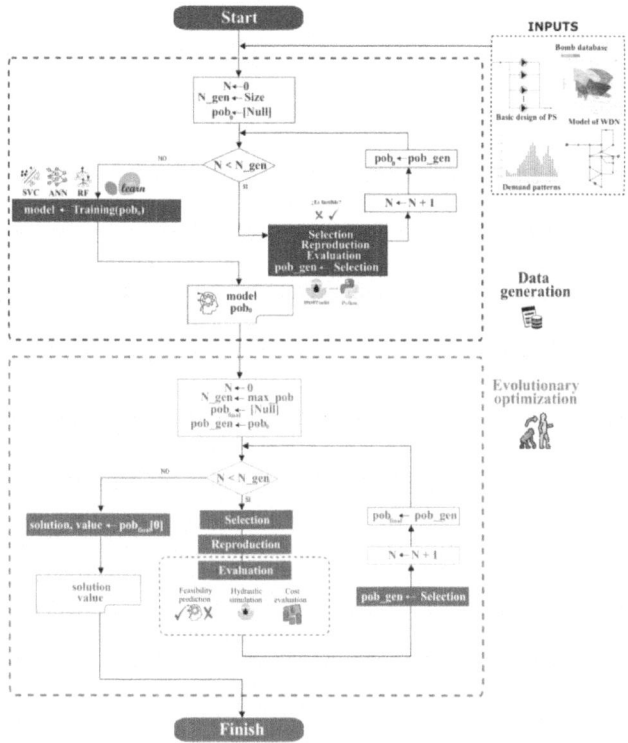

Fig. 2. Proposal of Data-Driven Evolutionary Algorithm.

2.1 Case Study

The study applies to a network called MTF, introduced by Gutiérrez-Bahamondes et al. (2021). Figure 3 illustrates the topology of the MTF network, which comprises 15 demand nodes, 3 pumping stations, and 25 pipes. The analysis spans 24 periods, each corresponding to one hour. The base demand is 100 L/s, and the minimum node pressure is 20 m. The problem resolution involves a pump catalog featuring 67 models.

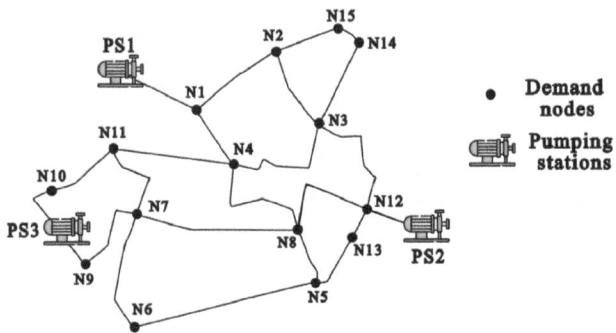

Fig. 3. MTF network.

2.2 Computational Implementation

A stopping criterion for the optimization processes is defined with a maximum of 50,000 solutions or 500 populations generated based on the convergence observed between 40,000 and 60,000 evaluations of the objective function (Gutiérrez-Bahamondes et al., 2021). The methodology compares three machine learning models (SVC, ANN, and RF) by testing different sizes of training datasets: 1,000, 5,000, and 10,000 solutions. One hundred experiments are conducted per scenario and fifty for the original method, totaling 950 experiments. In addition to optimal solutions, "good solutions" that vary up to 3% from the optimal value are considered (Daniel Mora Melia, 2012).

A stopping criterion for optimization is defined with a maximum of 50,000 solutions generated, based on observed convergence (Gutiérrez-Bahamondes et al., 2021). Three machine learning models (SVC, ANN, and RF) are compared, which will be trained with different datasets of 1000, 5000, and 10000 solutions. One hundred experiments are conducted for each mentioned variation and fifty for the original method. Additionally, the term "good solutions," which vary up to 3% from the optimal value found (Daniel Mora Melia, 2012), is considered.

3 Results

From the experimentation with the original method, an optimal value of €41,303 was obtained after 50,000 objective function evaluations (OFE). In the 9 additional experimental scenarios, different models with three database sizes were compared. Figure 4 shows the results with 1,000 training data, Fig. 5 with 5,000 data, and Fig. 6 with 10,000 data. The graphs highlight the optimal solution of the original method and the value of €42,543.04, which represents the threshold for what is considered good solutions (3% efficiency loss). Each figure also shows the optimal value found in the experimentation.

With 1,000 training data, it is €40,999.43, obtained by the Random forest DDEA in 35,661 OFE; with 5,000 training data, it is €41,116.66, obtained by the Random forest DDEA in 43,134 OFE; and with 10,000 training data, it is €40,670.49 (the optimum of the entire study), obtained by the Artificial neural network DDEA in 41,789 OFE.

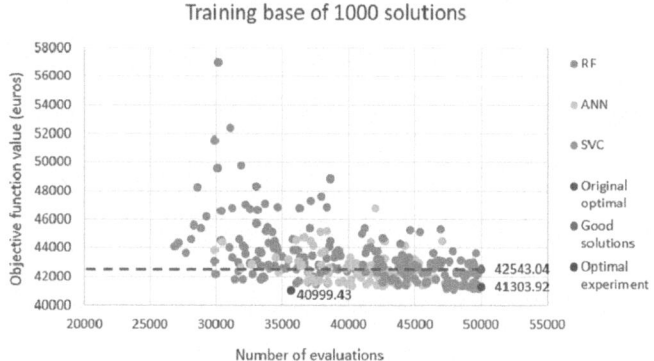

Fig. 4. Results with 1000 training data.

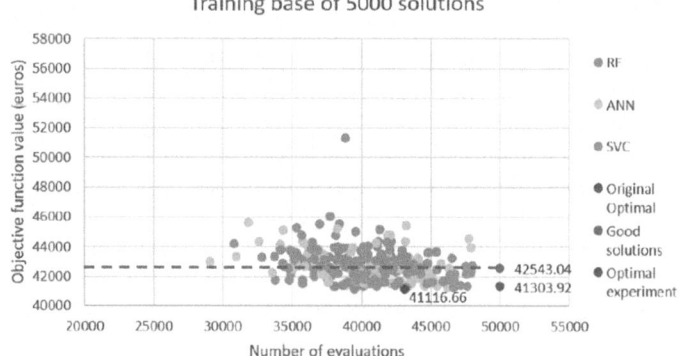

Fig. 5. Results with 5000 training data.

3.1 Discussion of Results

The results show a large number of good solutions in the different graphs, with a savings of up to 20,000 OFE. In some cases, solutions lower than the previously defined optimal value were achieved, with significant OFE savings. The lowest-cost solution, €40,670, was obtained with ANN and 10,000 training data after 41,789 evaluations, improving the previously set optimum and saving 16.42% of OFE. Compared to the optimal value of Gutiérrez-Bahamondes et al. (2021) of €40,259 with 100,000 OFE, the best solution found represents a higher cost by only 411 euros (1% loss in quality) but saves 58.21% of OFE. Good solutions are defined as those under €42,543.04, and the new proposed

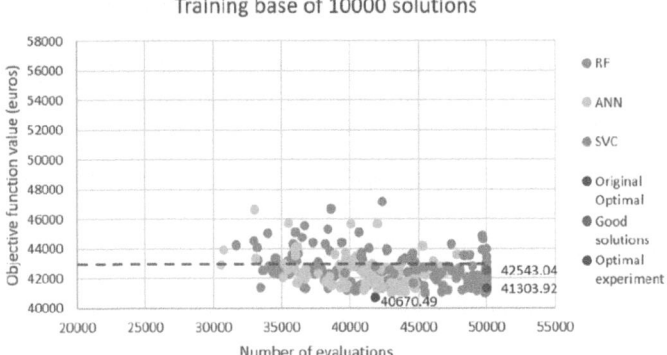

Fig. 6. Results with 10000 training data.

method allows obtaining them by evaluating approximately only 50% of the solutions under study.

The results presented show a large number of good solutions across the different graphs, with a clear savings of more than 20,000 objective function evaluations (OFE). In numerous cases, solutions were obtained that were better than the previously defined optimal value. This is evident in the optimal values from the experiments, where the DDEAs composed of Random Forest (RF) stood out by achieving the best solution with 1,000 and 5,000 training data, and the DDEA composed of Artificial neural networks (ANN) achieved the best solution with 10,000 training data. The latter was the lowest-cost solution in the entire study, with €40,670 in 41,789 evaluations, improving upon the previously established optimum and saving 16.42% in OFE. Another notable solution is the one obtained by the RF DDEA with 1,000 training data, which reduced the optimal value to €40,999.43 in 35,661 evaluations, achieving a 28.68% savings in OFE.

Considering the work of Gutiérrez-Bahamondes et al. (2021), which reports an optimal value of €40,259 in 100,000 OFE, the best solution found in this study (€40,670) represents an increase in cost of only €411 (1% loss in quality), but with a savings of 58.21% in OFE (41,789 evaluations). In the case of the solution obtained by the RF DDEA with 1,000 training data, the cost exceeded the optimal value by €740 (approximately 2% loss in efficiency), but with only 35,661 OFE, translating to a savings of 64.34% in OFE. This demonstrates the effectiveness of the proposed method in reducing the computational effort required for hydraulic simulations without significantly affecting the quality of the solutions, as "good solutions" (with values below €42,543.04) were still obtained, performing only 50% of the OFE needed in Gutiérrez's study.

Good solutions were obtained in all scenarios. The SVC trained with 10,000 solutions achieved the highest number of good solutions (56) and the best performance in finding solutions cheaper than the original optimum (13). On the other hand, the RF trained with 1,000 solutions obtained the fewest good solutions (22), in addition to showing greater value dispersion, indicating lower reliability. The SVC had the lowest average objective value (€42,315.5); although it did not achieve the lowest values, the majority remained within a range close to €42,000. In contrast, the RF had the highest average objective value (€44,304), as the greater dispersion in the results meant that, on some occasions, OFE savings resulted in a considerable loss of solution quality.

Regarding OFE savings, the RF stood out with an average savings of 31.55%, while the SVC had the lowest average (4.16%) and presented situations with no OFE savings. This indicates that, in some optimization processes, the model failed to correctly identify infeasible solutions. Overall, the graphs indicate that models trained with 1,000 data points show greater result dispersion, while when trained with 10,000 solutions, the SVC DDEA tends to perform more evaluations, even close to 50,000 OFE, showing practically no savings.

4 Conclusions

WDNs are a subject of constant research due to their high complexity. Currently, metaheuristic tools are used which, although effective, lose efficiency as the problem size increases, resulting in high costs and computational effort. This study proposes a two-stage DDEA to reduce the number of hydraulic simulations during optimization. Supervised learning classification models were used as a filter to label the solutions generated by the genetic algorithm. All tested models managed to reduce the number of OFE, reducing hydraulic simulations and computational demands, and in some cases, improving the original optimal solution. This results in a double benefit: cost reduction and fewer evaluations. Compared to other studies, the savings in time and resources are significant, especially when applied to real-scale studies. The new methodology has potential for future applications in optimizing other complex infrastructures and implementing online data.

References

1. Melia, D.M.: Diseño de redes de distribución de agua mediante algoritmos evolutivos. Análisis de eficiencia (2012)
2. Diao, K., Berardi, L., Laucelli, D.B., Ulanicki, B., Giustolisi, O.: Topological and hydraulic metrics-based search space reduction for optimal re-sizing of water distribution networks. J. Hydroinf. **24**(3), 610–621 (2022). https://doi.org/10.2166/HYDRO.2022.158
3. Gutiérrez-Bahamondes, J.H., Mora-Meliá, D., Iglesias-Rey, P.L., Martínez-Solano, F.J., Salgueiro, Y.: Pumping station design in water distribution networks considering the optimal flow distribution between sources and capital and operating costs. Water (Switzerland), **13**(21), 3098 (2021). https://doi.org/10.3390/w13213098
4. Hu, Z., Tan, D., Chen, B., Chen, W., Shen, D.: Review of model-based and data-driven approaches for leak detection and location in water distribution systems. In: Water Supply, vol. 21, no. 7, pp. 3282–3306. IWA Publishing (2021). https://doi.org/10.2166/ws.2021.101
5. Jin, Y., Wang, H., Chugh, T., Guo, D., Miettinen, K.: Data-driven evolutionary optimization: an overview and case studies. IEEE Trans. Evol. Comput. **23**(3), 442–458 (2019). https://doi.org/10.1109/TEVC.2018.2869001
6. Mora-Melià, D., Martínez-Solano, F.J., Iglesias-Rey, P.L., Gutiérrez-Bahamondes, J.H.: Population size influence on the efficiency of evolutionary algorithms to design water networks. Procedia Eng. **186**, 341–348 (2017). https://doi.org/10.1016/j.proeng.2017.03.209

7. Müller, T.M., Leise, P., Lorenz, I.S., Altherr, L.C., Pelz, P.F.: Optimization and validation of pumping system design and operation for water supply in high-rise buildings. Optim. Eng. **22**(2), 643–686 (2021). https://doi.org/10.1007/s11081-020-09553-4
8. Sangroula, U., Han, K.H., Koo, K.M., Gnawali, K., Yum, K.T.: Optimization of water distribution networks using genetic algorithm based SOP–WDN program. Water (Switzerland) **14**(6), 851 (2022). https://doi.org/10.3390/w14060851
9. Wang, B., et al.: A continuous pump location optimization method for water pipe network design. Water Resour. Manage. **35**(2), 447–464 (2021). https://doi.org/10.1007/s11269-020-02722-1
10. Jin, Y., Wang, H., Sun, C.: Data-Driven Evolutionary Optimization: Integrating Evolutionary Computation, Machine Learning and Data Science. Springer, Cham (2021). https://doi.org/10.1007/978-3-030-74640-7

Depth Map Completion Using a Specific Graph Metric and Balanced Infinity Laplacian for Autonomous Vehicles

Vanel Lazcano[(✉)]

Facultad de Ciencias, Ingeniería y Tecnología, Universidad Mayor, Santiago, Chile
vanel.lazcano@umayor.cl

Abstract. Depth map images are an essential source of information for various applications such as robotics, autonomous vehicles, 3D cinema post-production, video games, and many others. These maps can be acquired by sensors such as LiDAR or time-of-flight cameras. However, the acquired data present large areas without information (or holes) or data with low-confidence values. Filling in these holes is crucial for robotics and other applications, which helps robots avoid obstacles or plan paths. In order to complete the depth data, we use a method that involves solving the variation of the infinity Laplacian guided by a color reference image of the considered scene. We associate the image grid to a graph and given the rectangular image domain $\Omega \subset \mathbb{R}^2$, and a metric $d_{\mathbf{xy}}$ we created a manifold $\mathcal{M} = (\Omega, d_{\mathbf{xy}})$, where we solve the infinity Laplacian. Our GPU implementation significantly reduces processing time. The contribution of this work is three-fold: i) we use a graph-based approach, ii) we used suitable graph metrics, and ii) we used a variation of the infinity Laplacian. Our results show that our proposal outperforms other contemporary models and performs similarly to approaches based on infinity Laplacian. Our implementation is fast and easy to implement and represents a low-cost tool for many applications. Future work will explore using different metrics and a more contemporary interpolation model.

Keywords: Graph · Manifolds · Metric · Image reconstruction · Depth completion · GPU implementation

1 Introduction

An acquired depth map of a scene is an image whose pixel values represent the distance of the corresponding object to a fixed point (e.g., a Kinect sensor). Depth maps often contain holes due to sensor errors, reflections, and other factors. Depth maps have many applications in modern video games, 3D reconstruction, elevation models [1], robotics, and more. Filling in these holes in-depth maps is a fundamental task for these applications. For instance, a robotic vehicle must know the depth data at each point in one scene to plan a path trajectory.

In this paper, we associate a graph with the depth map and interpolate the available data in this setting using a suitable graph metric. To interpolate the data, we use a variation of the infinity Laplacian and solve it in this discrete setting.

The contribution of this paper is threefold: i) we use a graph-based approach, ii) we use a graph-suitable metric, and ii) we use a variation of the infinity Laplacian.

Section 2 reviews the literature to solve the problem of depth completion. Section 3 presents the proposed model to solve the problem of interpolation of depth maps. Section 5 shows the considered data set to estimate the models' parameters and evaluate the final performance. Section 6 shows our obtained results, and finally, we present in Sect. 7 our conclusions.

2 Related Works

Aronsson discovered the infinity Laplacian in the sixties [2,3] and revisited more recently in [4,13]. Recently, [5] applied the infinity Laplacian for inpainting and clusterings.

Variations of the infinity Laplacian were applied to depth completions in [9]. Also, this work tested different metrics and used them to improve the performance of the variations of the infinity Laplacian.

In [11], an efficient approximation of p-Laplacian is presented, demonstrating that the approximation preserves the local geometry of a manifold. P-Laplacian was applied to regression using kernels and to 2D and 3D signal filtering.

The proposal that uses a neural network has been stated to tackle the problem of depth map completion. In Yang et al. [7,10], authors proposed a convolutional neural network to recover depth maps. They diffuse the available data, controlling it by estimating the gradient. They assessed their proposal in the NYU_V2 [16] data set. The authors in [12] present a model to complete a depth map that simultaneously reconstructs a gray level of the scene the depth map. This setting lets the model learn complementary image features that help the model define interobject structures better.

In Raad, [15], an application of the infinity Laplacian is presented to complete optical flow in synthetic video sequences.

In [8,9] a study using the euclidean metric $\mathcal{L}^2$, $\mathcal{L}^1$ and $\mathcal{L}^{\frac{1}{2}}$ is presented. The metric used metric in those works is less flexible than the one we used in this work. Furthermore, in those, they compare points in order to interpolate the data, but in this work, we compare graph segments.

3 Model

We associated the depth map with a weight graph. The data is associated with vertices, and we created a connection between the data points given by edges. Following the notation on [5] let us define a graph $(\mathcal{G}, \mathcal{V} \times \mathcal{U}, \mathcal{E})$ where $\mathcal{V} \times \mathcal{U}$ are the vertices and $\mathcal{E}$ edges. Let us define a weight function $w : \mathcal{V} \times \mathcal{U} \rightarrow \mathbb{R}$ and $w(\mathbf{x}, \mathbf{y}) > 0$.

As shown in Fig. 1, we associated a graph with the complete depth map. The weights between edges are computed by:

$$\omega_{\mathbf{x},\mathbf{y}} = (\kappa_x \|\mathbf{x} - \mathbf{y}\|^p + \kappa_c \|\mathbf{x} - \mathbf{y}\|^r)^{\frac{1}{q}} \tag{1}$$

with $(\mathbf{x}, \mathbf{y}) \in \mathcal{V} \times \mathcal{U}$, κ_x, κ_y are positive parameters of the metric, p,r and q are exponents of the metric. Instead of using the classical $p = r = 2$ and $q = 1$, we use parameters p,q, and r estimated in the training stage. This fact is the difference concerning [9]. In that work, the authors use the weight as a distance between vertices.

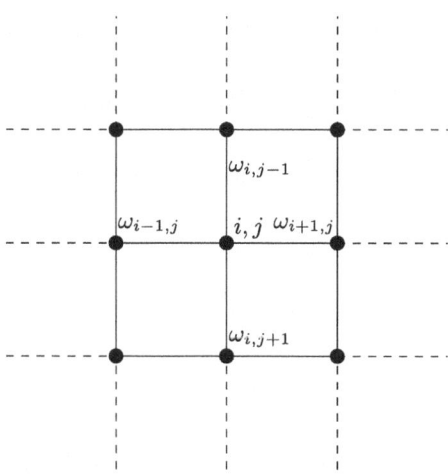

Fig. 1. Central vertex $\mathbf{x} = (i,j)$ and its surrounding edges and assigned weights $\omega_{i-1,j}$, $\omega_{i+1,j}$, $\omega_{i,j-1}$, and $\omega_{i1,j+1}$.

3.1 Interpolator

To interpolate the data, we solved the infinity Laplacian in the graph. We justify the use of the infinity Laplacian because it is simple to implement and easy to compute. According to Caselles et al., [4], the infinity Laplacian is the simplest interpolator that satisfies a set of axioms.

The infinity Laplacian equation is given by:

$$\Delta_{\infty,g} u = 0 \in \Omega \subset \mathbb{R}^2,$$

where g is the metric, u is the interpolated data, Δ_∞ is the infinity Laplacian operator. We solve the equation given the boundary condition: $u|_{\partial\Omega} = \theta$, where

θ is the available data. The discrete version of the infinity Laplacian numerical model is stated in [13] as:

$$\Delta_{\infty,g} = \frac{1}{2}\left(\|\nabla u(\mathbf{x})\|_x^+ + \|\nabla u(\mathbf{x})\|_x^-\right) = 0, \qquad (2)$$

where $\|\nabla u(\mathbf{x})\|_x^+$ and $\|\nabla u(\mathbf{x})\|_x^-$ are the discrete version of the positive eikonal and negative eikonal operator, respectively.

Let $\mathbf{y}$ be in a neighborhood centered in $\mathbf{x}$ of radius $Radius$, we write $\mathbf{y} \sim \mathbf{x}$, then infinity Laplacian can be written as:

$$\max_{\mathbf{y}\sim\mathbf{x}} \frac{u(\mathbf{y}) - u(\mathbf{x})}{d_{\mathbf{xy}}} + \min_{\mathbf{z}\sim\mathbf{x}} \frac{u(\mathbf{z}) - u(\mathbf{x})}{d_{\mathbf{xz}}} = 0. \qquad (3)$$

where $d_{\mathbf{xy}}$ is the metric between $\mathbf{x}$ and $\mathbf{y}$, and $d_{\mathbf{xz}}$ is the distance between $\mathbf{x}$ and $\mathbf{z}$.

3.2 Balanced Infinity Laplacian

The balanced infinity Laplacian is given by the model:

$$\frac{1}{2}\left(\gamma(\mathbf{x})\|\nabla u(\mathbf{x})\|_x^+ + (1-\gamma(\mathbf{x}))\|\nabla u(\mathbf{x})\|_x^-\right) = 0,$$

where $\gamma(x) : \mathbb{R}^2 \to [0,1]$ is a balance map between the positive eikonal operator and negative eikonal operator.

The $\gamma(\mathbf{x})$ map is given explicitly by the expression:

$$\gamma(\mathbf{x}) = \frac{1}{1 + e^{\beta(|\|\nabla u(\mathbf{x})\|_x^+| \; |\tau\|\nabla u(\mathbf{x})\|_x^-|)}}.$$

where β and $\tau \in \mathbb{R}^+$. The main idea behind $\gamma(\mathbf{x})$ is that if the absolute value of the positive eikonal operator is larger than the negative eikonal operator $|\|\nabla u(\mathbf{x})\|_x^+| \gg |\tau\|\nabla u(\mathbf{x})\|_x^-|$ means that the difference $|\|\nabla u(\mathbf{x})\|_x^+| - |\tau\|\nabla u(\mathbf{x})\|_x^-|$ should be positive and $e^{\beta(|\|\nabla u(\mathbf{x})\|_x^+| - |\tau\|\nabla u(\mathbf{x})\|_x^-|)}$ should be large and,

$$\gamma(\mathbf{x}) = \frac{1}{1 + e^{\beta|\|\nabla u(\mathbf{x})\|_x^+| - |\tau\|\nabla u(\mathbf{x})\|_x^-|}} \approx 0,$$

the model is solved using the negative eikonal operator. On the other hand, if $|\|\nabla u(\mathbf{x})\|_x^+| \ll |\tau\|\nabla u(\mathbf{x})\|_x^-|$, means that the difference $|\|\nabla u(\mathbf{x})\|_x^+| - |\tau\|\nabla u(\mathbf{x})\|_x^-|$ should be negative, and $e^{\beta(|\|\nabla u(\mathbf{x})\|_x^+| - |\tau\|\nabla u(\mathbf{x})\|_x^-|)}$ should be small,

$$\gamma(\mathbf{x}) = \frac{1}{1 + e^{\beta|\|\nabla u(\mathbf{x})\|_x^+| - |\tau\|\nabla u(\mathbf{x})\|_x^-|}} \approx 1,$$

the model is solved using the negative eikonal operator. Depending on the values of both eikonal operators, the $\gamma(\mathbf{x})$ map balances between one or another. If $\gamma(\mathbf{x}) = 0.5$, we recover the infinity Laplacian.

4 Numerical Model

Let us call Y and Z the value that maximizes and minimizes Eq. 3, respectively. Substituting in Eq. 2,

$$\frac{1}{2}\left(\gamma(\mathbf{x})\frac{Y-u(\mathbf{x})}{d_{\mathbf{xy}}} + (1-\gamma(\mathbf{x}))\frac{Z-u(\mathbf{x})}{d_{\mathbf{xz}}}\right) = 0, \qquad (4)$$

solving for $u(\mathbf{x})$ we obtain:

$$u(\mathbf{x}) = \frac{(1-\gamma(\mathbf{x}))Z d_{\mathbf{xy}} + \gamma(\mathbf{x})Y d_{\mathbf{xz}}}{(1-\gamma(\mathbf{x}))d_{\mathbf{xy}} + \gamma(\mathbf{x})d_{\mathbf{xz}}}. \qquad (5)$$

We used the iterated version of the infinity Laplacian,

$$u_k(\mathbf{x}) = \frac{(1-\gamma(\mathbf{x}))Z_{k-1} d_{\mathbf{xy}} + \gamma(\mathbf{x})Y_{k-1} d_{\mathbf{xz}}}{(1-\gamma(\mathbf{x}))d_{\mathbf{xy}} + \gamma(\mathbf{x})d_{\mathbf{xz}}} \text{ with } k = 1,..,n. \qquad (6)$$

where n is the number of iterations. We compute the distance between the vertex $\mathbf{x}$ and $\mathbf{y}$ comparing graph segments, and we will explain it in the following subsection.

4.1 Comparison of Subgraphs

We compare two subgraphs of the four-near-neighborhood around $\mathbf{x}$.

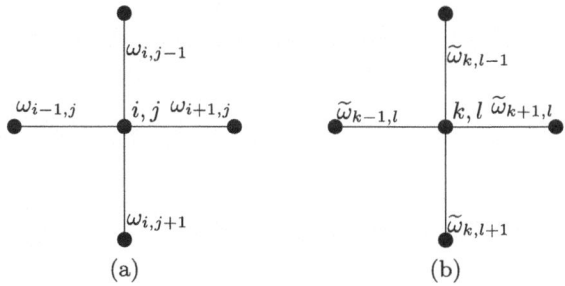

Fig. 2. Comparison of two segments of the graph. We compare the graph segment centered in (i,j) with the graph segment centered in (k,l).

We compare the edge weights around the vertex $\mathbf{x} = (i,j)$ and the vertex $\mathbf{y} = (k,l)$ as shown in Fig. 2. Computing the distance $d_{\mathbf{xy}}$ between this two graph segments:

$$d_{\mathbf{xy}} = \left\| \begin{pmatrix} 0 & \omega_{i,j-1} & 0 \\ \omega_{i-1,j} & 0 & \omega_{i+1,j} \\ 0 & \omega_{i,j+1} & 0 \end{pmatrix} - \begin{pmatrix} 0 & \widetilde{\omega}_{k,l-1} & 0 \\ \widetilde{\omega}_{k-1,l} & 0 & \widetilde{\omega}_{k+1,l} \\ 0 & \widetilde{\omega}_{k,l+1} & 0 \end{pmatrix} \right\|_p, \qquad (7)$$

we compute the distance using the norm:

$$d_{\mathbf{xy}} = |\omega_{i,j-1} - \widetilde{\omega}_{k,l-1}|^s + |\omega_{i-1,j} - \widetilde{\omega}_{k-1,l}|^s + |\omega_{i+1,j} - \widetilde{\omega}_{k+1,l}|^s + |\omega_{i,j+1} - \widetilde{\omega}_{k,l+1}|^s. \quad (8)$$

We tested many different values for s; best results were obtained for $s = 4$.

5 Parameter Estimation

The proposal considers many parameters we present in Table 1. We estimate

Table 1. Parameter of the model.

Name	Description	Number of parameter
Radius	Neighborhood size	1
p, q, r	weight's parameters	3
κ_x, κ_c	weight's parameters	2
n	number of iteration	1
β, τ	parameter of balanced AMLE	2
	Total	9

the parameters of the model using the Particle Swarm Optimization algorithm. We randomly created fifty instances μ_i ($i = 1, .., 50$), a vector containing the values of the parameters. We computed the $MSE + MAE$ (mean square and mean absolute error) for each instance in a subset of images and estimated the parameters. We minimized the functional:

$$J(\mu_i) = \sum_{i=1}^{N} MSE_i(\mu_i) + MAE_i(\mu_i) \quad (9)$$

Each instance μ_i evolves according to dynamics for position and velocity:

$$\nu_i^{t+1} = \alpha \nu_i^t + \varphi_g(\mu_i^t - \mu_g) + \varphi_b(\mu_i^t - \mu_b), \quad (10)$$

and,

$$\mu_i^{t+1} = \mu_i^t + \nu_i^t, \quad (11)$$

where $\alpha = 0.95$ is the evolution parameter for each solution candidate μ_i, $\varphi_g = 0.5$ and $\varphi_b = 1.0$ are positive weight parameters, ν_i is the velocity for each candidate solution, μ_b best instance or candidate solution of the current iteration and μ_b is the best candidate solution of all iterations. A saturation for ν_i is usually incorporated to avoid fast solution change. Practically, we used $\nu_{max} = 2$ and $\nu_{min} = -2$. The images used to estimate the model's parameters are described in Sect. 5.2.

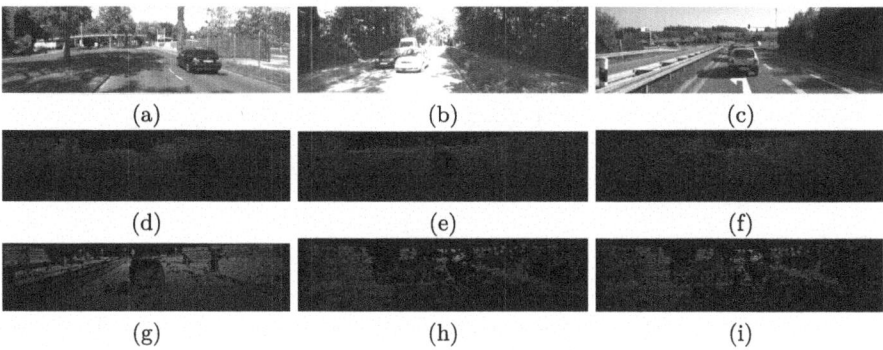

Fig. 3. Images used to estimate parameters. (a), (b), and (c) are color reference images. In (d), (e), and (f) incomplete depth maps. (g). (h), and (i) respective ground truth. (Color figure online)

5.1 Images Used to Estimate Model's Parameters

We used three images, three sparse depth maps, and their corresponding ground truth to estimate the model's parameters. We show these images in Fig. 3. Figure 3 (a), (b), and (c) show the color reference image. Figure 3 (d), (e), and (f) show the sparse depth map. Finally, Figs. 3 (g), (h), and (i) show the ground truth. A LiDAR sensor acquired the data. This data set will be explained in the following subsection. We color-coded the depth values using MATLAB colormap jet: dark red means low depth values, and yellow means large depth values.

5.2 Experiments

We considered the KITTI [6] data set to estimate the parameters of the model and validate the models. KITTI data set consists of 1000 reference color images and 1000 incomplete depth maps and their corresponding ground truth. A LiDAR obtained the incomplete depth map and the ground truth mounted on the top of a car. The car goes to the city, and the LiDAR scans the urban scene. An RGB camera simultaneously the scan takes photographs of the scene. Additionally, the ground truth was constructed by accumulating 11 LiDAR scans to increase the density of the acquired depth data. In this work, we used three of these images to estimate the parameters of the model and 997 images to assess the performance of the models. For each image, once completion was performed, we computed:

$$MSE = \frac{1}{N} \sum_{i=1}^{N_x} \sum_{j=1}^{N_y} (u(i,j) - gt(i,j))^2$$

where $N_x \times N_y$ is the number of pixels of the image, gt ground truth, u is the completed depth map, and:

$$MAE = \frac{1}{N} \sum_{i=1}^{N_x} \sum_{j=1}^{N_y} |u(i,j) - gt(i,j)|.$$

5.3 Implementation

We implemented our model in a parallel setting using a GPU, utilizing CUDA 12.1. The implementation was carried out on a Laptop computer, an HP Omen x16 i711800 with 16GB RAM and GeForce RTX 3070. We used a block configuration of 32 BLOCKX and 4 BLOCKY. As a graphical tool, we used MATLAB 2018a.

We measure the running time in a frame of 1216×352 pixels. This system yielded an average processing time per iteration of $\mu = 0.070$ ms with a $\sigma = 0,036$ ms variance. We performed 92 iterations per frame, i.e., the running time is $0.070 \times 92 = 6.45$ ms per frame, which enables us to process 155 images per second. We do not consider the time to upload images to the hard drive nor save images or results.

6 Results

Table 2 shows the performance of our model in the complete KITTI dataset. We evaluated the performance in 997 images. We also compare our results with contemporary models. As shown in Table 2, our proposal outperforms the bilateral

Table 2. Results obtained by our model in KITTI dataset.

Model	$RMSE$
Proposal	1.831
SocPar2020 [8]	1.801
Bilateral Filter	2.989
CNN [17]	2.100
DeepLidar [14]	0.687

filter, CNN [17] and performs similarly to the model in [8]. More complex models outperform our proposal in this data set. However, those models are complex, challenging to implement, and require many GPUs.

Figure 4 shows qualitative results of two completed depth maps. Figure 4 (a) and (d) shows a person riding a bike on the street. We color-coded the completed depth maps in (c) and (f). We clearly distinguish between people, cars, and traffic signals. Finally, (b) and (e) present the corresponding ground truth.

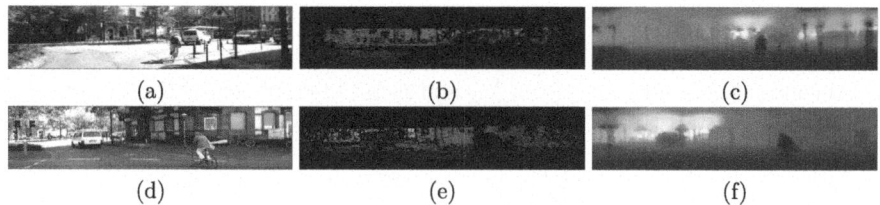

Fig. 4. Qualitative results. In (a) and (d), we show reference color images. In (b), (e) ground truth. In (c) and (f), the completed depth map is shown.

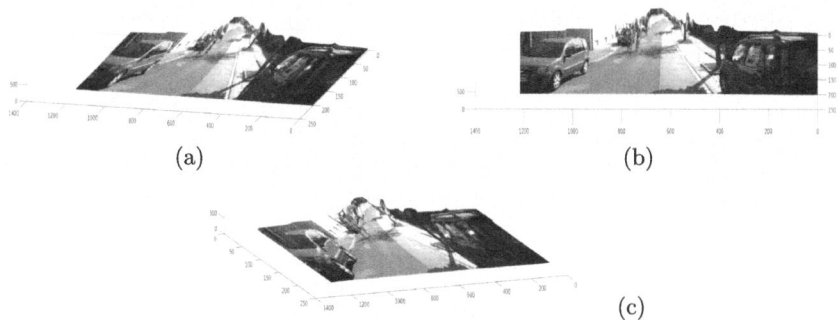

Fig. 5. Different views of the 3D reconstructed scene given the completed depth map and the color reference image.

Figure 5 presents a 3D reconstruction of one scene.

We show in Fig. 5 (a), (b), and (c) that the depth map is completed guided by the color reference image, and then is used to reconstruct the 3d scene. We observe the road from perspective but still present some errors, as seen in (a).

7 Conclusions

This paper presents a model to complete a depth map using a setting based on suitable graph metrics. The results show that our proposal performs similarly to contemporary models but is far from more complex models that use neural networks. The processing time allows the proposal to be implemented in real time, representing a low-cost implementation model for many computer vision applications, such as autonomous vehicles. Future work considers the use of a more contemporary interpolation model.

References

1. Almansa, A., Cao, F., Gousseau, Y., Rougé, B.: Interpolation of digital elevation models using AMLE and related methods. IEEE Trans. Geoci. Remote Sens. **40**(2), 314–325 (2002)
2. Aronsson, G.: Extension of functions satisfyng lipschitz conditions. Aktiv fuer Mathematik **6**(6), 551–561 (1967)
3. Aronsson, G.: On the partial differential equation $u_x^2 u_{,xx} + 2u_x u_y u_{xy} + u_y^2 u_{yy} = 0$. Aktiv fuer Matematik **7**(5), 395–425 (1968)
4. Caselles, V., Igual, L., Sander, O.: An axiomatic approach to scalar data interpolation on surfaces. Numer. Math. **102**(3), 383–411 (2006)
5. Ennaji, H., Quéau, Y., Elmoataz, A.: Tug of war games and PDEs on graphs with applications in image and high dimensional data processing. Sci. Rep. **6045** (2023). https://doi.org/10.1038/s41598-023-32354-5
6. Geiger, A., Lenz, P., Stiller, C., Urtasun, R.: Vision meets robotics: the KITTI dataset. Int. J. Robot. Res. (IJRR) (2013)
7. Lai, W.S., j. B. Huang, J.B., Ahuja, N., Yang, M.H.: Fast and accurate image super-resolution with deep Laplacian pyramid network. Trans. Pattern Anal. Mach. Intell. **41**(11), 2599–2613 (2019)
8. Lazcano, V., Calderero, F., Ballester, C.: Depth image completion using anisotropic operators. In: Abraham, A., et al. (eds.) Proceedings of the 12th International Conference on Soft Computing and Pattern Recognition (SoCPaR 2020), pp. 593–604. Springer International Publishing, Cham (2021). https://doi.org/10.1007/978-3-030-73689-7_57
9. Lazcano, V., Calderero, F., Ballester, C.: Comparing different metrics on an anisotropic depth completion model. Int. J. Hybrid Intell. Syst. **1**, 87–99 (2021)
10. Li, Y., j. B. Huang, Ahuja, N., Yang, M.H.: Joint image filtering with deep convolutional networks. Trans. Pattern Anal. Mach. Intell. **41**(8), 1909–1923 (2019). https://doi.org/10.1109/TPAMI.2018.2890623
11. Liu, W., Ma, X., Zhou, Y., Tao, D., Cheng, J.: p Laplacian regularization for scene recognition. IEEE Trans. Cybern. **49**(8), 2927–2940 (2019). https://doi.org/10.1109/TCYB.2018.2833843
12. Lu, K., Barnes, N., Anwar, S., Zheng, L.: From depth what can you see? Depth completion via auxiliary image reconstruction. In: Proceedings of the IEEE/CVF Conference on Computer Vision and Pattern Recognition (CVPR). IEEE (2020)
13. Manfredi, J., Oberman, A., Svirodov, A.: Nonlinear elliptic partial differential equations and p-harmonic functions on graphs. Differential Integral Equations **28**(12) (2012)
14. Qiu, J., et al.: DeepLiDAR: deep surface normal guided depth prediction for outdoor scene from sparse lidar data and single color image. In: The IEEE Conference on CVPR. IEEE (2019)
15. Raad, L., Oliver, M., Ballester, C., Haro, G., Meinhardt, E.: On anisotropic optical flow inpainting algorithms. IPoL: Image Processing on Line **10**, 78–104 (2020). https://doi.org/10.5201/ipol.2020.281

16. Silberman, N., Hoiem, D., Kohli, P., Fergus, R.: Indoor segmentation and support inference from RGBD images. In: Fitzgibbon, A., Lazebnik, S., Perona, P., Sato, Y., Schmid, C. (eds.) Computer Vision – ECCV 2012: 12th European Conference on Computer Vision, Florence, Italy, October 7-13, 2012, Proceedings, Part V, pp. 746–760. Springer, Berlin, Heidelberg (2012). https://doi.org/10.1007/978-3-642-33715-4_54
17. Uhrig, J., Schneider, N., Schneider, L., Franke, U., Brox, T., Geiger, A.: Sparsity invariant CNNs. In: International Conference on 3D Vision (3DV), pp. 11–20 (2017)

Beta Distribution Approach for Outlier Exposure in Multi-class Text Classification

Camilo Maldonado[1](✉) , Carlos Valle[2] , and Héctor Allende[1]

[1] Departamento de Informática, Universidad Federico Santa María, Valparaíso, Chile
camilo.maldonado@sansano.usm.cl, hallende@inf.utfsm.cl
[2] Departamento de Ciencia de Datos e Informática, Universidad de Playa Ancha, Valparaíso, Chile
carlos.valle@upla.cl

Abstract. In real-world environments, text classification faces limitations due to finite data and the closed-world assumptions combined with distributional identity in machine learning models. Leading to data with a distribution shifted from the learned one (in-distribution, ID), called out-of-distribution (OOD) data. OOD data present challenges related to safety, robustness, and the model's ability to discover new patterns. The literature on OOD detection mainly focuses on safety and robustness, leaving aside the discovery area where we can leverage the distinction between valuable OOD data, such as new classes, and anomalies, like outliers. To address this, we propose fine-tuning pre-training models with labeled OOD data and a regularization term for the outlier exposure method with samples of a mixture of beta distributions, forming a distribution with two peaks in the probability interval. This approach enhances the differentiation between two types of OOD data, near-OOD and far-OOD. Our experiments use RoBERTa, a BERT-based model, fine-tuned with LoRA, where the results show improved differentiation between near-OOD and far-OOD data without compromising performance on the principal task of multi-class classification and the base detection of ID and OOD. Using AUROC and FPR@95 metrics, our results indicate that this modified OOD detection with a beta approach for outlier exposure (OE) maintains good robustness and enhances the discovery capabilities of text classification models in real-world applications.

Keywords: Out-of-distribution · Natural language processing · Outlier exposure · Near-OOD · Far-OOD · Pre-trained transformer models

1 Introduction

The text classification problem is a fundamental task in natural language processing [1], typically operating under the assumptions of distributional identity and a closed-world scenario. However, maintaining these assumptions in real-world applications is challenging [2]. This problem generates out-of-distribution (OOD) data [3], which fall outside the learned distribution (ID) of a model. OOD data are vital for two main reasons: enhancing security by improving the model's robustness and reliability through filtering out data that do not meet established criteria [4], and facilitating discovery by enabling the model to identify potential new classes or patterns of interest [5]. This problem is significant because it occurs in various domains such as autonomous vehicle safety [6], robotics [7], medical applications [8], computer vision [3], task-oriented dialogue systems [9], and classic text classification [10].

The literature primarily addresses this challenge by distinguishing between in-distribution (ID) and out-of-distribution (OOD) data, further categorizing OOD data into near-OOD/easy-OOD and far-OOD/hard-OOD based on the type of distributional shift [11]. However, differentiation within the OOD group remains underexplored, and we consider it a significant matter for discovery. Therefore, we focus on this second-level detection without losing the ability to detect OOD data against ID, assuming that near-OOD data typically represent new classes or valuable information, while far-OOD data are more likely to be anomalies or non-useful information that falls outside the problem's context. For example, in a classification system for reporting major offenses, near-OOD might involve new types of crimes or emerging modus operandi, while far-OOD could include unrelated problems or minor infractions, like traffic violations. Additionally, the OOD detection problem intersects with related areas like novelty detection, anomaly detection, and open-set recognition, all of which share the challenge of distributional shifts from the trained data [12].

Starting from the premise that pre-trained transformer models help with the OOD data [13] and that the model can learn to deliver a uniform distribution for OOD data [14], we propose that if we seek a known and well-behaved distribution, we can guide a model based in BERT [15] and its distribution of OOD values toward one that facilitates binary classification between near-OOD and far-OOD while maintaining good performance in OOD detection and the main task of multi-class classification. For this research, we chose a mixture distribution generated by random sampling over beta distributions because the beta distribution offers moldability in shape and efficient integration into a model that works with multi-class classification and softmax.

In this work, we present in Sect. 2 the problem and the mathematical definitions formalized, besides the related work from the state of the art. In Sect. 3, we explain our proposed method and the ideas behind it. In Sect. 4, we present the methodology used in our experiments. In Sect. 5, we examine the results. Finally, in Sect. 6, we present our conclusions and explore directions for future research.

2 Problem and Formalization

In this section, we formalize the problem of OOD detection in the context of multi-class text classification models. We begin by defining the classification task and data representation. Next, we categorize the types of OOD data and discuss their characteristics. Subsequently, we introduce the OOD detector to distinguish between ID and OOD data, facilitating improved model robustness. Finally, we present a brief overview of related work in the field.

2.1 Classification and Representation

Given the context of utilizing deep models for a supervised multi-class text classification task in a real environment, we define the multi-class classification problem as follows: A set of examples and labels $(\{x_1, y_1\}, \ldots, \{x_j, y_j\})$, where $\mathbf{x} \in \mathbb{R}^n$ and $y \in K = \{1, 2, \ldots, K\}$. The task involves a parametric model $\mathcal{M}_\Theta : \mathcal{X} \to \mathcal{Y}$, with Θ being the trainable parameters and the $\mathcal{M}_\Theta$ can be represented by a composition of function $g \circ f(x)$ where $g(x)$ is the task-specific layer and $f(x)$, the features. The main task of the model is to learn to generalize the true joint distribution $P^*(x, y)$. The model approximates this joint distribution as $P(x, y)$.

We assume that for any representation vector $\phi(\mathbf{x})$, we can divide it into two independent and disjoint components [16], the **semantic features**, which are strongly related to the label, and the **background features**, which contain statistical information at the population level. Similar ideas exist under different names, such as decomposing into *content* (semantic) and *style* (background) [17]. Again, we use $\phi^*(\mathbf{x})$ to refer to the ideal theoretical representation, with the approximate learned representation being $\phi(\mathbf{x})$. We have:

$$\phi(x) = [\phi_s(\mathbf{x}); \phi_b(\mathbf{x})], \\ p^*(x) = p^*(\phi_s^*(\mathbf{x}))p^*(\phi_b^*(\mathbf{x})), \quad (1)$$

where $\forall y \in \mathcal{Y}$, it holds:

$$p^*(\phi_b^*(\mathbf{x})|y) = p^*(\phi_b^*(\mathbf{x})), \\ p^*(\phi_s^*(\mathbf{x})|y) \neq p^*(\phi_s^*(\mathbf{x})). \quad (2)$$

2.2 Types of Data

Considering the previous definitions and equations, we take an example $\mathbf{x}$ outside the training and validation sets to define the types of data and combine definitions with concepts from [3,11].

We have three types of OOD data derived from two types of shift distribution from the learned distribution (ID). First, we define the two distributional shifts:

- Covariate-shift/background-shift/domain-shift occurs when the domain differs from the distribution learned by the model.

$$\phi_b^*(\mathbf{x}) \neq \phi_b(\mathbf{x}). \quad (3)$$

- Semantic-shift/label-shift occurs when the example does not belong to the known classes of the problem.

$$\phi_s^*(\mathbf{x}) \neq \phi_s(\mathbf{x}). \tag{4}$$

The combination of these two shifts generates the three OOD types: generalized OOD, near-OOD, and far-OOD. The OOD generalization studies the covariate shift, which emphasizes robustness [18,19]. Near-OOD/hard-OOD occurs exclusively with semantic shift, while far-OOD/easy-OOD involves both shifts [3]. The last two types cover aspects related to the OOD detection problem.

With this, we have the near-OOD based on the assumption that it will be composed of possible new classes or topics related to the problem. On the other hand, far-OOD is supposed to contain values like anomalies or outliers with no relation to the context or topic. In this work, we do not study the full-spectrum [11] of the problem, where the setup considers all types of data, and the system works well in generalizing and detecting semantic shifts. We only study near-OOD and far-OOD to differentiate between valuable and non-useful data, considering the basis of OOD detection against ID data and the main task of multi-class classification.

2.3 OOD Detector

The problem is a binary classification that lies in differentiating whether an example $\mathbf{x}$ belongs to true distribution $P_\mathbf{x}^*$ or not, denoted as $Q_\mathbf{x}$ when it does not correspond to the distribution [20]. In our definitions, we will use the positive value 1 to detect values from $P_\mathbf{x}^*$.

To define a general detector based on the definition from the work [20], we first define a parameterized scoring function $s_\omega : \mathbb{R}^n \to \mathbb{R}$, whose goal is to quantify the confidence of the chosen class classified with parameters ω. Additionally, a decision threshold $\gamma \in \mathbb{R}$ and a detector $G : \mathbb{R}^n \to [0,1]$ are defined. We condition s_ω on the entire model $\mathcal{M}_\Theta$ because it is possible to use different layers of the model for s_ω not only the output of the model $g(x)$. Thus, the detector is:

$$G(\mathbf{x}|s_\omega, \mathcal{M}_\Theta, \gamma) = \mathbb{I}[s_\omega(\mathbf{x}\,|\,\mathcal{M}_\Theta) \geq \gamma] : \mathbf{x} \in \mathbb{R}^n. \tag{5}$$

With the detector defined, the theoretical detection error can be defined.

$$\mathcal{L}(P_\mathbf{x}^*, Q_\mathbf{x}; G) = \mathbb{E}_{\mathbf{x} \sim P_\mathbf{x}^*}[\mathbb{I}(G(\mathbf{x}) = 0)] + \mathbb{E}_{\mathbf{x} \sim Q_\mathbf{x}}[\mathbb{I}(G(\mathbf{x}) = 1)]. \tag{6}$$

The problem of OOD detection lies in choosing a scoring function, a type of training, and a model that helps to differentiate between $P_\mathbf{x}^*$ and $Q_\mathbf{x}$. Generally, the goal is only to differentiate between ID/OOD or ID/near-OOD; however, we also seek not only to ensure a good OOD detector but also to distinguish between near-OOD and far-OOD data.

2.4 Related Work

The work [21], which analyzes characters in computer vision, presents how deep models are more advantageous for OOD data than shallow models. Another important work that has led to various branches is the principal baseline in the current literature [22], which demonstrates that using the maximum value of the softmax output of a classification model with neuronal networks can be employed as a decision score when detecting OOD data. There are similarities with other problems that have a similar setup, where training occurs in one distribution and testing in another, such as anomaly detection, open set recognition, and model uncertainty, discussed in [3,12].

There are two primary techniques for differentiating OOD data: post-hoc methods [20,22–25], which are applied directly to the trained model, and training-based methods, which can be with or without labeled OOD data [14,26–28]. An additional critical aspect contributing to a more detailed taxonomy is the basis of decision scores, as highlighted in [3]. These scores can originate from various sources, including model outputs [22,29], distributions generated by the model [23], model gradients [30], or distances in a specific feature space [23,24].

The primary distinction between post-hoc methods and training-based methods lies in their ease of use. The post-hoc methods are model-agnostic and do not require retraining, but they can involve costly or unstable computations, such as inverse matrices [23]. In contrast, training-based methods face the challenge of labeling OOD data, which can be addressed by generating OOD data through algorithms or manual partitions [10] or by using existing manually labeled datasets [31].

While there are exhaustive benchmarks for computer vision [3], the NLP field lacks similarly detailed, varied, and extensive benchmarks [10]. Our approach focuses on text classification, marking a difference from existing benchmarks. We differentiate between near-OOD and far-OOD data by conducting experiments with a dataset from the literature already labeled with OOD data [31], as well as with other datasets generated through partitions above known classes [10].

3 Propose Method

In this section, we propose our method for OOD detection and enhance the differentiation of OOD data. The workflow of Fig. 1 shows the fine-tuning with OE and beta distributions and their context of use. Our principal target is to enhance the second filter of γ'.

The principal idea is to generate a two-peak distribution for OOD data, aiming to minimize the overlap between the beta distributions, as shown in Fig. 2. Simultaneously, the model outputs for ID distribution tend to separate, since the models already provide a solid baseline differentiation between ID and OOD, as demonstrated by the baseline work [22]. Consequently, this helps the model's distribution for ID data to trend towards higher values while giving a specific distribution to OOD values, which tend to produce lower scores.

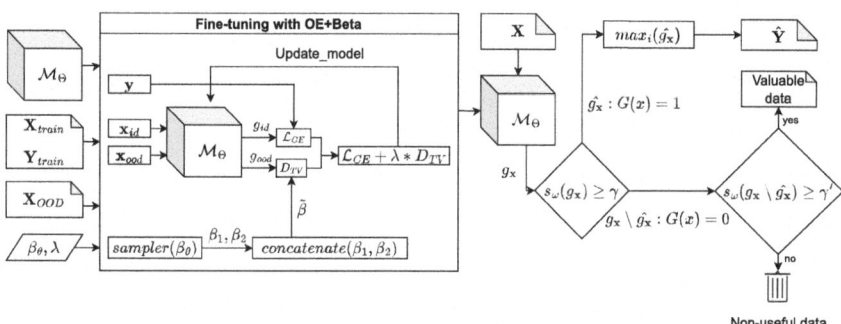

Fig. 1. Fine-tuning with OE and text classification with OOD detection.

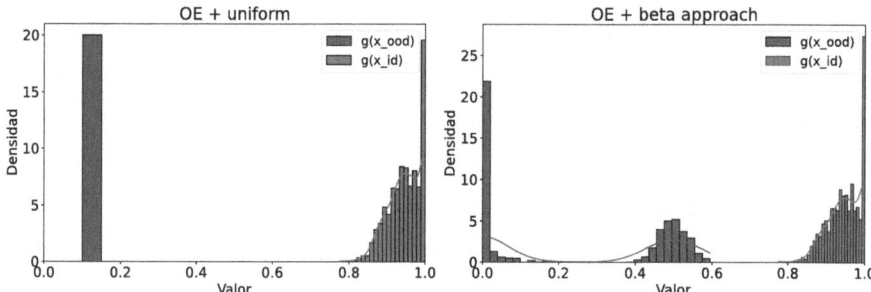

Fig. 2. The plots show the output values from a hypothetical model. In the left graph, OE with a normal distribution stacks all values towards a single point. In contrast, the right plot illustrates how our method introduces separation.

We chose to work with pre-trained transformers due to their superior performance in various natural language processing tasks, besides their improvement in OOD robustness [14] and use de MSP [22] score function in our experiments, which uses max softmax probability of the model's output. Besides, based on the previous section and formalization, we define the loss for OE with a beta distribution approach in equation (7). We represent de last layer of the model with $g_\Theta(x)$ and employ the standard cross-entropy loss $\mathcal{L}_{CE}$ for multi-class classification problems in the first part of the equation. Additionally, we incorporate a regularizer based on the total variation distance function D_{TV} for OOD data $\mathbf{x} \sim Q_{\mathbf{x}}$. We hypothesize that the total variation distance is more effective at mitigating the impact of distribution tails compared to the KL divergence used in other studies [14,35] because the model outputs tend to toward high scores with a small but widely distributed tail. This asymmetry complicates their comparison with distributions like two-peak beta, making the total variation distance a more suitable metric. Furthermore, we use a mixture of samples from two beta distributions, β_1 and β_2, denoted as $\tilde{\beta} = [\beta_1, \beta_2]$, ensuring that both beta distributions are normalized to sum 1.

$$\mathcal{L}(g_\Theta(\mathbf{x}), y) = \mathbb{E}_{\mathbf{x} \sim P_\mathbf{x}^*}[\mathcal{L}_{CE}(g_\Theta(\mathbf{x}), y)] + \lambda\, \mathbb{E}_{\mathbf{x} \sim Q_x}[D_{TV}(g_\Theta(\mathbf{x}), \tilde{\beta}))]. \tag{7}$$

We present a clear and comprehensive explanation of our proposed method, detailed in Algorithm 1, which incorporates fine-tuning with OE using Beta distributions.

Algorithm 1. Fine-tuning with OE and beta distribution

Require: $\mathbf{X}_{train}$: Training data, $\mathbf{Y}_{train}$: Training labels, $\mathbf{X}_{ood}$: Out-of-distribution data, λ: Regularization parameter, *epochs*: Number of epochs, $\mathcal{M}_\Theta$: Pre-trained model, β_θ: Beta parameters.
Ensure: $\mathcal{M}_\Theta$: Fine-tuned model
1: **for** each e in *epochs* **do**
2:    **for** each *batch* in $(\mathbf{X}_{train}, \mathbf{Y}_{train}, \mathbf{X}_{ood})$ **do**
3:      $\mathbf{x}_{id}, \mathbf{y}, \mathbf{x}_{ood} \leftarrow batch$
4:      $\beta_1, \beta_2 \leftarrow sampler(\beta_\theta)$
5:      $\tilde{\beta} \leftarrow concatenate(\beta_1, \beta_2)$
6:      $g_{id} \leftarrow \mathcal{M}_\Theta(\mathbf{x}_{id})$
7:      $loss \leftarrow \mathcal{L}_{CE}(\mathbf{y}, g_{id})$
8:      $g_{ood} \leftarrow \mathcal{M}_\Theta(\mathbf{x}_{ood})$
9:      $regularizer \leftarrow \lambda * D_{TV}(\tilde{\beta}, g_{ood})$
10:     $final\_loss \leftarrow loss + regularizer$
11:     Update *model* with $final\_loss$
12:    **end for**
13: **end for**
14: **return** *model*

In the algorithm, steps 4 and 5 generate two samples of betas and concatenate in one matrix, tailoring the desired distribution for OOD data with the conditions: they have the same size as a batch, and the vector of beta samples is the length equal to the last layer of the model. Next, Steps 6 and 7 involve performing standard prediction and loss calculations for classification models. Subsequently, steps 8 and 9 calculate the regularizer by employing a total variation function for the model outputs on OOD data against the sample mixture of Beta distribution. Finally, steps 10 and 11 involve fine-tuning the model with the two components of the loss function. Each epoch helps the model learn the classification task and also guides the model's outputs for OOD data towards a desirable distribution of our interest by penalizing model outputs on OOD data that deviate from the Beta mixture distribution.

4 Methodology

To assess our method, we will use a RoBERTa-base pre-trained model [36] and fine-tuning with LoRA [37] that significantly reduces the number of trainable parameters, implemented with the PEFT library [38], which shows advantages in training time and allows us to see its impact on state-of-the-art algorithms. It will be studied with three experimental runs for each experiment with 10 epochs of fine-tuning and a learning rate of 1e-4. On the other hand, we will use various text datasets, three experimental setups, and three techniques to compare: the OOD detection baseline Max Softmax Probability (MSP) [22], one of the best post hoc methods Virtual-Logit Matching (ViM) [24], and the comparison baseline for Outlier Exposure with a uniform distribution using KL divergence [14]. Both post-hoc methods, MSP and ViM, use a cross-entropy loss and LoRA technique. The code and experiments are available in a public repository[1].

4.1 Datasets

Our primary datasets are CLINC150 [31], a dataset on user intent classification distributed across 150 classes; NewsCategory (NC) [39], a collection of posts from a news website with the headline and a short description, comprising 42 classes, where our study only uses the headlines; the TREC dataset [40,41] is a question classification dataset, where we will use its coarse version with 6 classes to observe behavior with fewer classes. Lastly, the NewsGroups (NG) dataset [42] is a collection of newsgroup documents with 20 classes and long texts. As shown in the summary of each dataset each one is distinct. Each principal dataset has additional datasets used as far-OOD, and different ones are utilized for training/validation and testing. These are the YELP [43], SST2 [44], SNLI [45], Reddit [33], and IMDB [34] datasets, which are user reviews for products (YELP), sentences extracted from movie reviews (SST2), and human-written sentences (SNLI), posts from Reddit website (Reddit) and movie reviews (IMDB) respectively. For near-OOD data, only CLINC150 includes labeled OOD data (C150-OOD). According to the authors, these are potential new classes and texts related to the ID data, making it feasible to consider them labeled near-OOD. For the other datasets, there is the option to exclude certain classes at all stages and treat them as near-OOD, as demonstrated in the work [10]. Following the previously cited work, we exclude the least frequent classes and add another condition: the classes left as near-OOD in train and validation should not have classes that appear in testing as near-OOD. To refer to these partitions, we label the most frequent classes from the original dataset as TREC-I (ID), while the classes designated as OOD are labeled TREC-O. However, as mentioned earlier, we divide this into two groups with disjoint classes, TREC-O1 and TREC-O2. A summary of the specific experimental structures is presented in Table 1.

4.2 Metrics

In the experiments, we use two commonly used metrics from the literature. The first is AUROC, which measures the area under the ROC curve and can be

---

[1] https://github.com/cmaldona/BOE-Beta-distribution-for-Outlier-Exposure/tree/v1.0.

Table 1. Each value in the table indicates the type of data, either near-OOD or far-OOD, respectively. The first column represents the dataset as ID.

	Train	Validation	Testing
CLINC150	C150-OOD/SST2	C150-OOD/YELP	C150-OOD/NC
NC	NC-O1/SNLI	NC-O1/YELP	NC-O2/SST2
TREC	TREC-O1/SST2	TREC-O1/YELP	TREC-O2/SNLI
NG	NG-O1/Reddit-1	NG-O1/Reddit-2	NG-O2/IMDB

interpreted as the probability that the positive class scores higher than the negative class, considering different gap values, representing a more generic metric on the behavior of the methods [22]. For AUROC, we are interested in values close to 1. On the other hand, we consider a specific threshold metric, FPR@N, which measures the False Positive Rate (FPR) when a True Positive Rate (TPR) is N%, aiming for a value close to 0 to minimize false positives. We choose FPR@95 to place a higher demand on the methods and use ID values as positive, similar to our main comparison work in OE [14]. Therefore, we omit the AUPR metric that prioritizes the positive class because most data are ID and add little value to our study compared to metrics that prioritize strict detection and use OOD as the positive class.

4.3 Experiments

We analyze four experiments with three runs for each one, besides each one with a different ID dataset. These experiments follow this structure: a training set containing ID, near-OOD (labeled or generated depending on the dataset), and far-OOD data, represented by a different dataset. This same structure is applied to the validation and testing sets, ensuring that the far-OOD datasets of their classes do not repeat across stages. On the other hand, for generated near-OOD data, we opted to exclude certain classes and divide them into train, validation, and test sets. Considerate that classes from the test set do not appear in the train or validation set.

The techniques compared include three significant state-of-the-art (SOTA) works: MSP [22], ViM [24], and OE [14]. MSP, the simplest method, uses the max value of the softmax layer as a decision score. ViM combines class-dependent and class-agnostic information, such as outputs from intermediate layers, to generate a virtual output that serves as a decision score. ViM creates the virtual logit using information from the principal subspaces of the features. Both MSP and ViM do not require labeled OOD data. MSP is the baseline for post-hoc methods, while ViM, a more recent post-hoc method, demonstrates to be one of the best current post-hoc methods. Lastly, OE is compared as a baseline, using a regularizer and a uniform distribution for the OOD data with KL divergence in the loss function for training. Hyperparameters for ViM are tuned using validation data, and for

OE, the lambda parameter used is 1, following recommendations from their NLP studies.

An important aspect is the selective use of classes in the TREC, NC, and NG datasets. Less frequent are excluded from the ID set to create the data labeled as near-OOD. Specifically, we retained 4 out of 6 classes for TREC, 30 out of 42 for News Category as ID, and 15 of 20 for Newsgroups as ID. In contrast, for the CLINC150 dataset, all 150 classes are considered ID due to their fully labeled nature, making such exclusions unnecessary.

The studied beta combination is $\tilde{\beta} = [(0.1, 10), (100, 100)]$, chosen for their distinct shapes and the premise that near-OOD should approximate 0.5 values, while far-OOD should exhibit low probability values. These tuples represent the alpha and beta parameters of the Beta distribution, $B(\alpha, \beta)$. The generated mixed distribution combines random samples from both distributions, each with vector sizes of K, to produce a vector comparable to the model output. We employ a lambda value of 1.

5 Results

5.1 Robustness

This experiment compares the robustness of our method $OE(\tilde{\beta})$ against the other selected techniques, demonstrating that it retains its qualities as an OOD detector and classifier. However, as expected, it does not excel at distinguishing ID from OOD data. Due to the increased flexibility in the OOD data distribution, which can lead to a slight performance loss compared to methods like the OE baseline, $OE(U)$. Table 2 presents the average results from three experimental runs, including accuracy to indicate performance on the classification task and how the regularizer can slightly affect its performance. We focus on near-OOD data, as it is the most challenging, and including far-OOD data would dilute the metric's impact for the near-OOD by increasing the data volume.

Table 2. Each box represents the metrics ACC/AUROC/FPR@95. Only ID and near-OOD data are considered, with ID data treated as positive and accuracy calculated over the ID data. The best results are bolded.

	Clinc150	TREC4	NC30	NG15
MSP	**0.995**/0.992/0.03	**0.983**/0.955/0.271	**0.733**/0.744/0.849	**0.979**/0.804/0.517
ViM	**0.995**/0.996/0.01	0.968/**0.997**/**0.008**	**0.733**/0.758/0.78	**0.979**/0.893/0.222
OE(U)	0.987/**1**/**0**	**0.983**/0.994/0.028	0.698/**0.943**/**0.323**	0.923/**0.994**/0.03
OE($\tilde{\beta}$)	0.988/0.999/**0**	0.974/0.996/0.017	0.717/0.896/0.449	0.976/0.991/**0.02**

The accuracy helps us evaluate the impact of adding regularization, which is minimally affected and holds even when the performances are not as strong as in the NC30 experiment.

5.2 Discovery

This setup is the highlighting point of our proposal and study, aiming to improve its ability to differentiate the type of data between OOD data and enhance the second filter γ' from Fig. 1, while preserving the qualities of classifier and OOD detection viewed in the previous section. In this case, the improvement offered by the method in distinguishing between near-OOD and far-OOD compared to other methods can be observed in the Tables 3 and 4. However, contrary to expectations, the experiment with NC30, which initially appeared to have the worst result, does not perform the worst and can be for the relation between accuracy and overconfidence in deep models. When considering all techniques, the experiment with TREC4 shows an issue, and the difference with TREC4 compared to other datasets is the number of classes and the generality of those classes. While the TREC dataset allows for more granular classification, we use the coarse version to focus our experiments on the performance with fewer classes. This problem suggests a relationship between the number of classes and the model's ability to learn patterns, highlighting a crucial factor to consider in these methods.

Table 3. Table of AUROC values. Only near-OOD and far-OOD data are considered, with near-OOD being the positive. The mean results are shown, with the standard deviation in parentheses.

	Clinc150 ($\uparrow$)	TREC4 ($\uparrow$)	NC30 ($\uparrow$)	NG15 ($\uparrow$)
MSP	0.633(0.017)	0.438(0.031)	0.88(0.028)	0.657(0.033)
ViM(D=256)	0.628(0.022)	**0.685**(0.047)	0.55(0.08)	0.636(0.052)
OE(U)	0.971(0.008)	0.34(0.096)	0.996(0.001)	0.825(0.118)
OE($\tilde{\beta}$)	**0.993**(0.112)	0.619(0.14)	**0.999**(0)	**0.929**(0.056)

Table 4. Table of FPR values when TPR is 95%. Only near-OOD and far-OOD data are considered, with near-OOD being the positive. The mean results are shown, with the standard deviation in parentheses.

	Clinc150 ($\downarrow$)	TREC4 ($\downarrow$)	NC30 ($\downarrow$)	NG15 ($\downarrow$)
MSP	0.823(0.032)	0.978(0.008)	0.482(0.067)	0.804(0.054)
ViM(D = 256)	0.943(0.015)	**0.831**(0.012)	0.981(0.018)	0.827(0.087)
OE(U)	0.133(0.074)	0.964(0.02)	0.013(0.004)	0.775(0.259)
OE($\tilde{\beta}$)	**0.02**(0.026)	0.871(0.182)	**0.004**(0.001)	**0.4348**(0.364)

An additional point of interest is that OE(U) demonstrates a quality not studied in their work [14], with the method helping to differentiate valuable data from anomalies. In previous studies on OE, there was no differentiation between near-OOD and far-OOD in the data. In both OE versions, we provide near-OOD and far-OOD and do not distinguish their labels in the loss functions.

6 Conclusions

We have shown in this paper that outlier exposure methods can offer additional classification and differentiation beyond traditional OOD detection. Our proposal incorporates a favorable and flexible distribution, such as beta distributions, which led to improved discernment between valuable data with new classes (near-OOD) and anomalies from other datasets (far-OOD). The beta distribution provided a desirable shape that helped in this task. Lastly, remark the presented method has better results but involves an extra hyperparameter, the beta distribution parameters, compared to the OE baseline, which also has a lambda value for the regularizer. We recommend $\tilde{\beta} = [(0.1, 10), (100, 100)]$ and $\lambda = 1$, as these values have been studied and consistently demonstrated strong. Future research interests include the explicit separation of near and far values in the loss function and automatic generation or augmentation of these data for training since near-OOD data are more challenging to obtain than far-OOD.

Acknowledgments. This work was funded by ANID PIA/APOYO AFB230003 and in part by Project DGIIP-UTFSM PI-LIR23-13. The work of Carlos Valle was supported by the National Agency for Research and Development (ANID) Chile through the National Fund for Scientific and Technological Development (FONDECYT) Iniciación Project under Grant 11230351.

Disclosure of Interests. The authors declare no competing interests.

References

1. Li, Q., et al.: A survey on text classification: from traditional to deep learning. ACM Trans. Intell. Syst. Technol. **13**(2), 31 (2022)
2. Parmar, J., Chouhan, S., Raychoudhury, V., Rathore, S.: Open-world machine learning: applications, challenges, and opportunities. ACM Comput. Surv. **55**(10), 205 (2023)
3. Yang, J., et al.: OpenOOD: benchmarking generalized out-of-distribution detection. In: Proceedings of the 36th Conference on Neural Information Processing Systems (NeurIPS 2022), Track on Datasets and Benchmarks (2022)
4. Mohseni, S., Wang, H., Xiao, C., Yu, Z., Wang, Z., Yadawa, J.: Taxonomy of machine learning safety: a survey and primer. ACM Comput. Surv. **55**(8), 157 (2022)
5. Uhlemeyer, S., Lienen, J., Hüllermeier, E., Gottschalk, H.: Detecting novelties with empty classes. arXiv preprint arXiv:2305.00983 (2023)
6. Koopman, P., Wagner, M.: Autonomous vehicle safety: an interdisciplinary challenge. IEEE Intell. Transp. Syst. Mag. **9**(4), 90–96 (2017)

7. Mantegazza, D., Giusti, A., Gambardella, L.M., Guzzi, J.: An outlier exposure approach to improve visual anomaly detection performance for mobile robots. IEEE Robot. Autom. Lett. **7**, 11354–11361 (2022)
8. Ulmer, D., Meijerink, L., Cinà, G.: Trust issues: uncertainty estimation does not enable reliable OOD detection on medical tabular data. In: Alsentzer, E., McDermott, M.B.A., Falck, F., Sarkar, S.K., Roy, S., Hyland, S.L. (eds.) Proceedings of the Machine Learning for Health NeurIPS Workshop, vol. 136, pp. 341–354. PMLR (2020)
9. Zhang, H., Xu, H., Lin, T.-E.: Deep open intent classification with adaptive decision boundary. In: AAAI Conference on Artificial Intelligence (2020)
10. Baran, M., Baran, J., Wójcik, M., Zięba, M., Gonczarek, A.: Classical out-of-distribution detection methods benchmark in text classification tasks. In: Padmakumar, V., Vallejo, G., Fu, Y. (eds.) Proceedings of the 61st Annual Meeting of the Association for Computational Linguistics (Volume 4: Student Research Workshop), pp. 119–129. Association for Computational Linguistics, Toronto, Canada, July 2023
11. Zhang, J., et al.: OpenOOD v1.5: enhanced benchmark for out-of-distribution detection. arXiv preprint arXiv:2306.09301 (2023)
12. Salehi, M., Mirzaei, H., Hendrycks, D., Li, Y., Rohban, M.H., Sabokrou, M.: A unified survey on anomaly, novelty, open-set, and out-of-distribution detection: solutions and future challenges. Trans. Mach. Learn. Res. 2835–8856 (2022)
13. Hendrycks, D., Liu, X., Wallace, E., Dziedzic, A., Krishnan, R., Song, D.: Pretrained transformers improve out-of-distribution robustness. In: Jurafsky, D., Chai, J., Schluter, N., Tetreault, J. (eds.) Proceedings of the 58th Annual Meeting of the Association for Computational Linguistics, pp. 2744–2751. Association for Computational Linguistics, Online (2020)
14. Hendrycks, D., Mazeika, M., Dietterich, T.: Deep anomaly detection with outlier exposure. In: Proceedings of the International Conference on Learning Representations (ICLR), pp. 1–10. New Orleans, LA, USA (2019)
15. Devlin, J., Chang, M. W., Lee, K., Toutanova, K.: BERT: pre-training of deep bidirectional transformers for language understanding. In: Proceedings of the North American Chapter of the Association for Computational Linguistics (NAACL) (2019)
16. Ren, J., et al.: Likelihood ratios for out-of-distribution detection. In: Wallach, H., Larochelle, H., Beygelzimer, A., d'Alché-Buc, F., Fox, E., Garnett, R. (eds.) Advances in Neural Information Processing Systems, vol. 32. Curran Associates, Inc. (2019)
17. Fu, Z., Tan, X., Peng, N., Zhao, D., Yan, R.: Style transfer in text: exploration and evaluation. In: Proceedings of the AAAI Conference on Artificial Intelligence, vol. 32, no. 1 (2018)
18. Hendrycks, D., Dietterich, T.: Benchmarking neural network robustness to common corruptions and perturbations. In: International Conference on Learning Representations (ICLR 2019), New Orleans, USA (2019)
19. Hendrycks, D., et al.: The many faces of robustness: a critical analysis of out-of-distribution generalization. In: 2021 IEEE/CVF International Conference on Computer Vision (ICCV), pp. 8320–8329 (2021)
20. Liang, S., Srikant, R.: Enhancing the reliability of out-of-distribution image detection in neural networks. In: Proceedings of the International Conference on Learning Representations (ICLR 2018), Vancouver, Canada (2018)

21. Bengio, Y., et al.: Deep learners benefit more from out-of-distribution examples. In: Gordon, G., Dunson, D., Dudík, M. (eds.) Proceedings of the Fourteenth International Conference on Artificial Intelligence and Statistics, Proceedings of Machine Learning Research, vol. 15, pp. 164–172. PMLR, Fort Lauderdale, FL, USA, 11–13 April 2011
22. Hendrycks, D., Gimpel, K.: A baseline for detecting misclassified and out-of-distribution examples in neural networks. In: Proceedings of the International Conference on Learning Representations (ICLR), pp. 1–10. Toulon, France (2017)
23. Lee, K., Lee, K., Lee, H., Shin, J.: A simple unified framework for detecting out-of-distribution samples and adversarial attacks. In: Proceedings of the 32nd Conference on Neural Information Processing Systems (NIPS 2018), Montréal, Canada (2018)
24. Wang, H., Li, Z., Feng, L., Zhang, W.: ViM: out-of-distribution with virtual-logit matching. In: Proceedings of the IEEE Conference on Computer Vision and Pattern Recognition (CVPR), pp. 4911–4920. IEEE (2022)
25. Sun, Y., Ming, Y., Zhu, X., Li, Y.: Out-of-distribution detection with deep nearest neighbors. In: Proceedings of the 39th International Conference on Machine Learning (ICML 2022), vol. 162. PMLR, Baltimore, Maryland, USA (2022)
26. Tack, J., Mo, S., Jeong, J., Shin, J.: CSI: novelty detection via contrastive learning on distributionally shifted instances. In: Proceedings of the 34th Conference on Neural Information Processing Systems (NeurIPS 2020), Vancouver, Canada (2020)
27. Wei, H., Xie, R., Cheng, H., Feng, L., An, B., Li, Y.: Mitigating neural network overconfidence with logit normalization. In: Proceedings of the 39th International Conference on Machine Learning (ICML 2022), vol. 162. PMLR, Baltimore, Maryland, USA (2022)
28. Yang, J., et al.: Semantically coherent out-of-distribution detection. In: Proceedings of the IEEE International Conference on Computer Vision (ICCV 2021). IEEE (2021)
29. Liu, W., Owens, J.D., Wang, X., Li, Y.: Energy-based out-of-distribution detection. In: 34th Conference on Neural Information Processing Systems (NeurIPS), Vancouver, Canada (2020)
30. Huang, R., Geng, A., Li, Y.: On the importance of gradients for detecting distributional shifts in the wild. In: 35th Conference on Neural Information Processing Systems (NeurIPS 2021). Department of Computer Sciences, University of Wisconsin-Madison, Madison, WI, USA (2021)
31. Larson, S., et al.: An evaluation dataset for intent classification and out-of-scope prediction. In: Proceedings of the 2019 Conference on Empirical Methods in Natural Language Processing and the 9th International Joint Conference on Natural Language Processing, pp. 1311–1316, Hong Kong, China, 3–7 November 2019. Association for Computational Linguistics (2019)
32. Lang, K.: NewsWeeder: learning to filter netnews. In: Prieditis, A., Russell, S. (eds.) Machine Learning Proceedings 1995, pp. 331–339. Morgan Kaufmann, San Francisco, CA (1995). https://doi.org/10.1016/B978-1-55860-377-6.50048-7
33. Baumgartner, J., Zannettou, S., Keegan, B., Squire, M., Blackburn, J.: The pushshift reddit dataset. In: Proceedings of the International AAAI Conference on Web and Social Media, vol. 14, pp. 830–839 (2020)
34. Maas, A.L., Daly, R.E., Pham, P.T., Huang, D., Ng, A.Y., Potts, C.: Learning word vectors for sentiment analysis. In: Proceedings of the 49th Annual Meeting of the Association for Computational Linguistics: Human Language Technologies, pp. 142–150 (2011)

35. Jiang, X., et al.: Detecting out-of-distribution data through in-distribution class prior. In: Proceedings of the 40th International Conference on Machine Learning, vol. 202, pp. 15067–15088. PMLR (2023)
36. Liu, Y., et al.: RoBERTa: a robustly optimized BERT pretraining approach. arXiv:1907.11692 (2019)
37. Hu, E.J., et al.: LoRA: low-rank adaptation of large language models. In: International Conference on Learning Representations (2022)
38. Mangrulkar, S., Gugger, S., Debut, L., Belkada, Y., Paul, S., Bossan, B.: PEFT: state-of-the-art parameter-efficient fine-tuning methods (2022)
39. Misra, R.: News Category Dataset (2018)
40. Hovy, E., Gerber, L., Hermjakob, U., Lin, C.-Y., Ravichandran, D.: Toward semantics-based answer pinpointing. In: Proceedings of the First International Conference on Human Language Technology Research (2001)
41. Li, X., Roth, D.: Learning question classifiers. In: COLING 2002: The 19th International Conference on Computational Linguistics (2002)
42. Lang, K.: NewsWeeder: learning to filter netnews. In: Prieditis, A., Russell, S. (eds.) Machine Learning Proceedings 1995, páginas 331–339. Morgan Kaufmann, San Francisco (1995)
43. Zhang, X., Zhao, J., LeCun, Y.: Character-level convolutional networks for text classification. In: Advances in Neural Information Processing Systems (NIPS 2015), vol. 28 (2015)
44. Socher, R., et al.: Recursive deep models for semantic compositionality over a sentiment treebank. In: Proceedings of the 2013 Conference on Empirical Methods in Natural Language Processing, páginas, pp. 1631–1642. Association for Computational Linguistics, Seattle, Washington, USA (2013)
45. Young, P., Lai, A., Hodosh, M., Hockenmaier, J.: From image descriptions to visual denotations: new similarity metrics for semantic inference over event descriptions. Trans. Assoc. Comput. Linguist. páginas **2**, 67–78, Cambridge, MA. MIT Press (2014)

Impact of Quantization on Large Language Models for Portuguese Classification Tasks

Danilo Samuel Jodas, Gabriel Lino Garcia(✉), Pedro Henrique Paiola, João Renato Ribeiro Manesco, and João Paulo Papa

School of Sciences, São Paulo State University (UNESP), Bauru, Brazil
{danilo.jodas,gabriel.lino,pedro.paiola,joao.r.manesco,
joao.papa}@unesp.br

Abstract. Large Language Models have emerged as transformative agents in the frequently evolving landscape of artificial intelligence, reshaping the world towards a disruptive and modern technological era. This paradigm stresses their crucial role in extending the generative capabilities in the context of natural language processing. Generative Artificial Intelligence, an innovative and cutting-edge research topic, is critical to unlocking remarkable opportunities in our era of unparalleled technological progress. Despite the remarkable progress made in language model architectures, their exponential growth still raises pertinent concerns regarding their deployment and the associated costs for retraining efforts tailored to specific tasks. We present a study achieving a detailed analysis of the impact resulting from the application of diverse quantization methodologies on an open-source large language model tailored for Portuguese classification tasks, aka Bode. Our research thoroughly evaluates the performance nuances introduced by various quantization strategies, thus providing valuable insights into the constant concerns surrounding the optimization of large language models, aiming for enhanced efficiency and effectiveness in growing applications for the Portuguese community.

Keywords: Quantization · Large Language Models · Natural Language Processing · Generative Artificial Intelligence · Bode

1 Introduction

Large Language Models (LLMs) have recently emerged to reshape the cognitive aspects of human society and promote artificial intelligence (AI) generative capabilities. Generative AI is a novel and cutting-edge research field that unlocks promising opportunities in the modern era of remarkable technological progress, providing the ability to support the creation of natural content imitating the human creative process. In this context, Generative Pretrained Transformers (GPT) have been the firstborn milestone in the new era towards a wide range of novel generative natural language text and multimedia applications, exhibiting a noteworthy performance and allowing disruptive research beyond the human knowledge frontier in unexpected ways.

ChatGPT, OpenAI's first GPT model, was designed to encompass a massive deep neural network architecture comprising 175 billion parameters in its third generation. Despite the outstanding and ever-evolving capabilities, the dramatic increase in the LLM architectures has raised several concerns about the cost of further improvements and retraining such models applied to a specific task, thus leading to the urging of more studies focused on building efficient and smaller generative architectures.

Recent progress has unlocked novel strategies to focus on building Small Language Models (SLMs) exhibiting similar capabilities in the context of natural content generation. MetaAI has recently revealed its own open-source Large Language Model Meta AI (LLaMa) architecture exhibiting outstanding generative capabilities that outperform the GPT-3 model across a collection of models with different setups ranging from 7 billion to 65 billion parameters [17,18]. Further, small open-source architectures have also been designed as competitive language models to cope with commercial counterparts with a massive number of parameters. Mistral [10], Falcon [1], Alpaca[1], and Vicuna[2] [2] are the most famous examples of open-source models whose generative capacities resulted in competitive performance, thus promoting further applications in generative AI.

Apart from the impressive innovation carried by LLMs and their generalist capabilities in a wide range of applications, they also positively impact the Portuguese community, offering transformative potential in various aspects of language processing and communication. LLMs enable innovative applications and tools tailored to Portuguese speakers by leveraging state-of-the-art natural language generation and understanding capabilities. These models facilitate advances in sentiment analysis [6], text summarization [5], and content generation [15], among other areas, improving accessibility to digital content and services for Portuguese speakers. Furthermore, LLMs contribute to preserving and promoting the Portuguese language by providing resources for language learning, cultural preservation, and linguistic research. As LLM technology continues to evolve, its impact on the Portuguese-speaking community will expand, thus pushing further innovation in developing and applying quantization methods across multiple applications.

Despite the best research efforts focused on optimizing the language model parameters, training deep neural networks, including LLMs, becomes challenging since the process demands even more expensive hardware. Optimizing very large language architectures becomes especially critical when such models are required to perform in low-cost hardware. Further, several studies have focused on the impact of lower-precision operations on the model's performance, including converting the high-precision values of the models' weights to a lower-precision data type targeting the models' size reduction. In this context, the term *quantization* has been introduced to address such challenges by providing an effective approach to reduce the weight's precision of the model while maintaining its performance and generative capabilities in the best possible way.

---

[1] https://crfm.stanford.edu/2023/03/13/alpaca.html.
[2] https://lmsys.org/blog/2023-03-30-vicuna/.

This paper provides a comprehensive analysis of the impact derived from the application of several post-training quantization methodologies on leading open-source LLMs applied to the Portuguese language. We specifically evaluated the quantization effects on a recent LLM proposed by Garcia et al. [6]. They developed the Bode model based on the LLaMa 2 architecture, which underwent fine-tuning using the Portuguese version of the Alpaca dataset. This model was chosen because it achieved superior performance on various Portuguese classification tasks, overcoming Mistral, Falcon, and OpenCabrita 3B models. Therefore, the proposed study is specifically tailored to assess their implications in the context of the classification task, providing an extensive evaluation of the performance nuances introduced by different quantization strategies when the models are trained with Portuguese content.

The contributions of the work are threefold:

- To provide a comprehensive analysis of the impact of LLM quantization when applied to the Portuguese language;
- To assess different quantization techniques applied to the Bode model and the LLaMa 2 architecture, both configured with 7 billion and 13 billion parameters; and
- To promote quantized LLMs for Portuguese classification tasks on resource-limited hardware.

The remainder of this paper is organized as follows: Sect. 2 provides a theoretical background regarding the quantization of LLMs. Section 3 describes the related work in this field, while Sect. 4 presents the proposed approach for LLM quantization. Section 5 describes the research methodology, while Sect. 6 discusses the experiments and results. At last, conclusions and future works are stated in Sect. 7.

2 Theoretical Background

In the context of LLMs, Quantization refers to a technique used to reduce the memory and computational requirements while maintaining their performance [12,20]. It involves characterizing the model's parameters and activation values using fewer bits than the standard floating-point format typically used. This reduction in precision allows for more efficient storage and computation.

The quantization process splits into *weight quantization* and *activation quantization*. Weight quantization involves converting the model's parameters (weights) from floating-point numbers to lower-precision fixed-point numbers, thus reducing their precision. Activation quantization reduces the precision of the intermediate activation values produced during the forward pass of inference in a neural network. These activations are the outputs of each layer in the network and serve as inputs to the subsequent layers.

In addition, LLM quantization can be applied to the level of the key-value (KV) cache of the attention mechanism in transformed-based models [4,7,19]. In these models, the attention mechanism involves calculating attention scores between query vectors (typically derived from the input tokens) and key vectors (typically derived from the context or memory). Quantization of key-value caches typically involves reducing the precision of the key and value vectors stored in the cache. These vectors are used repeatedly across multiple layers and tokens during attention calculation. By quantizing the key-value caches, the memory footprint of the model can be reduced, allowing for more efficient storage and retrieval of these vectors during inference. This reduction in precision can be achieved using techniques similar to weight and activation quantization, where the vectors are represented using lower-precision fixed-point numbers or integer formats.

LLM quantization is categorized into *Quantization Aware Training* (QAT) and *Post-Training Quantization* (PTQ). While the former incorporates quantization into the training process of LLMs, the latter approach involves applying quantization to a previously already trained model, which simplifies the quantization procedure but exposes it to harm in precision. PTQ includes methods like Generalized Post-Training Quantization (GPTQ) and GPT-Generated Unified Format (GGUF) quantization. GPTQ was primarily proposed for GPU-driven quantization, while GGUF is a new approach designed to focus on models that run on CPU devices.

Quantization can significantly reduce LLMs' memory requirements and computational costs, making them more practical for deployment on resource-constrained devices such as mobile phones and edge devices. However, quantization may also introduce some loss in model performance due to the reduced precision. Therefore, fine-tuning and calibration are often employed to mitigate this performance degradation.

3 Related Works

The quantization of LLMs has been a research topic addressed in numerous recent studies. Novel strategies have been proposed to improve the prediction capabilities and speedup of the models on mobile and edge resource-limited devices at lower-bit precision [13]. In addition, reducing the bit scaling while upholding the performance during prediction is crucial for pushing the quantization of LLMs even further to 4 or lower bit precision [8]. For example, Shang et al. [16] proposed the Partially Binarized LLM (PB-LLM), an efficient method for compressing the network's weights to 1-bit precision. They adopted a strategy to preserve high-bit precision for a small set of the most relevant weights while the other weights are quantized to a 1-bit scale. A similar approach was proposed by Huang et al. [8], in which the most salient weights, i.e., the ones with importance for the LLM, are selected based on the Hessian Matrix and binarized through the residuals of the original values and their binarized version.

Liu et al. [14] enhance LLM quantization by integrating knowledge distillation to overcome resource constraints and limited data acquisition for QAT-based methods. They generate data for fine-tuning using a pre-trained model and apply quantization at various architectural levels, including weights, activations, and the KV cache. This approach refines model performance without additional data and mitigates resource-intensive training. The method combines knowledge distillation and targeted quantization for efficient LLM optimization. They found interesting results using different precision level combinations, including remarkable results for 8-bit and 4-bit quantization at the weight level.

Jin et al. [11] addressed the challenges regarding the lack of comprehensive benchmarks of the LLM quantization in the application to specific tasks or general knowledge capabilities across multiple domains. In addition, they claimed that there was a lack of protocol evaluation concerning unbiased, reliable responses and resource usage, such as memory and inference speed. They adopted three critical aspects: knowledge and capacity, alignment, and efficiency. Their outcomes provided important insights among three quantization methods applied to Qwen-chat models at 8, 4, 3, and 2-bit precision levels, demonstrating similar performance of their non-quantization counterpart versions and the 8-bit and 4-bit quantization equivalents.

In a recent study, Huang et al. [9] assessed the effects of post-training quantization methods on the newly released LLaMa 3 model across various bit precision levels. They found remarkable outcomes even at weights reduced to a 4-bit precision, with results retaining the accuracy of the non-quantized version of LLaMa 3 with 8 billion (8B) and 70 billion (70B) parameters. In addition, they contrasted the quantization performance with the fine-tuning accuracy yielded from the quantized version of the LLaMa 3 8B with 4-bit precision. This comparison demonstrates the effectiveness of the PTQ without any fine-tuning procedure, revealing the necessity for a robust dataset for further training of the model over its quantized counterpart.

Despite the remarkable performance achieved by the different LLM quantization techniques, the previous attempts only focused on the impact of the quantization performance on English-based models. However, only recent studies have focused on expanding the application of LLMs to the Portuguese language. Garcia et al. [6] explored the utilization of LLMs for Portuguese classification tasks by employing the second version of the LLaMa 2 architecture to propose the Bode model. They introduced a trained variant of the LLaMa 2 architecture by using the Low-Rank Adaptation (LoRA) method to fine-tune the model (Hu et al., 2021; Larcher et al., 2023). Their investigation was conducted using the Portuguese version of the Alpaca dataset, providing insights into the effectiveness of LLMs in handling different classifications using three Portuguese datasets.

Our work is the first attempt to evaluate the quantized models' performance across multiple Portuguese tasks. To accomplish this purpose, we opted for the Bode model due to its remarkable performance over different classification tasks, outperforming models like Cabrita and OpenCabrita on three distinct Portuguese datasets [6].

4 Proposed Approach

Figure 1 depicts the pipeline of the proposed strategy for post-training quantization. We adopted a series of methods using different quantization levels to target the models' inference over CPU devices. In the wake of the recent improvements provided by GGUF, we employed this technique to address this need since it focuses on CPU devices while loading some layers on the GPU for acceleration improvements. Then, we compared their performance through the zero-shot approach using only Portuguese datasets.

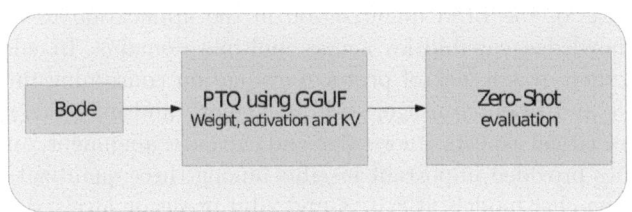

Fig. 1. Diagram of the proposed quantization approach.

In the first step, the models undergo the GGUF method for the network quantization utilizing different precision levels. The models are fed using their original weights provided by previously pre-training the architectures over publicly available datasets. Regarding the Bode model, we adopted the original architecture whose weights were computed using a fine-tuning procedure over the Portuguese version of the Alpaca dataset. In the second stage, we perform the networks' quantization using 16, 8, and 4 bits of precision using the following approaches[3]:

- **f16:** It represents the IEEE 754 standard for 16-bit floating-point arithmetic, which has been widely adopted for representing and performing arithmetic on floating-point numbers in computing. The 16-bit floating-point precision, or half-precision, is a specific format within this standard. It is designed to balance range, precision, and storage requirements;
- **q8_0:** It represents the eight-bit quantization with only one scaling factor. It works by dividing the weights within a block of values. Afterwards, each weight within the block is scaled to an 8-bit precision. Since it needs a high amount of memory, q8_0 is suitable for hardware with significant RAM;
- **q5_km:** It represents a new strategy encompassed by the k-quants quantization, in which some weights are mostly quantized to a higher-bit precision. In contrast, others are assigned to a lower precision level. The q5_km applies quantization to half of attention and feed-forward layer weights to a 6-bit precision, while the other weights are represented as 5-bit;
- **q4_km:** It applies 6-bit precision to half of attention and feed-forward layer weights, but other weights are assigned to 4-bit precision;
- **q4_0:** It is the same as q8_0, but each weight is scaled to a 4-bit precision.

After the models are quantized, we provide prompts using a zero-shot approach to assess the prediction capabilities by asking questions concerning a specific task in the Portuguese language. We prompt the models without providing

---
[3] https://huggingface.co/TheBloke/Llama-2-7B-GGML.

any response examples to accurately evaluate their generalist performance. It's important to note that no additional fine-tuning occurs during the quantization and inference stages. In addition, only Portuguese prompts are provided during the models' assessment, thus meeting the standards of the proposed research.

5 Methodology

This section presents the methodology for performing the experiments on three renowned datasets and describes the setup for conducting the experiments.

5.1 Datasets

Experiments were performed over three datasets incorporating only Portuguese texts:

- **Sentiment analysis:** It comprises a set of Portuguese texts collected from Twitter posts that are used to predict users' sentiment expressions. This dataset includes 5,000 test samples categorized into three classes of sentiments: neutral, positive, and negative. All test samples were used in the models' inference stage;
- **News categorization:** It is a dataset comprising a compilation of Portuguese news articles organized into four news categories: sports, finance, economy, and technology. The test set comprises 7,600 samples;
- **FakeRecogna:** It is a dataset for binary classification assembled from Portuguese genuine and fake news articles collected from nine respected fact-checking Brazilian agencies. It comprises a set of 52,800 new articles evenly distributed into real and fake news categories. However, owing to insufficient GPU VRAM to process the entire dataset, only 7,000 samples were used for the models' inference, thus meeting the maximum input capacity of the hardware.

For each dataset, we only used the test sets to align with the maximum capacity of the GPU devices.

5.2 Experimental Setup

We employed the Bode model and the LLaMa 2 architecture, both configured with 7 billion and 13 billion parameters, to assess the impact of the quantization methods. To perform quantization, we adopted the GGUF implementation included in the LLaMa.cpp framework[4].

Experiments were performed over the Kaggle[5] platform using two Graphic Processing Units (GPU) NVIDIA© T4 model with 16 GB of Virtual RAM. Quantization was performed by loading all the layers into the GPU for both

---
[4] https://github.com/ggerganov/llama.cpp.
[5] https://www.kaggle.com.

tested models to accelerate the process. Regarding the inference step, we set the batch size to 512 and the context size to 2,048 to meet the maximum input of both models. In addition, we set the temperature to 0.1 to prevent excessive hallucinations. It produces different responses since the temperature is set to a non-zero value. However, the model performs at a small generative variation to avoid taking risks of modifying the probability of the next word, thus preventing the random sampling of low-probability words. Furthermore, we select the top 10 most probable words to compose the next sentence.

6 Experiments and Results

This section provides results and detailed discussions regarding each quantization method applied to the baseline LLMs. The obtained scores are given using a graphical presentation encompassing bar and line plots to represent the accuracy and the F1-score values, respectively.

Figure 2 shows the average scores received from each LLM after applying the quantization methods.

Regarding the models with 7 billion parameters, one can notice an improvement in the BODE 7B in most quantization approaches compared to the scores presented by the LLaMa 2 7B model, particularly for f16, q8_0, q5_km and q4_km. For sentiment analysis, depicted in Fig. 2a), the average accuracies attained by f16, q8_0, q5_km, q4_km and q4_0 are 60.13%, 60.05%, 59.57%, 58.81% and 55.77% for the Bode 7B model, respectively. Likewise, the LLaMa 2 7B showed accuracies of 57.51%, 57.83%, 58.75%, 57.19% and 57.97% for the same quantization methods. Despite the lower accuracy while applying the q4_0 method over the Bode 7B, the model overcame the LLaMa 2 7B in terms of accuracy in the other quantization methods. Similar results are given for the other datasets, with the News Categorization achieving accuracies of 70.76%, 70.87%, 71.76%, 71.48% and 67.95% for f16, q8_0, q5_km, q4_km and q4_0 applied to the Bode 7B model, respectively. Regarding the LLaMa 2 7B, the attained accuracies are 68.30%, 68,37%, 68,43%, 65,88% and 67,42%. Again, Bode 7B showed the highest accuracy in all quantization methods. The same behaviour is shown for the FakeRecogna dataset, with Bode 7B attaining remarkable performance with the highest accuracy compared to LLaMa 2 7B.

Regarding the models with 13 billion parameters, one can notice the outstanding achievements provided by Bode 13B in the Sentiment Analysis and FakeRecogna datasets. For Sentiment Analysis, the model attained accuracies of 64.80%, 64.70%, 64.00%, 64.70% and 64.50% for f16, q8_0, q5_km, q4_km and q4_0, respectively. In contrast, LLaMa 13 B showed accuracies of 58.49%, 58.55%, 58..67%, 57.57% and 56.21%. The remarkable performance lies in the FakeRecogna dataset, in which the Bode 13B attained 66.24%, 66.34%, 64.86%, 65.34% and 58.40% for f16, q8_0, q5_km, q4_km and q4_0, respectively. For the sake of comparison, LLaMa 2 13B exhibited accuracies of 25.04%, 24.48%, 21.80%, and 22.25% for q8_0, q5_km, q4_km, and q4_0. Regarding f16 quantization, the accuracy could not be computed since the workload for the model

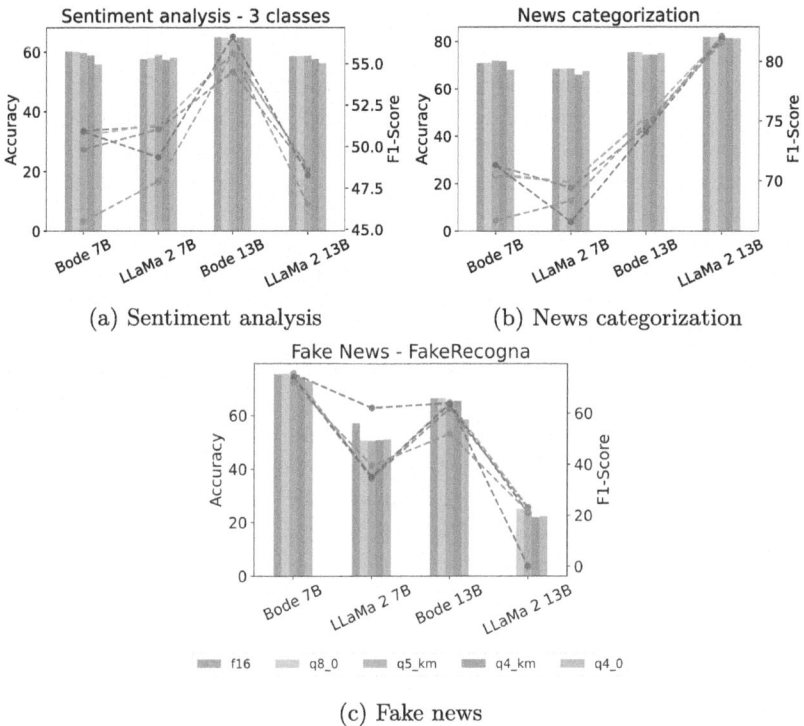

Fig. 2. Accuracy (bars) and F1-Score (Dotted lines) obtained by each quantization method applied to the LLMs.

did not meet the hardware requirements. Regardless, this shows the effectiveness of the quantized Bode 13B model in predicting fake news in the Portuguese language.

Regarding the F1-Score, the models showed different behaviours in terms of performance across all quantization methods. For the sentiment analysis, both versions of Bode, i.e. 7B and 13B, exceeded the LLaMa 2 performance only in the q4_km quantization method. In contrast, Bode 13B showed remarkable results regardless of the applied quantization method, exhibiting F1-Score values of 55.60%, 55.50%, 54.50%, 56.60% and 55.60%, while LLaMa 2 13B attained 48.75%, 48 79%, 48.57%, 48.27% and 46.53% for f16, q8_0, q5_km, q4_km and q4_0, respectively. For news categorization, the Bode 7B model exhibited consistently higher F1-score values than LLaMa 2 7B across most quantized versions, except for q4_0. However, the performance declined upon applying the quantization methods to the Bode 13B version. Similar to accuracy, Bode 13B showed impressive results when applied to the FakeRecogna dataset.

Our findings are in line with previous studies conducted to compare the performance of different LLM quantization levels. Liu et al. [14] obtained similar accuracy between the non-quantized LLMs and their equivalent quantized versions at lower-bit precision. Despite being applied at different levels of the structural parts of the LLM, they reported remarkable results for the 8-bit and 4-bit quantization levels applied to the weights of the LLaMa 7B, 13B, and 30B models. Their findings showed higher accuracies for those quantization levels and comparable performance with the non-quantized models. In addition, similar findings were reported in the benchmark conducted by Jin et al. [11]. In their study, the quantized models with 8-bit and 4-bit precision levels showed equivalent accuracies with the non-quantized Qwen-Chat model. In addition, the quantized models exhibited similar perplexity with the non-quantized counterparts [11]. Another study by Dettmers and Zettlemoyer [3] demonstrated the efficacy of 4-bit level scaling using a comprehensive benchmark encompassing the analysis of several LLMs. In our study, similar outcomes are shown in the context of the Portuguese language, with remarkable results attained by the Bode model in almost all datasets, thus revealing its practical implementation using its quantization equivalent for inference in Portuguese classification tasks.

For comparison purposes, we contrasted the quantized model outcomes against the experiments conducted by Garcia et al. [6] over the same datasets using the original models' architecture. Table 1 shows the accuracies obtained by the original models in their study and the ones yielded from the corresponding quantized versions.

Table 1. Comparison among the quantized LLM accuracies and the ones obtained from the original models' results reported in Garcia et al. [6]

Model	Version	Dataset		
		Sentiment analysis	News categorization	FakeRecogna
Bode 7B	Original	**60.30%**	69.10%	69.40%
	f16	60.13%	70.76%	75.33%
	q8_0	60.05%	70.87%	**75.53%**
	q5_km	59.57%	**71.76%**	74.03%
	q4_km	58.81%	71.48%	74.30%
	q4_0	55.77%	67.95%	72.66%
Bode 13B	Original	64.30%	**79.50%**	65.90%
	f16	**64.80%**	75.40%	66.24%
	q8_0	64.70%	75.39%	**66.34%**
	q5_km	64.00%	74.30%	64.86%
	q4_km	64.70%	74.30%	65.34%
	q4_0	64.50%	74.85%	58.40%

*The highest scores are highlighted in bold for each dataset.

Regarding the Bode 7B model, the results reported in their study revealed accuracies of 60.30%, 69.10% and 69.40% for sentiment analysis, news categorization and FakeRecogna datasets, respectively. For the 13 billion parameter version, the model attained accuracies of 64.30%, 79.50% and 65.90% over the same datasets. In addition, they reported F1-score values of 55.40%, 69.20% and 69.50% achieved by the Bode 7B model and 55.20%, 79.60% and 64.20% for Bode 13B. These scores demonstrate comparability when contrasted with the models after applying the quantization methods. Table 2 summarizes the F1-score values obtained from the original and the quantized Bode models. Likewise, it reveals comparable behaviour as demonstrated by accuracies in Table 1.

Table 2. Comparison among the quantized LLM F1-score values and the ones obtained from the original models' results reported in Garcia et al. [6]

Model	Version	Dataset		
		Sentiment analysis	News categorization	FakeRecogna
Bode 7B	Original	**55.40%**	69.20%	69.50%
	f16	50.94%	70.48%	75.40%
	q8_0	50.68%	70.56%	**75.59%**
	q5_km	49.82%	**71.29%**	73.99%
	q4_km	50.89%	71.28%	74.12%
	q4_0	45.47%	66.67%	71.92%
Bode 13B	Original	55.20%	**79.60%**	**64.20%**
	f16	55.60%	75.30%	63.84%
	q8_0	55.50%	75.33%	64.06%
	q5_km	54.50%	74.41%	61.73%
	q4_km	**56.60%**	74.09%	63.51%
	q4_0	55.60%	74.83%	51.79%

*The highest scores are highlighted in bold for each dataset.

Additionally, the quantized models at the 16B configuration exhibited superior performance against the original version with 7B parameters in almost all datasets. For sentiment analysis, Bode 7B attained 60.30% in its non-quantized version. In contrast, q8_0 and q4_km quantization applied to Bode 13B attained both 64.70% of accuracy. In the news categorization dataset, the non-quantized 7B model achieved an accuracy of 69.10%, while the Bode 13B with f16 quantization resulted in a higher accuracy of 75.40%. In addition, almost all quantized versions achieved higher scores than the non-quantized model. The worst performance is shown for the FakeRecogna dataset, in which the accuracy deteriorates after applying the quantization methods over the 13B model configuration. Nevertheless, such findings agree with Liu et al. [14] and Jin et al. [11]. They indicate improved performance when employing low-bit scaling on large model configurations compared to solely relying on models with fewer parameters and high-bit precision levels for inference.

The incidence of hallucinations is still a challenge during the LLM inference since they can produce nonsensical responses that disconnect from the intake prompt. In the context of LLM, hallucinations can arise from a combination of factors, such as the model's intrinsic probabilistic response, overfitting, and training with insufficient data. Identifying this aspect becomes critical to measuring the LLM performance and correct responses. We evaluate how the LLM quantization can impact the models' accuracy in terms of hallucinations compared to the counterpart non-quantized version. We computed the hallucinations produced by each quantized Bode and LLaMa 2 version for each tested dataset (Fig. 3).

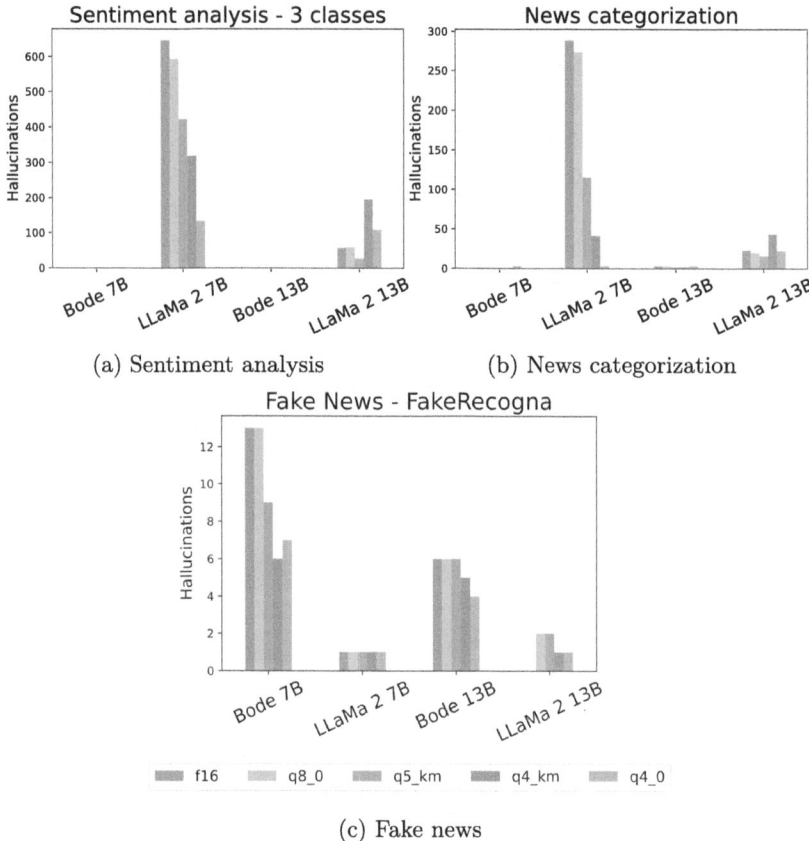

Fig. 3. Hallucinations achieved by each LLM at different quantization levels.

Regarding sentiment analysis, the Bode model's performance over the quantized LLaMa 2 7B and 13B is evident and remarkable, demonstrating confidence in its capabilities through a few hallucinations. In contrast, the LLaMa 2 models hallucinate often, showing many incorrect responses in sentiment analysis and news categorization. The only exception is the FakeRecogna dataset, in which the quantized Bode models revealed more hallucinations than LLaMa 2 in both versions, 7 billion and 13 billion parameters. However, it is particularly small compared to sentiment analysis and news categorization. Although LLaMa 2 models hallucinated less in the FakeRecogna dataset, the former revealed lower accuracy than the Bode model in both parameter configurations, demonstrating its tendency to offer incorrect responses regarding the authenticity of the news.

Finally, the proposed strategy considered the time required by each quantization method when applied to each LLM. The computed times are displayed in Fig. 4. Remarkably, the methods exhibited similar performance in terms of quantization time. This outcome is somehow expected since the Bode model is designed under the same LLaMa 2 architecture. The only exception is observed for the FakeRecogna dataset, where the computational cost of quantizing the LLaMa 2 architecture is notably higher compared to Bode. Note the significant computational time the LLaMa 2 7B requires for the FakeRecogna dataset. The Kaggle instability or scheduling in the processing allocation somehow explains this behaviour. However, Bode 13B and LLaMa 2 13B provided similar execution times.

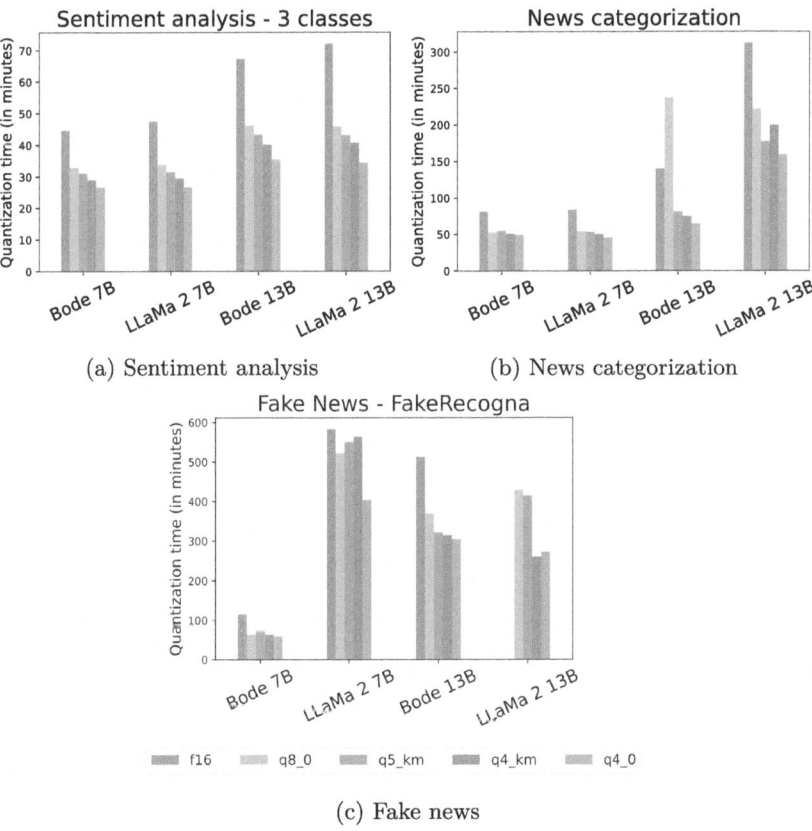

Fig. 4. Computational cost for each quantization method and dataset.

7 Conclusions

The influence of LLMs is rapidly expanding and pushing innovation and advancement across a wide range of domains in the Portuguese-speaking community. In this sense, this paper presented a novel investigation regarding the impact of several post-training quantization methods applied to Large Language Models in the context of Portuguese classification tasks. The purpose of this study was to evaluate the performance of quantized language models using three Portuguese datasets for classification tasks. The proposed strategy revealed a remarkable performance while adopting simple quantization methods without requiring further training during the quantization process. In addition, the research findings are in step with previous studies reporting equivalent results between quantized models and the non-quantized counterparts. Such analysis showed a positive impact concerning the application of quantization to a Portuguese-based LLM model, overcoming its counterpart version in different scenarios.

In future works, we aim to incorporate additional LLMs to offer a comprehensive evaluation of the quantization effects across a diverse set of language models. In addition, we intend to incorporate further Portuguese datasets in future experiments, aiming to investigate the impact of the content variability and prediction capabilities in lower bit precision.

Acknowledgements. The authors are grateful to the Brazilian National Council for Scientific and Technological Development (CNPq) grants 308529/2021-9 and 400756/2024-2, to the São Paulo Research Foundation (FAPESP) grants 2013/07375-0, 2019/07665-4, 2023/14427-8, 2023/10823-6, 2023/03726-4, 2023/01374-3, 2023/14354-0, and 2024/00789-8, to Unesp-IEPe-RC-#06/2023 PROPe grant, and to the Petrobrás Brazil grant #2017/00285-6 for their financial support.

References

1. Almazrouei, E., et al.: The falcon series of open language models. arXiv preprint arXiv:2311.16867 (2023)
2. Chiang, W.L., et al.: Vicuna: an open-source chatbot impressing GPT-4 with 90%* ChatGPT quality (2023). https://lmsys.org/blog/2023-03-30-vicuna/
3. Dettmers, T., Zettlemoyer, L.: The case for 4-bit precision: k-bit inference scaling laws. In: International Conference on Machine Learning, pp. 7750–7774. PMLR (2023)
4. Dong, S., Cheng, W., Qin, J., Wang, W.: QAQ: quality adaptive quantization for LLM KV cache. arXiv preprint arXiv:2403.04643 (2024)
5. Garcia, G.L., Paiola, P.H., Jodas, D.S., Sugi, L.A., Papa, J.P.: Text summarization and temporal learning models applied to portuguese fake news detection in a novel Brazilian corpus dataset. In: Proceedings of the 16th International Conference on Computational Processing of Portuguese, pp. 86–96 (2024)
6. Garcia, G.L., et al.: Introducing bode: a fine-tuned large language model for Portuguese prompt-based task. arXiv preprint arXiv:2401.02909 (2024)
7. Hooper, C., et al.: KVQuant: towards 10 million context length LLM inference with KV cache quantization. arXiv preprint arXiv:2401.18079 (2024)

8. Huang, W., et al.: BiLLM: pushing the limit of post-training quantization for LLMs. arXiv preprint arXiv:2402.04291 (2024)
9. Huang, W., et al.: How good are low-bit quantized LLaMA3 models? An empirical study. arXiv preprint arXiv:2404.14047 (2024)
10. Jiang, A.Q., et al.: Mistral 7b. arXiv preprint arXiv:2310.06825 (2023)
11. Jin, R., et al.: A comprehensive evaluation of quantization strategies for large language models. arXiv preprint arXiv:2402.16775 (2024)
12. Li, S., et al.: Evaluating quantized large language models. arXiv preprint arXiv:2402.18158 (2024)
13. Lin, J., Tang, J., Tang, H., Yang, S., Dang, X., Han, S.: AWQ: activation-aware weight quantization for LLM compression and acceleration. arXiv preprint arXiv:2306.00978 (2023)
14. Liu, Z., et al.: LLM-QAT: data-free quantization aware training for large language models. arXiv preprint arXiv:2305.17888 (2023)
15. Lopes, R., Magalhães, J., Semedo, D.: GlórIA-A generative and open large language model for Portuguese. arXiv preprint arXiv:2402.12969 (2024)
16. Shang, Y., Yuan, Z., Wu, Q., Dong, Z.: PB-LLM: partially binarized large language models. arXiv preprint arXiv:2310.00034 (2023)
17. Touvron, H., et al.: LLaMA: open and efficient foundation language models. arXiv preprint arXiv:2302.13971 (2023)
18. Touvron, H., et al.: Llama 2: open foundation and fine-tuned chat models. arXiv preprint arXiv:2307.09288 (2023)
19. Yue, Y., Yuan, Z., Duanmu, H., Zhou, S., Wu, J., Nie, L.: WKVQuant: quantizing weight and key/value cache for large language models gains more. arXiv preprint arXiv:2402.12065 (2024)
20. Zhu, X., Li, J., Liu, Y., Ma, C., Wang, W.: A survey on model compression for large language models. arXiv preprint arXiv:2308.07633 (2023)

GemBode and PhiBode: Adapting Small Language Models to Brazilian Portuguese

Gabriel Lino Garcia[1]( ), Pedro Henrique Paiola[1], Eduardo Garcia[2], João Renato Ribeiro Manesco[1], and João Paulo Papa[1]

[1] School of Sciences, São Paulo State University (UNESP), Bauru, SP, Brazil
{gabriel.lino,pedro.paiola,joao.r.manesco,joao.papa}@unesp.br
[2] Institute of Informatics, Federal University of Goiás, Goiânia, GO, Brazil

Abstract. Recent advances in generative capabilities provided by large language models have reshaped technology research and human society's cognitive abilities, bringing new innovative capacities to artificial intelligence solutions. However, the size of such models has raised several concerns regarding their alignment with hardware-limited resources. This paper presents a comprehensive study on training Portuguese-focused Small Language Models (SLMs). We have developed a unique dataset for training our models and employed full fine-tuning, as well as PEFT approaches for comparative analysis. We used Microsoft's Phi and Google's Gemma as base models to create our own, named PhiBode and GemBode. These models range from approximately 1 billion to 7 billion parameters, with a total of ten models developed. Our findings provide valuable insights into the performance and applicability of these models, contributing significantly to the field of Portuguese language processing. This research is a step forward in understanding and improving the performance of SLMs in Portuguese. The comparative analysis of the models provides a clear benchmark for future research in this area. The results demonstrate the effectiveness of our training methods and the potential of our models for various applications. This paper significantly contributes to language model training, particularly for the Portuguese language.

Keywords: Small Language Models · Portuguese · Bode · Generative Artificial Intelligence · Natural Language Processing

1 Introduction

Recent Natural Language Processing (NLP) advances have been primarily driven by developing increasingly sophisticated and powerful language models [30]. Large Language Models (LLMs), in particular, are known for their ability to understand and generate text with great precision, opening ways for a wide range of applications, including automatic translation, text summarization, and question and answer, among others [21].

Despite their impressive effectiveness, these models usually come paired with a substantial cost in computing resources, often requiring huge amounts of memory and processing power, unattainable to several researchers and developers with limited infrastructure [18]. To tackle this issue, Small Language Models (SLMs) have emerged in the literature as a way to minimize the computational resource requirement. These models allow language models to be deployed in resource-constrained environments with limited memory and processing power.

The underlying concept behind SLMs is to offer a more affordable and resource-efficient alternative to various NLP applications. However, a fundamental issue faced when dealing with SLMs is their performance trade-off compared to large-scale models, as demonstrated by [30]. When trained with more data, these smaller models can match or even outperform their larger counterparts in certain benchmarks. Still, given their size and training capacity limitations, SLMs often fail to achieve the same accuracy and generalization capabilities as their larger counterparts in many other applications. Nonetheless, several prominent works have emerged in this direction, including Microsoft's Phi-1.5 [18], Phi-2[1] and Phi-3 [1], as well as Gemma models, based on Gemini, made available by Google [29].

As Portuguese is an important language spoken by more than 250 million people worldwide, it is paramount to address the necessity of developing language models for that scenario. As such, several works have already been proposed as a way to adapt LLMs to the Brazilian Portuguese (BP) language, such as Sabiá [22], a model developed by Maritaca AI, explicitly designed for BP, which has the first model based on the LLaMA-1 and Sabiá-J architecture, and a second version under a proprietary license, with specifics regarding parameter count and architectural design not publicly disclosed.

With the release of the LLaMa open-weight models, several models that enhance its capabilities for Brazilian Portuguese have been proposed, such as Canarim [10] and Bode [14]. The Canarim model was obtained by pretraining an LLaMa2-7B model using the Portuguese subset of Common Crawl solely on the language modeling task. Conversely, Bode employs low-rank adaptation (LoRA) to fine-tune and improve the LLaMa-2 model using the Alpaca dataset in prompt-based tasks for Portuguese instruction-following response.

Another notable model is the openCabrita [17], a 3B parameter model based on the OpenLLaMa, designed to provide a more cost-effective solution for Portuguese language processing. It employs a language-specific tokenizer adaptation approach to optimize token usage and reduce inference time, offering tangible benefits for real-world applications. Following a similar approach, the TeenyTinyLLaMa [6] model was proposed, presenting a compact model under the permissive Apache 2.0 license, with only 460 million parameters, following the Chinchilla scaling laws.

Despite the progress in language models for Brazilian Portuguese, there is still a need for more inclusive and versatile language approaches that prioritize efficiency and are based on modern SLM architectures. This work aims to

---

[1] Available at: https://huggingface.co/microsoft/phi-2 Accessed on: April 18th, 2024.

enhance the adaptation of linguistic models, specifically focusing on the Phi and Gemma models provided by Microsoft and Google, respectively. We introduce the adapted versions of PhiBode (1.5B, 2B, and 3B) and GemBode (2B, 2B-it, 7B, and 7B-it). These models are designed to bridge the gap in developing linguistic resources for Brazilian Portuguese, with the aim of improving performance across various Natural Language Processing (NLP) tasks. Our ultimate goal is to make these models more resource-efficient and enable a broader range of researchers and developers to benefit from these technologies. As such, the main contributions of this work are described as follows:

- **Comparison of Tuning Techniques:** This work compares the effectiveness of fully fine-tuned models with others fine-tuned through parameter-efficient tuning techniques (PEFT), aiming to evaluate not only models with smaller parameter sizes but also the impact of using efficient tuning techniques.
- **Introduction of UltraAlpaca Dataset:** A new multi-task dataset named UltraAlpaca, has been created by aggregating multiple English datasets translated into Portuguese.
- **In-depth Evaluation of Small Language Models:** We conduct a thorough evaluation of small language models specifically within the context of the Portuguese language.

This paper is organized as follows: Sect. 2 introduces the methodology employed by the proposed method, technical details of their architectures, the proposed dataset, and information on the tuning process. Section 3 discusses the experimental setup configurations used for evaluating the models on the benchmarks. Section 4 outlines the results, and Sect. 5 states conclusions.

2 Methodology

This section describes the proposed method and the foundational models used as the basis for our approach, along with details on our proposed dataset and the conditions under which it was created. Moreover, we describe the tuning techniques employed for evaluation over different scenarios.

2.1 Proposed Method

In this study, we have chosen the linguistic models Phi and Gemma, provided by Microsoft and Google, respectively, as the foundation for our investigation. Despite their sophistication, they were not specifically trained in Portuguese. Nevertheless, it is reasonable to assume that these models may have been exposed to Portuguese data during their training. To optimize the performance of these models for Portuguese, we have decided to use the UltraAlpaca dataset for further training.

The UltraAlpaca dataset, a unique collection of exclusively Portuguese samples, is a rare resource exceptionally suited to enhancing language models for this language. Leveraging this dataset, we conducted both PEFT and full fine-tuning of the Phi and Gemma models.

By performing PEFT and full fine-tuning based on UltraAlpaca, we aimed to adapt the Phi and Gemma models to yield more accurate and relevant results in Portuguese. This approach will allow us to maximize the potential of these pre-trained models, adapting them to the nuances and peculiarities of the Portuguese language. We hope that the results obtained with this methodology will significantly contribute to the advancement of natural language processing in Portuguese and for specific applications of this language.

2.2 Foundation Models

Two families of SLMs, Phi and Gemma, developed by Microsoft and Google, respectively, were chosen as the baseline for developing our Portuguese language models. The decision to use these models was motivated by their effectiveness in Portuguese and English, especially their performance per number of parameters. Both are capable of achieving performance comparable to that of large-scale models. This section will introduce these models, with details of their architectures discussed in more depth.

Introduced by Microsoft, the Phi [18] family of language models represents a turning point in the research of SLMs. The first iteration of the Phi model family focused on training a compact model using high-quality textbook-like texts and synthetic data from GPT-3.5, achieving significant results with only 1.3B parameters. Its successor expanded to 2.7 billion parameters, trained on an augmented dataset from the first model, delivering comparable performance to much larger models. The latest Phi-3-mini iteration of the family [1] is made of a 3.8 billion parameter SLM trained on 3.3 trillion tokens, rivaling larger models while being mobile-friendly. This generation also introduced configurations with 7B and 14B parameters, capable of achieving even more significant results.

Gemma, another staple in the SLM state-of-the-art, comprises a family of open models based on Google's Gemini [29]. These models are trained on up to 6 trillion tokens, featuring variants with parameter counts ranging from 2 billion to 7 billion. Like the Phi model, Gemma adopts a transformer decoder architecture with a context length of 8,192 tokens. However, Gemma incorporates several enhancements built upon previous works, such as multi-query attention, RoPE embeddings, GeGLU Activations, and RMSNorm. Although inspired by the Gemini architecture, the Gemma model was trained primarily on English data and did not demonstrate multimodal capabilities.

2.3 UltraAlpaca

In our research, we used models provided by Microsoft and Google and trained them using a dataset created by our team called UltraAlpaca. This dataset is based on several English databases translated into Portuguese. The datasets used to compose UltraAlpaca were:

- **Alpaca** [28]: A dataset of 52,000 instruction-following samples generated by OpenAI's text-davinci-003 engine. We used the translated version of Alpaca [17] for training the first version of Cabrita.
- **UltraChat** [9]: Self-refinement dataset that comprises 1.47 million multi-turn dialogues generated by GPT-3.5-TURBO, spanning 30 topics and 20 distinct types of text material. From this dataset, 70,000 samples were selected and translated to compose UltraAlpaca.
- **Aya** [27]: A multilingual instruction fine-tuning dataset curated by an open-science community via the Aya Annotation Platform from Cohere For AI. It contains 204,000 human-annotated prompt-completion pairs, along with annotator demographics data. For this study, we filtered the Portuguese samples from this dataset.
- **OpenAssistant Conversations (OASST1)** [16]: Multilingual corpus of 161,443 assistant-style messages in 35 languages, annotated with 461,292 quality ratings, forming over 10,000 fully annotated conversation trees. The dataset was created through a global crowd-sourcing effort with over 13,500 volunteers. For UltraAlpaca, the Portuguese samples from this dataset were filtered.
- **Code Alpaca** [5]: A dataset of 20,000 samples built similarly to Alpaca, focusing on code generation. The dataset was fully translated to be integrated into UltraAlpaca.
- **MetaMathQA-40K-PTBR** [32]: Specialized dataset designed to enhance mathematical reasoning capabilities in LLMs comprising 395,000 samples. We used 40,000 previously translated samples from this dataset.

The choice of these datasets to compose UltraAlpaca resulted from an attempt to write a comprehensive database comprising several tasks. Some works in the literature [19,33] point to the benefits of using multi-task datasets for training LLMs, increasing their chances of gaining emerging capabilities.

After translating these databases, we combined them to create UltraAlpaca. It is important to note that syntactic and semantic problems may occur even with translations, as translation does not guarantee high-quality data. However, in our experiments, these translated datasets improved the performance of the original models, as later discussed. This work represents a significant leap forward in developing LLMs, particularly for languages other than English. While our approach has challenges, it offers a promising direction for future research.

2.4 Efficient Tuning Techniques

As a route to adapt already existing language models to Portuguese, it is necessary to fine-tune the model in a language-specific dataset. One way to do that is called *full fine-tuning*, in which a pre-trained model is loaded, and all model weights are adjusted based on the calculated error. Although heavy in computer resource usage, this approach usually has the potential to provide the best results since it is capable, in theory, of working on the finer details of the model.

On the other hand, to use fewer computer resources while still maintaining acceptable model accuracy, fine-tuning can be performed on only the most relevant weights of the model. Certain techniques such as Low-rank Adaptation (LoRA) [15] perform *Parameter Efficient Fine-Tuning (PEFT)* by introducing two low-rank decomposition matrices within the dense layers, which are then fine-tuned to represent an update to the original model's weights.

QLoRA [8] refers to a more efficient alternative of LoRA, in which, by integrating quantization techniques, fewer computer resources are used for fine-tuning. QLoRA applies quantization to 4-bit precision on a pre-trained model to reduce space usage. After that, non-quantized decomposition matrices are calculated using LoRA to fine-tune the model properly. This work compares the fully fine-tuned techniques with the PEFT-based fine-tuning, in which QLoRA was employed.

3 Experimental Setup

This section discusses the experimental setups used on the developed models for each experiment and provides detailed information on the benchmark and techniques.

3.1 Open Portuguese LLM Leaderboard

Resources like the Hugging Face Open LLM Leaderboard [3] offer valuable insights for English language models. Still, they need more coverage for Portuguese, creating a significant gap in understanding the capabilities of LLMs for this language. Existing Portuguese-centric benchmarks, such as the Poeta benchmark [22], do not have a fully open reproducible evaluation script and incorporate machine-translated datasets, potentially introducing biases and limiting their reliability for assessing proper language understanding.

The Open Portuguese LLM Leaderboard [13] is a fully open and reproducible benchmark designed to evaluate the performance of Large Language Models. It was created using the EleutherAI-Language Model Evaluation Harness [12], a unified framework for testing generative language models on various evaluation tasks. The leaderboard ensures consistent and standardized evaluation across diverse tasks, enabling meaningful comparisons between different models and fostering transparency in performance assessment. The leaderboard currently includes nine benchmarks:

- **ENEM** [20,23,26]: The *Exame Nacional do Ensino Médio* (ENEM) is a standardized Brazilian national exam for high school students, covering various subjects such as natural sciences, human sciences, languages, and mathematics. This benchmark comprises 1,430 questions from exams between 2010 and 2018, as well as 2022 and 2023, excluding questions that require image understanding. For utilizing this dataset in the Open Portuguese LLM Leaderboard, 3 examples were used for each prompt, following a few-shot approach. The performance was evaluated based on accuracy.

- **BLUEX** [2]: The Brazilian Leading Universities Entrance eXams (BLUEX) dataset comprises entrance exams from two top Brazilian universities, spanning from 2018 to 2024. This benchmark includes 724 questions that do not necessitate image understanding. For this dataset's application in the Open Portuguese LLM Leaderboard, 3 examples were employed for each prompt, and this performance was assessed based on accuracy.
- **OAB Exams** [7]: The Order of Attorneys of Brazil (OAB) Exams are professional certification exams for lawyers in Brazil, evaluating their legal knowledge and reasoning skills. This benchmark includes over 2,000 questions from exams between 2010 and 2018. This dataset was used with 3 examples for each prompt. The performance metric used was accuracy.
- **ASSIN2 RTE** [24]: The ASSIN2 Recognizing Textual Entailment (RTE) task assesses a model's ability to determine whether a given text entails another text. This benchmark uses a dataset of sentence pairs annotated with human judgments for RTE. For this dataset, 15 examples were used for each prompt and were evaluated using the macro F1 score as the performance metric.
- **ASSIN2 STS** [24]: The ASSIN2 Semantic Textual Similarity (STS) task measures a model's capability to determine the degree of semantic similarity between two sentences. This benchmark utilizes the same dataset as ASSIN2 RTE but with annotations for STS. This dataset used 15 examples for each prompt, and the performance metric used was the Pearson Correlation Coefficient.
- **FAQUAD NLI** [25]: The FAQUAD Natural Language Inference (NLI) task is derived from the FAQUAD question-answering dataset and evaluates a model's ability to perform textual entailment between a question and its possible answers. This benchmark contains 900 questions about 249 reading passages. In this dataset, we used 15 examples for each prompt. The performance metric used was the macro F1 score.
- **HateBR** [31]: The HateBR dataset is a collection of 7,000 Brazilian Instagram comments annotated for hate speech and offensive language detection. For this dataset's application in the Open Portuguese LLM Leaderboard, 25 examples were employed for each prompt. Performance was assessed based on a macro F1 score.
- **PT Hate Speech** [11]: The Portuguese Hate Speech dataset consists of 5,668 tweets labeled for hate speech detection in Portuguese. This dataset used 25 examples for each prompt, and the performance metric used was macro F1 score.
- **TweetSentBR** [4]: The tweetSentBR dataset is a corpus of 15,000 Brazilian Portuguese tweets annotated for sentiment analysis. For this dataset, 25 examples were used for each prompt and were evaluated using the macro F1 score as the performance metric.

4 Results and Discussion

In this section, we present and discuss the results obtained by our models on each dataset. The models were categorized based on the number of parameters they possess, resulting in three distinct subsections: Models around 1B parameters, Models around 3B parameters, and Models around 7B parameters. For this analysis, the best result will be displayed in gold, the second best in silver, and the third best in bronze.

4.1 Models Around 1B Parameters

In this subsection, we present an analysis of the evolution from the base model to the specific model we have developed, with an emphasis on optimization for the Portuguese language. This analysis aims to assess how our model compares with other models of similar dimensions (approximately 1 billion parameters) that have also been trained for Portuguese.

PhiBode-1.5 vs Phi-1.5 Comparison: Table 1 presents a detailed comparison between PhiBode-1.5 and its base model, Phi-1.5. Each dataset's results and final average are provided, allowing for a comprehensive evaluation of the models' performance.

Table 1. Comparison results between the baseline model and our Portuguese model of around 1 billion parameters.

Model	ENEM	BLUEX	OAB	Assin2RTE	Assin2STS	FAQUAD NLI	HateBR	PTHateSpeech	tweetSentBR	Average
PhiBode-1.5	**23.58%**	20.72%	**24.87%**	**69.07%**	04.94%	43.97%	**34.94%**	41.23%	24.19%	**31.95%**
Phi-1.5	21.62%	**23.50%**	23.92%	33.33%	**13.02%**	43.97%	22.20%	41.23%	**32.88%**	28.41%

The results indicate that the PhiBode-1.5 model outperforms the Phi-1.5 model across various metrics. For instance, the PhiBode-1.5 model achieved a score of 34.94% on the dataset involving the HateBR, while the Phi-1.5 model scored 22.20%. Similarly, on the Assin2RTE dataset, the PhiBode-1.5 model reached 69.07%, whereas the Phi-1.5 model scored 33.33%. However, when considering the Assin2STS dataset, the PhiBode-1.5 model did not perform as well as the Phi-1.5 model. This discrepancy may be attributed to the training data, as some samples were artificially created. Additionally, the translation process into Portuguese may have resulted in the loss of semantic information.

Despite this, the overall findings demonstrate that the PhiBode-1.5 model is capable of surpassing the original model in several tasks, achieving an average percentage increase of more than 3% compared to the base model. This suggests that the PhiBode-1.5 model holds promise for future applications, even though there are areas where further optimization could enhance performance.

Benchmarking PhiBode-1.5 vs Other Portuguese Models: Table 2 provides a comprehensive comparison between the PhiBode-1.5 model and other Portuguese models that have approximately 1B parameters. The objective is to benchmark the performance of these models against each other.

Table 2. Comparison between Portuguese models with approximately 1B parameters.

Model	ENEM	BLUEX	OAB	Assin2RTE	Assin2STS	FAQUAD	NLI	HateBR	PTHateSpeech	tweetSentBR	Average
PhiBode-1.5	23.58%	20.72%	24.87%	69.07%	04.94%	43.97%	34.94%	41.23%	24.19%	31.95%	
TeenyTinyLlama	20.15%	25.73%	27.02%	53.61%	13.00%	46.41%	33.59%	22.99%	17.28%	28.86%	
TinyLLama1.1B	17.07%	21.56%	23.23%	43.90%	00.52%	43.97%	33.33%	42.33%	28.04%	28.22%	
gp2-small-pt	19.31%	21.42%	03.14%	33.59%	03.44%	43.97%	33.33%	22.99%	13.62%	21.65%	
Samba1.1b	10.22%	08.07%	15.03%	33.33%	01.30%	17.78%	35.79%	27.26%	03.27%	16.89%	
Glria1.3B	01.89%	03.20%	05.19%	00.00%	02.32%	00.26%	00.28%	23.52%	00.19%	04.09%	

When comparing the PhiBode-1.5 model with other models of similar size, such as the TeenyTinyLlama and TinyLLama1.1B, the PhiBode-1.5 model continues to stand out. For instance, in the HateBR dataset, the PhiBode-1.5, TeenyTinyLlama, and TinyLLama1.1B models achieved scores of 34.94%, 33.59%, and 33.33%, respectively, demonstrating that the PhiBode-1.5 model is competitive even against other models. However, it is important to note that there is still room for improvement. For example, in the BLUEX dataset, the PhiBode-1.5 model scored 20.72%, while the TeenyTinyLlama model scored 25.73%. This suggests that future iterations of the PhiBode-1.5 model could benefit from further adjustments and optimizations.

4.2 Models Around 3B Parameters

In this subsection, we embark on a comprehensive analysis of the progression from the base model to the specific model we have developed, with a distinct focus on optimization for the Portuguese language. The purpose of this analysis is to evaluate how our model stands in comparison with other models of similar dimensions (approximately 3 billion parameters) that have also been trained for Portuguese. This analysis includes models such as PhiBode-3B, PhiBode-2B, Gembode-2B-it, and Gembode-2b. This will allow us to understand where our model stands regarding performance and effectiveness relative to these models.

PhiBode and GemBode vs Base Models: Table 3 presents a detailed comparison between the PhiBode-3B and PhiBode-2B models and their base models, Phi-3B-mini and Phi-2B, respectively. These models are part of Microsoft's Phi family and have approximately 3 billion parameters. In addition, we also compare the Gembode-2B and Gembode-2B-it models and their base models, Gemma-2B and Gemma-2B-it, which are based on the Gemma family from Google. The results for each dataset are provided, along with the final average, allowing for a comprehensive evaluation of the performance of these models.

Table 3. Comparison of PhiBode and GemBode models against baseline models.

Model	ENEM	BLUEX	OAB	Assin2RTE	Assin2STS	FAQUAD NLI	HateBR	PTHateSpeech	tweetSentBR	Average
PhiBode-3-3B	56.12%	40.75%	38.50%	88.56%	69.63%	50.65%	82.19%	68.10%	51.67%	60.69%
Phi-3-mini-4k	**65.22%**	**53.96%**	**45.69%**	**90.64%**	**73.60%**	**56.06%**	**84.34%**	**70.82%**	**57.40%**	**66.41%**
PhiBode-2B	**38.85%**	25.17%	**29.61%**	**45.39%**	**24.43%**	**43.97%**	54.15%	**54.59%**	**43.34%**	**39.89%**
Phi-2	34.99%	**26.98%**	28.29%	38.38%	8.87%	43.92%	**59.63%**	51.23%	36.37%	36.52%
Gembode-2B	**34.71%**	25.87%	**31.71%**	**71.37%**	34.08%	**60.09%**	47.01%	**57.04%**	49.37%	**45.69%**
Gemma-2B	26.45%	**28.37%**	28.34%	63.53%	**36.35%**	44.75%	**77.82%**	36.81%	**54.25%**	44.07%
GemBode-2B-it	21.69%	25.31%	26.83%	52.71%	**16.28%**	52.95%	**67.52%**	**24.22%**	37.54%	**36.12%**
Gemma-2B-it	**28.76%**	**25.87%**	**28.38%**	**57.17%**	5.51%	**55.20%**	44.55%	23.59%	**39.20%**	34.25%

The PhiBode-3B model performed less effectively than its base model, Phi-3B-mini, indicating that the recently released base model from Microsoft already has a strong proficiency in Portuguese, showing a significant improvement when compared to the Phi-2B, moving from an average of 36.52% to 66.41%. This also suggests that our training dataset, UltraAlpaca, may contain grammatical and syntactical errors, and its translations may not have been satisfactory. In contrast, the PhiBode-2B showed an improvement of approximately 3% compared to the Phi-2B, achieving a general average of 39.89%, demonstrating that its training was effective.

Regarding the Gemma family from Google, the overall results improved compared to the base model but did not reach a 3% improvement in the general average. This could also be related to the training data despite slightly improving overall performance.

PhiBode and GemBode vs Other Portuguese Models: Table 4 presents a comprehensive comparison between the PhiBode and GemBode models and other Portuguese models. It provides a detailed view of these models' performance across various tasks, allowing for a complete evaluation of their effectiveness and accuracy in comparison to other models.

Table 4. Comparison between Portuguese models with approximately 3B parameters.

Model	ENEM	BLUEX	OAB	Assin2RTE	Assin2STS	FAQUAD NLI	HateBR	PTHateSpeech	tweetSentBR	Average
PhiBode-3-3B	56.12%	40.75%	38.50%	88.56%	69.63%	50.65%	82.19%	68.10%	51.67%	60.69%
Zephyr-3B	41.64%	36.02%	33.03%	87.57%	61.89%	67.10%	67.08%	54.25%	40.75%	54.37%
RecurrentGemma-2B-it	32.68%	31.43%	30.34%	79.42%	35.66%	28.02%	75.47%	60.20%	57.86%	47.92%
Gembode-2B	34.71%	25.87%	31.71%	71.37%	34.08%	60.09%	47.01%	57.04%	49.37%	45.69%
PhiBode-2B	38.85%	25.17%	29.61%	45.39%	24.43%	43.97%	54.15%	54.59%	43.34%	39.89%
GemBode-2B-it	21.69%	25.31%	26.83%	52.71%	16.28%	52.95%	67.52%	24.22%	37.54%	36.12%
Luana-2B	24.42%	24.34%	27.11%	70.86%	01.51%	43.97%	40.05%	51.83%	30.42%	34.95%
Periquito-3B	17.98%	21.14%	22.69%	42.01%	08.92%	43.97%	50.46%	41.19%	47.96%	33.04%
OpenCabrita-3B	17.98%	21.14%	22.69%	43.01%	08.92%	43.97%	50.46%	41.19%	47.96%	33.04%

Compared to other 3B models explicitly trained for the Portuguese language, the PhiBode-3-3B shows a significant performance improvement, beating the second-best model by a margin of 6.32%. It also greatly dominates almost all the other benchmarks, except for FAQUADNLI and tweetSentBR, in which the model was bested by Zephyr and RecurrentGemma, respectively. The PhiBode-3 model also excels in specific tasks, such as HateBR and PTHateSpeech, with a significant margin compared to other models, indicating the model's effectiveness when dealing with problems related to hate speech. Regarding the Phibode-2 model, although it can maintain a relevant result on average compared to other Portuguese-based models, it still needs to improve in terms of efficacy by a large amount to the more robust models, including the Gembode-2B.

The other developed models, Gembode-2B and GemBode-2B-it, although displaying a gap in effectiveness in comparison to the PhiBode-3 model and falling behind other models such as Zephyr and RecurrentGemma, are still able to outperform all the other Portuguese language models, including Luana-2B, which also belongs to the Gemma Family and Periquito and OpenCabrita, from the OpenLLaMA family. These models also display excellent results in the FAQUADNLI task, with Gembode-2B even surpassing the PhiBode-3 model, indicating that the proposed dataset may contribute to enhancing the model's capabilities to deal with textual entailment and question-answering scenarios.

4.3 Models Around 7B Parameters

In this subsection, we analyze the efficacy of our models of around 7 billion parameters by comparing them to their base models and other models of similar size from the literature. This comparison enables us to properly understand the capabilities and limitations of our models in various Portuguese language tasks.

GemBode X Base Model: Table 5 displays a comparison between the GemBode-7B and GemBode-7B-it models and their respective base models, Gemma-7B and Gemma-7B-it, as listed on the Open Portuguese LLM Leaderboard. It is important to note that it was impossible to perform full tuning on the 7B models, meaning the models presented are trained using PEFT.

Table 5. GemBode: comparison against baseline models.

Model	ENEM	BLUEX	OAB	Assin2RTE	Assin2STS	FAQUADNLI	HateBR	PTHateSpeech	tweetSentBR	Average
GemBode-7B	66.90%	**57.16%**	**45.47%**	**86.61%**	**71.39%**	67.40%	79.81%	**63.75%**	65.49%	**67.11%**
Gemma-7b	67.04%	56.47%	42.87%	81.34%	64.28%	69.23%	**85.69%**	42.51%	68.19%	64.18%
GemBode-7B-it	49.34%	36.58%	34.76%	79.09%	64.95%	64.67%	86.27%	63.61%	66.17%	60.60%
Gemma-7b-it	36.60%	30.32%	27.70%	**81.44%**	60.84%	57.36%	72.73%	55.99%	23.53%	49.61%

One can observe that the GemBode-7B model, when compared to its base model Gemma-7B, achieved an overall improvement of nearly 3% in the benchmark, demonstrating satisfactory performance across various datasets. The GemBode-7B-it model, on the other hand, exhibited a substantial performance increase compared to its base model Gemma-7B-it, achieving an increase of more than 10% in the general average of the benchmark. Consequently, its performance improved significantly in almost all the analyzed datasets, indicating its successful adaptation and acquisition of information in Portuguese.

Comparative Analysis of GemBode-7B with Portuguese SLMs: Table 6 provides a comparative analysis of the GemBode-7B model with other Portuguese models of equivalent size.

Table 6. Comparison between Portuguese models with approximately 7B parameters.

Model	ENEM	BLUEX	OAB	Assin2RTE	Assin2STS	FAQUAD NLI	HateBR	PTHateSpeech	tweetSentBR	Average
Llama-3-8B-Dolphin	67.39%	53.82%	45.38%	92.97%	75.99%	80.17%	88.26%	58.45%	69.56%	70.22%
Internlm-ChatBode-7B	63.05%	51.46%	42.32%	91.33%	80.69%	79.80%	87.99%	68.09%	61.11%	69.54%
GemBode-7B	66.90%	57.16%	45.47%	86.61%	71.39%	67.40%	79.81%	63.75%	65.49%	67.11%
Zephyr-7B-alpha	56.26%	51.04%	40.27%	90.11%	72.33%	69.63%	85.26%	65.29%	65.49%	66.19%
Llama-3-8B	67.46%	57.30%	48.11%	89.51%	76.65%	43.97%	75.63%	62.43%	70.50%	65.73%
InternLM2-7B	60.18%	51.88%	39.86%	88.20%	81.15%	60.07%	67.98%	68.24%	66.23%	64.87%
Mistral-7B-Instruct	58.92%	52.16%	39.23%	90.52%	74.32%	66.10%	81.21%	70.26%	50.57%	64.81%
GemBode-7B-it	49.34%	36.58%	34.76%	79.09%	64.95%	64.67%	86.27%	63.61%	66.17%	60.60%
Vicuna-7B	50.59%	41.31%	36.08%	79.57%	50.76%	59.38%	69.81%	65.87%	59.89%	57.03%
Bode-7B	34.36%	28.93%	30.84%	79.83%	43.47%	67.45%	85.06%	65.73%	43.25%	53.21%
Llama-2-7b	31.91%	31.29%	35.44%	67.02%	31.10%	53.87%	75.16%	55.26%	59.06%	48.90%
Canarim-7B	25.96%	29.76%	31.48%	75.74%	12.08%	43.92%	79.57%	64.01%	66.00%	47.36%
Sabi-7B	55.07%	47.71%	41.41%	46.68%	01.89%	58.34%	61.93%	64.13%	46.64%	47.09%

The Gembode-7B, despite being only third in the overall ranking of models, still displays impressive performance, achieving a 67.11% accuracy on average, being the best-scorer in certain benchmarks such as BLUEX and OAB test scores. It is still relevant to notice that despite the top-scorer being tuned for the Portuguese language on the much more recent LLaMA-3 model (70.22%), the difference between them and our Gembode-7B model regarding the average score is only 3.11%.

An interesting analysis is that Gembode-7B can beat the average score of the base versions of the top scorers. For example, it surpasses the base Llama-3-8B's average of 65.73%, despite Llama-3-8B undergoing significant finetuning to reach its higher performance. Gembode-7B also performs better than the InternLM2-7B model, which scores an average of 64.87%.

The Gembode-7B-it model, however, despite improving the performance of the Gemma-7B-it, as shown in Table 5, and being capable of competing with certain 7B models, still lags behind the top-scorers, especially the Gembode-7B model of the same family, indicating some limitation inherited from its base model. Still, the Gembode-7B-it can outperform several models in certain tasks, such as in the case of the HateBR dataset, in which it achieves the third-best score. In the case of the tweetSentBR, on top of achieving the third-best score, it also beats the top two average scorers, indicating that this model may be more adequate for certain types of domain-specific tasks.

4.4 Full Tuning X PEFT

In addition to the training on models with approximately 3B parameters involving full tuning, we have incorporated an analysis of the training of these models using the same dataset but with a PEFT approach. This allows us to explore the extent to which these training approaches can influence the final outcome of the model. Table 7 presents the results and comparisons of outcomes between the types of training, focusing on the use of the PhiBode-2B models and the GemBode-2B and GemBode-2B-it models.

Table 7. Comparison between fine-tuned and full-tuned models. In this case, ∗ indicates full-tuning and † PEFT.

Model	ENEM	BLUEX	OAB	Assin2RTE	FAQUAD	NLI	HateBR	PTHateSpeech	tweetSentBR	Average
PhiBode-2B*	**38.35%**	25.17%	**29.61%**	45.39%	**43.97%**	54.15%	54.59%	43.34%		39.89%
PhiBode-2B-PEFT†	33.94%	**25.31%**	28.56%	**68.10%**	**43.97%**	**60.51%**	**54.60%**	**46.78%**		**43.59%**
GemBode-2B*	31.77%	24.20%	27.84%	69.51%	**55.55%**	53.18%	**64.74%**	50.18%		**45.25%**
GemBode-2B-PEFT†	24.14%	20.31%	25.56%	**69.75%**	52.63%	33.33%	41.65%	19.15%		32.30%
GemBode-2b-it*	**34.71%**	**25.87%**	**31.71%**	**71.31%**	**60.09%**	**47.01%**	**57.04%**	49.37%		**45.69%**
GemBode-2b-PEFT†	32.05%	21.56%	27.47%	33.33%	43.00%	36.41%	34.22%	**51.79%**		31.19%

The results indicate that in the case of PhiBode, its full-tuning training did not yield the expected impact, falling short compared to its PEFT training alternative. This raises the question of whether the data may not have had the desired impact and resulted in optimal performance during the model's training. In contrast, for GemBode, the results were quite significant, achieving more than a 10% increase in overall performance.

5 Conclusion and Future Works

This paper thoroughly investigates the training of Portuguese-centric Small Language Models (SLMs). A unique dataset was curated for training the models, and both full-tuning and PEFT methodologies were employed for comparative evaluation. Base models from Microsoft's Phi and Google's Gemma were used to construct our models, namely PhiBode and GemBode, which span from approximately 1 billion to 7 billion parameters. Ten models were developed in total.

Most models exhibited enhancements over their base models, except PhiBode-3B. This could be attributed to the significant advancements in the Portuguese language by its base model, Phi-3k, compared to Phi-2B. Notably, the dataset used for our training, termed UltraAlpaca, could have adversely influenced this training. The training data for Phi-3k could be more refined and devoid of translation errors, which are potential issues in our dataset.

Moreover, our dataset's size is relatively small compared to other Portuguese training datasets, which could also impact the training. However, it is crucial to highlight that we achieved superior results in all other models, with improvements ranging from 3% to over 10% of the general average of the Open Portuguese LLM Leaderboard benchmark.

Our research is a significant stride in the development of open-source models in Portuguese. We contribute a new dataset for training large models, novel training methodologies, and a comprehensive discussion of the results obtained through this training. This not only enhances the performance of full fine-tuning training compared to PEFT but also enriches the understanding and application of language model training, particularly for the Portuguese language.

Looking ahead, our research sets a promising trajectory for the field. We aspire to train a 7B parameter model with full fine-tuning, refine our dataset for even more impressive results, and release new models to the Brazilian scientific community. This will not only promote and boost Brazilian models for the scientific community but also for companies and individuals to use without additional costs. Our research significantly contributes to the field of language model training, particularly for the Portuguese language, and paves the way for exciting future advancements.

Acknowledgments. This study was funded by the São Paulo Research Foundation (FAPESP) grants 2013/07375-0, 2019/07665-4, 2023/14427-8, 2024/00789-8, and 2024/01336-7 as well as the National Council for Scientific and Technological Development (CNPq) grants 308529/2021-9 and 400756/2024-2.

References

1. Abdin, M., et al.: Phi-3 technical report: a highly capable language model locally on your phone. arXiv preprint arXiv:2404.14219 (2024)
2. Almeida, T.S., Laitz, T., Bonás, G.K., Nogueira, R.: Bluex: a benchmark based on Brazilian leading universities entrance exams (2023)
3. Beeching, E., et al.: Open LLM leaderboard (2023). https://huggingface.co/spaces/HuggingFaceH4/open_llm_leaderboard
4. Brum, H., das Graças Volpe Nunes, M.: Building a sentiment corpus of tweets in Brazilian Portuguese. In: Proceedings of the Eleventh International Conference on Language Resources and Evaluation (LREC 2018). European Language Resources Association (ELRA), Miyazaki, Japan (2018)
5. Chaudhary, S.: Code alpaca: an instruction-following llama model for code generation (2023). https://github.com/sahil280114/codealpaca
6. Corrêa, N.K., Falk, S., Fatimah, S., Sen, A., de Oliveira, N.: Teenytinyllama: open-source tiny language models trained in Brazilian Portuguese. arXiv preprint arXiv:2401.16640 (2024)
7. Delfino, P., Cuconato, B., Haeusler, E.H., Rademaker, A.: Passing the Brazilian OAB exam: data preparation and some experiments. In: Legal Knowledge and Information Systems, pp. 89–94. IOS Press (2017)

8. Dettmers, T., Pagnoni, A., Holtzman, A., Zettlemoyer, L.: Qlora: efficient finetuning of quantized LLMs. In: Advances in Neural Information Processing Systems, vol. 36 (2024)
9. Ding, N., et al.: Enhancing chat language models by scaling high-quality instructional conversations. arXiv preprint arXiv:2305.14233 (2023)
10. Domingues, M.: canarim-7b (2023). https://doi.org/10.57967/hf/1356
11. Fortuna, P., Rocha da Silva, J., Soler-Company, J., Wanner, L., Nunes, S.: A hierarchically-labeled Portuguese hate speech dataset. In: Proceedings of the 3rd Workshop on Abusive Language Online (ALW3), pp. 94–104. Association for Computational Linguistics (2019). https://doi.org/10.18653/v1/W19-3510
12. Gao, L., et al.: A framework for few-shot language model evaluation (2023). https://doi.org/10.5281/zenodo.10256836
13. Garcia, E.A.S.: Open Portuguese LLM leaderboard (2024). https://huggingface.co/spaces/eduagarcia/open_pt_llm_leaderboard
14. Garcia, G.L., et al.: Introducing bode: a fine-tuned large language model for Portuguese prompt-based task (2024)
15. Hu, E.J., et al.: LoRA: low-rank adaptation of large language models. arXiv preprint arXiv:2106.09685 (2021)
16. Köpf, A., et al.: Openassistant conversations – democratizing large language model alignment (2023)
17. Larcher, C., Piau, M., Finardi, P., Gengo, P., Esposito, P., Caridá, V.: Cabrita: closing the gap for foreign languages (2023)
18. Li, Y., Bubeck, S., Eldan, R., Giorno, A.D., Gunasekar, S., Lee, Y.T.: Textbooks are all you need ii: phi-1.5 technical report (2023)
19. Minaee, S., et al.: Large language models: a survey (2024)
20. Nunes, D., Primi, R., Pires, R., Lotufo, R., Nogueira, R.: Evaluating GPT-3.5 and GPT-4 models on Brazilian university admission exams (2023)
21. OpenAI, Achiam, J., Adler, S., Agarwal, S., Ahmad, L., et al: GPT-4 technical report (2024)
22. Pires, R., Abonizio, H., Almeida, T.S., Nogueira, R.: Sabiá: Portuguese large language models. In: Naldi, M.C., Bianchi, R.A.C. (eds.) BRACIS 2023. LNCS, vol. 14197, pp. 226–240. Springer, Cham (2023). https://doi.org/10.1007/978-3-031-45392-2_15
23. Pires, R., Almeida, T.S., Abonizio, H., Nogueira, R.: Evaluating GPT-4's vision capabilities on Brazilian university admission exams (2023)
24. Real, L., Fonseca, E., Gonçalo Oliveira, H.: The ASSIN 2 shared task: a quick overview. In: Quaresma, P., Vieira, R., Aluísio, S., Moniz, H., Batista, F., Gonçalves, T. (eds.) PROPOR 2020. LNCS (LNAI), vol. 12037, pp. 406–412. Springer, Cham (2020). https://doi.org/10.1007/978-3-030-41505-1_39
25. Sayama, H.F., Araujo, A.V., Fernandes, E.R.: Faquad: reading comprehension dataset in the domain of Brazilian higher education. In: 2019 8th Brazilian Conference on Intelligent Systems (BRACIS), pp. 443–448 (2019)
26. Silveira, I.C., Mauá, D.D.: University entrance exam as a guiding test for artificial intelligence. In: Proceedings of the 6th Brazilian Conference on Intelligent Systems, pp. 426–431. BRACIS (2017)
27. Singh, S., et al.: Aya dataset: an open-access collection for multilingual instruction tuning (2024)
28. Taori, R., et al.: Alpaca: a strong, replicable instruction-following model. Stanford Center Res. Found. Models 3(6), 7 (2023)
29. Team, G., et al: Gemma: open models based on Gemini research and technology (2024)

30. Touvron, H., et al: Llama: open and efficient foundation language models (2023)
31. Vargas, F., Carvalho, I., de Góes, F.R., Pardo, T., Benevenuto, F.: HateBR: a large expert annotated corpus of Brazilian Instagram comments for offensive language and hate speech detection. In: Proceedings of the Thirteenth Language Resources and Evaluation Conference, pp. 7174–7183 (2022)
32. Yu, L., et al.: Metamath: bootstrap your own mathematical questions for large language models (2024)
33. Zhao, W.X., et al.: A survey of large language models. arXiv preprint arXiv:2303.18223 (2023)

Hate Speech Detection in Portuguese Using BERTimbau

João Otávio Rodrigues Ferreira Frediani, Gabriel Lino Garcia(✉), Pedro Henrique Paiola, Leandro Aparecido Passos, João Paulo Papa, and Aparecido Nilceu Marana

School of Sciences, São Paulo State University (UNESP), Bauru, Brazil
{j.frediani,gabriel.lino,pedro.paiola,leandro.passos,joao.papa, nilceu.marana}@unesp.br

Abstract. Hate speech refers to language expressions that attack individuals or groups based on specific characteristics associated with their identities, causing lasting damage. Social networks have become a pertinent environment for hate speech proliferation since they allow anonymity and maintain a safe distance from aggressors and assaulted victims. With the amount of data published every minute, automatic identification of hate speech using machine learning gathered much attention from academic and industrial researchers. However, as with many natural language processing tasks, the efforts mainly focused on English, and languages like Portuguese remain less explored. Therefore, this paper aims to experiment with different techniques to deal with the challenges associated with low-resource languages in automatic hate speech detection. It evaluates whether knowledge transferred from offensive speech detection as a source task can be effective for hate detection and if the unbalanced data poses an obstacle for a Portuguese pre-trained BERT model, BERTimbau. Experimental results show that transferring learning between tasks does not improve performance and that using balanced data leads to better F1 scores and Cohen's Kappa.

Keywords: Hate Speech · Natural Language Processing · Machine Learning · Undersampling · Portuguese Language

1 Introduction

Usually, the development of new technologies aims to provide humankind with more comfort and quality of life. In this context, social networks successfully connected 5 billion people worldwide by the start of April 2024 [15], encouraging constructive communication among users and allowing them to share information and express opinions. Notwithstanding, it is common to identify vicious applications of such approaches in destructive practices and illegal activities. Such misbehaviour can be exemplified by sharing abusive language and encouraging hatred against individuals or groups [18].

In this context, the work of Fortuna and Nunes [6] provides a review of hate speech from computer science point of view, defining hate speech as a language that attacks or diminishes individuals or groups, inciting violence and hate based on specific characteristics such as physical appearance, religion, descent, national or ethnic origin, sexual orientation, gender identity or other, which can occur with different linguistic styles, even in subtle forms or when humour is used.

Social networks fostered the proliferation of this type of discourse due to the speed at which information spreads and the possibility of anonymity. It caused an escalation in violence around the world [18], whose victims can suffer lasting consequences due to the potentially traumatic experience, resembling physical crimes, such as assault and domestic violence [30].

Among such consequences, one can cite anxiety, stress, and depressive thoughts in victims. Further, it may encourage polarization, reinforce stereotypes about minorities, and deteriorate the relationship between groups [20]. In extreme cases, violence carried out over the internet presents a concerning likelihood of leading to suicide [10].

Manual analysis of hostile content denotes a prohibitive task due to the massive amount of information shared on social networks [23]. Thus, academic and industrial researchers devoted considerable efforts toward automatic approaches to tackle the problem using machine learning [18]. Nevertheless, most essays are conducted considering English language speech only [6]. Recently, new studies considered Arabic [9,19,21], Turkish [4,13], Dutch [17], Spanish [2,27], and German [12] languages, while Portuguese, Korean, and Chinese continue with few publications in the area [11].

Performing hate speech detection in these languages becomes difficult due to the lack of labelled data, which can be associated with the problem of dataset creation, which denotes a challenging task itself. Hate speech is a vague and abstract concept comprising subtle nuances with which the human agent in charge of labelling the data may disagree. Further, the definition correlates to complex social and cultural structures in constant change [6]. Moreover, a hate speech dataset must gather information representing the actual environment. Thus, it is usually collected from social networks, mainly from X (formerly Twitter) and Facebook. Besides, it demands some care to assure the users' security and anonymity [11].

Another concerning issue in data extracted from social networks is the imbalanced class distribution, i.e., instances comprising hate speech are considerably rarer than standard publications, thus negatively affecting the classification process by leading to a bias towards the most common class [22,24].

Therefore, some strategies should be considered when dealing with such low-resource languages and unbalanced information to provide a fair evaluation. Regarding the former, this paper employs a transfer learning-based approach that fine-tunes BERTimbau [25], a transformers-based approach for natural language processing (NLP) in Portuguese, using offensive language detection as a source task and afterward repeats the fine-tuning process aiming to detect hate speech as the target task. Concerning the latter, it conducts several experiments using random undersampling to tackle the imbalance problem. Additionally, due to the low resources available in the Portuguese language, this paper also assembles multiple offensive language and hate speech datasets in Portuguese to assess the effects of increasing the amount of training data. Hence, the main contributions of this paper are described as follows:

- Providing an automated approach for hate speech detection in the Portuguese language using BERTimbau;
- Using transfer-learning and fine-tuning to evaluate BERTimbau in the context of hate speech detection; and
- Employing data imbalance solutions to tackle bias towards the majority class, i.e., non-hate speeches.

The remainder of this work is organized as follows. Section 2 introduces the related work concerning hate speech detection, while Sect. 3 describes the material, the methods and the experimental setup. Further, Sect. 4 presents the results. Finally, Sect. 5 states the conclusions and future work.

2 Related Work

Recently, many outstanding studies have addressed the problem of hate speech detection. Nevertheless, most such works are conducted over English speech datasets, leaving languages like Portuguese almost unexplored [11].

Regarding the hate speech detection in the Portuguese language, we highlight the recent works of Silva and Roman [23], Assis et al. [3], Firmino et al. [5], Da Silva and Rosa [24] and Aluru et al. [1].

Silva and Roman [23] compared four different machine-learning models over the *Fortuna* dataset [7], concluding that the effectiveness of these models is drastically affected by the text representation used and that classical models like Support Vector Machines (SVM) and Random Forests (RF) can achieve good results.

Assis et al. [3] compared the performance of BERT-based models against prompting techniques for generative models, concluding that BERT performs better in most cases.

Firmino et al. [5] explored the usage of Cross-lingual Learning on Pre-trained Language Models. They used one corpus in English and another in Italian as source languages to evaluate if the knowledge obtained from these texts can help to solve the problem for Portuguese by testing on the OffComBr-2 dataset. The experiments indicated that the use of other languages with more resources could be helpful when solving the problem in Portuguese, achieving an F1 score of 0.92 with both source languages.

Da Silva and Rosa [24] compared the classification performances of eleven distinct methods on seven different hate speech detection datasets, three of them in Portuguese. Experimental results show that BERT obtained the best results, followed closely by Linear Regression (LR) and the SVM.

Aluru et al. [1] perform a analysis of multi-lingual hate speech detection, using deep learning to detect hate speech on 16 different datasets written in 9 languages, including portuguese. Overall, observing high resource models, BERT-based models were the most effective.

3 Material and Methods

This section presents the datasets used in this work, containing offensive language and hate speech in Portuguese. It also presents the methods used for resampling and transfer learning. Then, it describes the experimental setup.

3.1 Datasets

This work employs two offensive language (ToLDBR and OlidBR) and two hate speech detection (Fortuna Dataset and HateBR) datasets in Portuguese, described as follows.

ToLDBR: The Toxic Language Dataset for Brazilian Portuguese (ToLDBR) is an offensive-language dataset, created by Leite et al. [16], that comprises 21k posts collected from Twitter, categorized into seven toxic language categories. The data was gathered using the GATE Cloud's Twitter Collector[1] between July and August 2019 using two strategies. The first consists of searching for terms and hashtags usually present in offensive tweets. The second avoids generalization by selecting around 50 influential users and scraping tweets that mentioned them.

The authors extracted the posts from a pre-selected set of 10 million tweets. These texts were sent to 42 demographically distributed volunteers to represent distinct cultural perceptions. Each text was annotated by three volunteers who classified it into seven categories: LGBTQ+phobia, obscene, insult, racism, misogyny or xenophobia, and black if it was non-toxic.

OlidBR: The Offensive Language Identification Dataset for Brazilian Portuguese (OlidBR), an offensive language dataset created by Trajano et al. [26], contains 6,354 texts collected from Twitter, Youtube comment section and other related sources. The data was classified into three levels. The first is a simple binary toxic classification, the second splits the toxic comment into eleven categories, and the third regards the target group.

Fortuna Dataset: The Fortuna dataset [7] consists of posts written in Portuguese collected from Twitter using the Twitter API between January and March 2017. The corpus comprises 5,668 texts from 1,156 authors.

The data was gathered in two ways: i) using a keyword filter and ii) using a user filter aimed at users known for propagating hate speech. Then, each text was classified by 18 native speakers, and a majority voting strategy was applied to determine the label.

The authors also proposed a hierarchical distribution of different hate speech categories, e.g., hate against lesbians, as a subcategory of homophobia. Such an approach enables the comprehension of relationships between various types of

---

[1] Available at https://cloud.gate.ac.uk/info/about/twitter.html.

hate speech by introducing a larger set of behaviors, thus allowing non-disjoint classes to participate in more generic categories.

HateBR: The HateBr dataset, proposed by Vargas et al. [28], consists of a corpus of Portuguese snippets gathered from political figures' comment sections on Instagram. Six public accounts were chosen, three belonging to conservative and three to liberal parties. Five hundred comments were extracted from publications posted in the second semester of 2019, resulting in 15,000 texts.

The data was then cleaned by removing links, characters without semantic value, and comments that presented only emoticons, laughs, or mentions, resulting in 7,000 snippets. These texts were classified into three levels: (i) binary offensive language classification, (ii) offensive texts classified as highly, moderately, or weakly offensive, and (iii) offensive texts containing hate speech or not.

3.2 Resampling

Data imbalance is a problem related to datasets in which one or more classes comprise a much greater number of examples than the others [8]. Training a classifier on an imbalanced dataset usually leads to biased estimations towards the majority class while neglecting the minority, leading to an unsatisfying performance [24].

The bias effect is usually undertaken through resampling techniques. Resampling is a technique that balances the number of examples for each class in a dataset and can be done in different ways, e.g., oversampling, generating synthetic samples of the minority classes, and undersampling, which prunes majority class instances [14].

This work tackles the problem through random undersampling, which removes random instances from the majority classes until reaching a predefined ratio. Despite the simple concept, random sampling can provide impactful improvements in the final results [24].

3.3 Transfer Learning

Transfer learning is a concept first introduced in educational psychology and is defined as the result of the generalization of knowledge. Suppose a person who knows how to play the violin, for instance. Learning to play a new instrument, like the piano, may be easier for him than most people since both violin and piano are musical instruments and share similar knowledge. Likewise, someone who knows how to ride a bike will have an easier time learning to ride a motorcycle.

While transfer learning holds promise, it's important to note that knowledge transfer is not always straightforward. Significant similarities between the domains are required, and even in cases where domains are similar, the resemblances can be misleading. For example, a Portuguese speaker learning English may make several mistakes due to false cognates, which are pairs of similar words with disconnected meanings [31].

In artificial intelligence, transfer learning seeks to generalize knowledge in one domain to be applied in another, being especially useful in cases in which the objective domain has little data available, and in multitask learning, a technique in which a machine learning model is trained to solve multiple tasks simultaneously. Figure 1 represents the learning process in a traditional supervised machine learning model (left), in which the training and testing sets come from the same domain, and for transfer learning (right), in which the training set comes from, at least partially, a different domain than the test set [29].

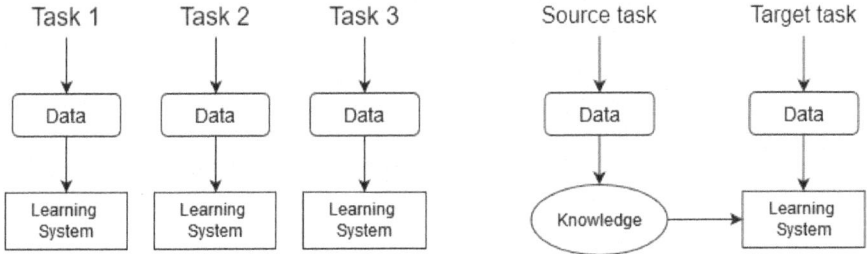

Fig. 1. Learning process of traditional machine learning (left) and transfer learning (right).

3.4 Experimental Setup

This section describes the overall experimental setup, such as parameters, data pre-processing, and machine specifications.

Hyperparameters. The experiments were conducted using BERTimbau[2]. The fine-tuning process was performed with a leaning rate of 2.5×10^{-5} with a weight decay of 0.01, on 5 epochs with a batch size of 16.

Data Pre-processing. As the task explored is a binary classification, for both souce and target task, all features except text and label were removed.

In this work, it was applied a random undersampling, using the Imbalanced Learn library[3], on the hate speech detection datasets. Table 1 shows details of these datasets before and after the undersampling.

For easier comprehension, the junction of both hate speech detection datasets (Fortuna and HateBr) will be referred to as JoinedHatePT from now on.

The data was split into three sets: (i) the training set with 75% of the instances, (ii) the validation set with 12.5% of the instances, and the test set with 12.5% of the instances.

---

[2] Available in Hugging Face at: https://huggingface.co/neuralmind/bert-base-portuguese-cased.
[3] Available at: https://imbalanced-learn.org.

Table 1. Number of instances and ratio for every dataset used.

	Total of instances	Positive instances	Negative instances	Ratio
Before Undersampling				
Fortuna Dataset	5,668	1,788	3,880	31.55/68.45
HateBR	7,000	702	6,298	10.02/89.98
JoinedHatePT	12,668	2,490	10,178	19.66/80.34
After Undersampling				
Fortuna Dataset	3,576	1,788	1,788	50/50
HateBR	1,404	702	702	50/50
JoinedHatePT	4,980	2,490	2,490	50/50

Hardware Specifications. All the experiments were performed on Google Colab[4], a service that provides access to computing resources using the cloud. The selected machine comprises 51 GB of system RAM, 15 GB of GPU RAM, and 201 GB of available disk storage.

4 Experiments and Results

The first experiment comprises the BERTimbau performance trained on the source task, i.e., offensive speech detection, whose results are presented in Table 2. The model is fine-tuned in multiple datasets to explore whether the amount of training data influences the model's performance on the target task.

Table 2. Performance of BERTimbau on the offensive speech detection task.

	Test Set			Validation Set		
Dataset	Accuracy	Cohen's Kappa	F1	Accuracy	Cohen's Kappa	F1
ToLDBR+OlidBR	75.35%	0.5074	0.7429	79.53%	0.5902	0.7811
ToLDBR	77.53%	0.5535	0.7713	76.41%	0.5299	0.7515
OlidBR	86.14%	0.5018	0.9169	83.08%	0.4885	0.8932

Afterwards, Table 3 shows the model's performance, fine-tuned on hate detection datasets, i.e., Fortuna, HateBR, and JoinedHatePT datasets. Moreover, it also presents the results after conducting a pre-fine-tuning step over offensive language datasets, namely OlidBR and ToLDBR. Regarding the latter, the first column states the standard model (BERTimbau), pre-fine-tuned on the OlidBR (BERTimbau OlidBR), on the OlidBR (BERTimbau ToLDBR), and both datasets (BERTimbau ToLDBR+OlidBR).

The findings delineated in Table 3 suggest that the source task did not significantly influence the model's performance, notwithstanding the apparent similarity. The BERTimbau model, devoid of prior fine-tuning, demonstrated superior

---
[4] https://colab.google/.

accuracy in two datasets. At the same time, it secured the second-best accuracy for the Fortuna dataset, albeit with values akin to the other models.

The most commendable results in Table 3 were achieved utilizing the HateBR dataset. However, this dataset is notably imbalanced, with a mere 10% of instances being positive. Following this, the JointHatePT dataset emerges, which attains superior accuracy and Cohen's Kappa for each model compared to the Fortuna Dataset, but records a lower F1 score for two models. It is crucial to underscore that, despite the disparities in the results, all variations fall within an acceptable margin, thereby indicating the robustness of the models across diverse scenarios.

Table 3. Performance of the BERTimbau model using transfer learning on the task of hate speech detection on the original datasets (unbalanced).

Model	Test Set			Validation Set		
	Accuracy	Cohen's Kappa	F1	Accuracy	Cohen's Kappa	F1
Fortuna Dataset						
BERTimbau	76.49%	0.4055	0.5603	**77.54%**	0.4245	0.5737
BERTimbau ToLDBR+OlidBR	74.29%	0.3887	0.5725	73.87%	0.3970	0.5879
BERTimbau ToLDBR	75.23%	0.4301	**0.6117**	74.71%	**0.4317**	**0.6215**
BERTimbau OlidBR	**76.59%**	**0.4406**	0.6070	74.57%	0.3983	0.5794
HateBR						
BERTimbau	**94.14%**	**0.6976**	**0.7304**	93.82%	0.5915	0.625
BERTimbau ToLDBR+OlidBR	93.39%	0.5883	0.6236	**94.97%**	0.6346	0.6615
BERTimbau ToldBR	93.46%	0.6278	0.6634	93.71%	**0.6398**	**0.6745**
BERTimbau OlidBR	92.54%	0.5332	0.5729	94.05%	0.6115	0.6438
JoinedHatePT						
BERTimbau	**86.65%**	0.5089	0.5856	**87.55%**	**0.5438**	**0.6175**
BERTimbau ToldBR+olidBR	84.88%	**0.5138**	**0.6074**	84.20%	0.5034	0.6019
BERTimbau ToldBR	86.12%	0.4648	0.5423	85.21%	0.4602	0.5411
BERTimbau OlidBR	86.31%	0.5106	0.5915	85.41%	0.5040	05897

Table 4 introduces the results concerning the model fine-tuned on the undersampled hate speech datasets. Although the accuracy decreased compared to Table 3, Cohen's Kappa and F1 achieved considerably better results in most cases, which is expected and suggests that the unbalanced data influenced the model to create a bias towards the majority class.

The best performance found in Table 4 were also achieved using the HateBR dataset, however in this case the dataset is considerably small. Compared to Table 3 there was a loss of accuracy but the Cohen's Kappa were slightly better, while the F1 score was significantly better, achieving an improvement of 0.31 for the BERTimbau OlidBR.

It is also noteworthy that joining both datasets leads to a better performance with or without undersampling when comparing to the Fortuna Dataset, but worse comparing to the HateBR dataset. The best results being achieved with the HateBR dataset could be linked to its imbalance or its small size after the undersampling, which could indicate that increasing the amount of training data could help the model perform better.

Table 4. Performance of the BERTimbau model using transfer learning on the task of hate speech detection on the undersampled datasets (balanced).

Model	Test Set			Validation Set		
	Accuracy	Cohen's Kappa	F1	Accuracy	Cohen's Kappa	F1
Fortuna Dataset						
BERTimbau	72.24%	0.4460	0.7269	72.48%	0.4475	0.7050
BERTimbau ToLDBR+OlidBR	72.64%	0.4526	0.7131	73.15%	0.4630	0.7357
BERTimbau ToLDBR	74.35%	0.4870	0.7458	**76.51%**	**0.5297**	**0.7732**
BERTimbau OlidBR	**75.14%**	**0.5028**	**0.7523**	70.69%	0.4081	0.6749
HateBR						
BERTimbau	86.19%	0.7162	0.8816	80.00%	0.5997	0.7853
BERTimbau ToLDBR+OlidBR	83.33%	0.6648	0.8448	84.00%	0.6778	0.8526
BERTimbau ToldBR	82.35%	0.6466	0.8297	**86.28%**	**0.7190**	**0.8812**
BERTimbau OlidBR	**88.15%**	**0.7627**	**0.8858**	78.85%	0.5486	0.8310
JoinedHatePT						
BERTimbau	73.97%	0.4799	0.7383	81.83%	**0.6351**	**0.8295**
BERTimbau ToldBR+olidBR	76.32%	0.5260	0.7482	**78.45%**	0.5704	0.7853
BERTimbau ToldBR	76.06%	0.5211	0.7486	77.65%	0.5501	0.7565
BERTimbau OlidBR	**76.80%**	**0.5355**	**0.7757**	76.52%	0.5305	0.7614

5 Conclusions

The main objective of this paper was to explore methods to improve the performance of the BERTimbau in detecting hate speech in Portuguese, a low-resource language that suffers from a lack of labeled datasets and unbalanced data.

It explored the possibility of transferring knowledge learned of offensive language detection tasks to hate speech detection. It concluded that prior finetuning the model in the source task did not help the model gather generic knowledge to improve the target task, thus providing similar results in experiments with and without prior fine-tuning.

Further, it employed resampling techniques to tackle the problem of biased results towards the majority class using a random undersampler. Although providing lower accuracies, a virtually insignificant metric to imbalanced datasets, the procedure achieved considerably better Cohen's Kappa and F1 scores, leading to more trustworthy results.

Moreover, the models performed better on the JoinedHatePT than in the Fortuna Dataset, indicating that increasing the training data implies positive effects.

In future work, we aim to explore diverse resampling techniques since the amount of data is demonstrated to influence model performance directly. We also aim to investigate the behaviour of oversampling techniques in the process.

Acknowledgment. This study was financed in part by the Coordination of Superior Level Staff Improvement - Brazil (CAPES) - Finance Code 001, the São Paulo Research Foundation (FAPESP) grants 2013/07375-0, 2019/07665-4, 2023/14427-8, and 2024/01336-7 as well as the National Council for Scientific and Technological Development (CNPq) grants 308529/2021-9 and 400756/2024-2.

References

1. Aluru, S.S., Mathew, B., Saha, P., Mukherjee, A.: Deep learning models for multilingual hate speech detection. arXiv preprint arXiv:2004.06465 (2020)
2. Plaza-del Arco, F.M., Molina-González, M.D., Urena-López, L.A., Martín-Valdivia, M.T.: Comparing pre-trained language models for Spanish hate speech detection. Expert Syst. Appl. **166**, 114120 (2021)
3. Assis, G., Amorim, A., Carvalho, J., de Oliveira, D., Vianna, D., Paes, A.: Exploring Portuguese hate speech detection in low-resource settings: lightly tuning encoder models or in-context learning of large models? In: Proceedings of the 16th International Conference on Computational Processing of Portuguese, pp. 301–311 (2024)
4. Beyhan, F., Çarık, B., Arın, I., Terzioğlu, A., Yanikoglu, B., Yeniterzi, R.: A Turkish hate speech dataset and detection system. In: Proceedings of the Thirteenth Language Resources and Evaluation Conference, pp. 4177–4185 (2022)
5. Firmino, A.A., de Souza Baptista, C., de Paiva, A.C.: Improving hate speech detection using cross-lingual learning. Expert Syst. Appl. **235**, 121115 (2024)
6. Fortuna, P., Nunes, S.: A survey on automatic detection of hate speech in text. ACM Comput. Surv. (CSUR) **51**(4), 1–30 (2018)
7. Fortuna, P., da Silva, J.R., Wanner, L., Nunes, S., et al.: A hierarchically-labeled Portuguese hate speech dataset. In: Proceedings of the Third Workshop on Abusive Language Online, pp. 94–104 (2019)
8. Haixiang, G., Yijing, L., Shang, J., Mingyun, G., Yuanyue, H., Bing, G.: Learning from class-imbalanced data: review of methods and applications. Expert Syst. Appl. **73**, 220–239 (2017)
9. Hassan, S., Samih, Y., Mubarak, H., Abdelali, A.: Alt at Semeval-2020 task 12: Arabic and English offensive language identification in social media. In: Proceedings of the Fourteenth Workshop on Semantic Evaluation, pp. 1891–1897 (2020)
10. Hinduja, S., Patchin, J.W.: Bullying, cyberbullying, and suicide. Arch. Suicide Res. **14**(3), 206–221 (2010)
11. Jahan, M.S., Oussalah, M.: A systematic review of hate speech automatic detection using natural language processing. Neurocomputing, 126232 (2023)
12. Jaki, S., De Smedt, T.: Right-wing German hate speech on Twitter: analysis and automatic detection. arXiv preprint arXiv:1910.07518 (2019)

13. Karayiğit, H., Akdagli, A., Aci, Ç.İ: Homophobic and hate speech detection using multilingual-BERT model on Turkish social media. Inf. Technol. Control **51**(2), 356–375 (2022)
14. Kaur, H., Pannu, H.S., Malhi, A.K.: A systematic review on imbalanced data challenges in machine learning: applications and solutions. ACM Comput. Surv. (CSUR) **52**(4), 1–36 (2019)
15. KEPIOS: Global social media statistics (2024). https://datareportal.com/social-media-users
16. Leite, J.A., Silva, D.F., Bontcheva, K., Scarton, C.: Toxic language detection in social media for Brazilian Portuguese: new dataset and multilingual analysis. arXiv preprint arXiv:2010.04543 (2020)
17. Markov, I., Gevers, I., Daelemans, W.: An ensemble approach for Dutch cross-domain hate speech detection. In: Rosso, P., Basile, V., Martínez, R., Métais, E., Meziane, F. (eds.) NLDB 2022. lncs, vol. 13286, pp. 3–15. Springer, Cham (2022). https://doi.org/10.1007/978-3-031-08473-7_1
18. Mozafari, M., Farahbakhsh, R., Crespi, N.: A BERT-based transfer learning approach for hate speech detection in online social media. In: Cherifi, H., Gaito, S., Mendes, J.F., Moro, E., Rocha, L.M. (eds.) COMPLEX NETWORKS 2019. SCI, vol. 881, pp. 928–940. Springer, Cham (2020). https://doi.org/10.1007/978-3-030-36687-2_77
19. Mubarak, H., Darwish, K., Magdy, W.: Abusive language detection on Arabic social media. In: Proceedings of the First Workshop on Abusive Language Online, pp. 52–56 (2017)
20. Obermaier, M., Schmuck, D., Saleem, M.: I'll be there for you? Effects of Islamophobic online hate speech and counter speech on Muslim in-group bystanders' intention to intervene. New Media Soc. 14614448211017527 (2021)
21. Omar, A., Mahmoud, T.M., Abd-El-Hafeez, T.: Comparative performance of machine learning and deep learning algorithms for arabic hate speech detection in OSNs. In: Hassanien, A.-E., Azar, A.T., Gaber, T., Oliva, D., Tolba, F.M. (eds.) AICV 2020. AISC, vol. 1153, pp. 247–257. Springer, Cham (2020). https://doi.org/10.1007/978-3-030-44289-7_24
22. Passos, L.A., Jodas, D.S., Ribeiro, L.C., Akio, M., De Souza, A.N., Papa, J.P.: Handling imbalanced datasets through optimum-path forest. Knowl.-Based Syst. **242**, 108445 (2022)
23. Silva, A., Roman, N.: Hate speech detection in Portuguese with Naïve Bayes, SVM, MLP and logistic regression. In: Anais do XVII Encontro Nacional de Inteligência Artificial e Computacional, pp. 1–12. SBC (2020)
24. da Silva, R.C.C., Rosa, T.C.: Combining data transformation and classification approaches for hate speech detection: a comparative study. Thierson, Combining Data Transformation and Classification Approaches for Hate Speech Detection: A Comparative Study
25. Souza, F., Nogueira, R., Lotufo, R.: BERTimbau: pretrained BERT models for Brazilian Portuguese. In: Cerri, R., Prati, R.C. (eds.) BRACIS 2020. LNCS (LNAI), vol. 12319, pp. 403–417. Springer, Cham (2020). https://doi.org/10.1007/978-3-030-61377-8_28
26. Trajano, D., Bordini, R.H., Vieira, R.: Olid-BR: offensive language identification dataset for Brazilian Portuguese. Lang. Resourc. Eval. 1–27 (2023)
27. del Valle, E., de la Fuente, L.: Sentiment analysis methods for politics and hate speech contents in Spanish language: a systematic review. IEEE Lat. Am. Trans. **21**(3), 408–418 (2023)

28. Vargas, F.A., Carvalho, I., de Góes, F.R., Benevenuto, F., Pardo, T.A.S.: HateBR: a large expert annotated corpus of Brazilian Instagram comments for offensive language and hate speech detection. arXiv preprint arXiv:2103.14972 (2021)
29. Weiss, K., Khoshgoftaar, T.M., Wang, D.D.: A survey of transfer learning. J. Big Data **3**(1), 1–40 (2016). https://doi.org/10.1186/s40537-016-0043-6
30. Williams, M.: Hatred behind the screens: A report on the rise of online hate speech (2019)
31. Zhuang, F., et al.: A comprehensive survey on transfer learning. Proc. IEEE **109**(1), 43–76 (2020)

An Effective Approach to Text Detection and Recognition in Degraded Historical Documents

Percy Maldonado-Quispe and Helio Pedrini

Institute of Computing, University of Campinas, Campinas, SP 13083-852, Brazil
helio@ic.unicamp.br

Abstract. In this paper, we present a method for improving text detection and recognition in historical and degraded document images by integrating Document Layout Analysis (DLA) with advanced Image Enhancement (IE) techniques and Optical Character Recognition (OCR). Our method utilizes the state-of-the-art Vision Grid Transformer for precise detection of text regions, followed by a series of enhancement processes including skew correction, Gaussian blur, binarization, and mathematical morphology to enhance the clarity and readability of the text. Evaluated on a custom dataset of scanned historical books from Project Gutenberg, our approach demonstrates superior performance in terms of Word Error Rate (WER) compared to existing methods, highlighting its effectiveness in accurately recovering text from complex and degraded document images.

Keywords: Document Layout Analysis · Image Enhancement · Optical Character Recognition

1 Introduction

Document Layout Analysis (DLA) [1,9,14,22] is a foundational aspect of document image processing, crucial for the efficient extraction of structural and semantic information from digitized documents. The accurate identification and processing of text regions within these documents are vital for numerous applications, including digital archiving, automated document retrieval, and historical document preservation. However, the variability in document layouts, especially in historical and degraded documents, poses significant challenges that require innovative solutions.

In this paper, we introduce a DLA methodology that integrates two primary components: the detection of text regions and the enhancement of the content within these regions. The first component employs advanced technique to accurately detect text regions in document images. By utilizing the state-of-the-art method, this approach effectively localizes text areas, ensuring robustness and precision even in the presence of noise and varying document layouts. This initial step is critical for delineating the areas that require further processing and improvement.

The second and principal contribution of our work lies in the enhancement of content within the detected text regions. This component focuses on improving the quality and legibility of the text, addressing issues such as noise, low contrast, and haze that are common in historical documents. Our enhancement techniques incorporate state-of-the-art image processing methods, including contrast adjustment, deblurring, and haze removal. These techniques collectively enhance the clarity and readability of the text, facilitating more accurate text recognition and interpretation.

Our approach not only improves the detection of text regions but also significantly enhances the content within these regions, setting our work apart from existing approaches. By prioritizing content enhancement, we ensure that the extracted text is not only accurately identified but also presented in a manner that is more legible and usable. This dual focus on detection and enhancement addresses both structural and quality-related challenges in DLA, providing a comprehensive solution for the analysis of complex and degraded document images.

We validate the effectiveness of our method through extensive experiments on own dataset of Books scanned and historical Peruvian newspaper images. The results demonstrate that our approach significantly outperforms classical methods in text detection and content enhancement. The integration of these advanced techniques into a unified framework offers a robust and reliable solution for DLA, paving the way for more accurate and accessible digitization and analysis of historical documents.

The paper is organized as follows. Section 2 reviews related work in the field of DLA and pre-processing techniques. Section 3 details our approach, including the specific algorithms used. Section 4 presents experimental results and discusses the impact of our enhancements. Finally, Sect. 5 concludes the paper with a summary of our contributions and potential directions for future work.

2 Related Work

The fields of Document Layout Analysis (DLA), Image Enhancement (IE), and Optical Character Recognition (OCR) are crucial for the efficient digitization and interpretation of document images. We provide an overview in this fields.

2.1 Document Layout Analysis

DLA [1,9,14,22] has evolved significantly, driven by the increasing need to digitize and organize large volumes of documents. Traditional approaches, such as the XY-Cut algorithm [8] and the Docstrum method [16], focus on geometric and structural properties to segment document images into text blocks, images, and tables. However, these methods often struggle with complex and heterogeneous layouts, especially in historical documents.

Recent advancements leverage machine learning and deep learning techniques to improve DLA accuracy and robustness. Models such as LayoutLM [23], which combine visual and textual information using transformer architectures, have set new benchmarks in layout analysis. Additionally, the creation of large annotated datasets such as PubLayNet [24] and DocBank [12] has significantly contributed to the development and evaluation of advanced DLA methods. The recent work Vision Grid Transformer [4] becomes the state of the art for the detection of layouts, where they propose Grid Transformer (GiT).

2.2 Image Enhancement

Traditional IE methods [19], such as histogram equalization and Gaussian filtering, have been used extensively in handling document issues [18]. Mathematical morphology techniques are essential for shape analysis and noise reduction. Erosion and Dilation are used to remove small noise elements and fill gaps, respectively. Opening and Closing operations, combinations of erosion and dilation, smooth contours and remove small unwanted elements [6]. The Distance Transform [2] is used to separate connected components and improve text block segmentation. Top-Hat and Black-Hat Transformations enhance elements that are lighter or darker than their surroundings, useful for correcting non-uniform illumination.

The advent of deep learning [7,10] has introduced more sophisticated approaches, such as GANs and convolutional autoencoders, which have shown superior performance in tasks such as image super-resolution and deblurring [11]. Recent models such as SwinIR [13], based on the Swin Transformer architecture, have demonstrated state-of-the-art results in image enhancement, effectively improving the visual quality and preserving textual information.

2.3 Optical Character Recognition

OCR [3,15,20] has undergone significant transformations, evolving from early template matching techniques to advanced deep learning-based methods. Traditional OCR systems such as Tesseract [21] rely on pattern recognition and feature extraction, which are effective for clean and structured documents but struggle with noisy and complex layouts. Commercial OCR systems, including Google's Vision API and ABBYY FineReader, incorporate these advanced techniques, benefiting from large annotated datasets that provide diverse training data for improved accuracy and robustness.

On the other hand, Okamato et al. [17] proposed a transformer-based methodology for visual document understanding, in addition to a weakly supervised framework for cost-efficient training. In contrast, Dell et al. [5] presented a pipeline specifically designed for text extraction from English newspaper images.

3 Methodology

Our approach consists of three main parts: first, we utilize the state-of-the-art Vision Grid Transformer (ViGT) to accurately detect text regions. Second, we apply a series of image enhancement techniques and finally detect and extract the text. This comprehensive approach ensures high precision in text detection and significantly improves the readability and usability of the processed document images.

Figure 1 presents a flowchart illustrating the three main modules of our approach. The process starts with an input image, which is first processed by the DLA module, where all regions present in the image are recognized, focusing specifically on the text regions components. From the DLA result, a detected region text is selected and sent to the next module, IE. In this module, image enhancement techniques such as angle correction, text readability enhancement and binarization are applied. Finally, the improved result is processed by the OCR module, which is responsible for extracting and recognizing the text present in the previously improved region text box.

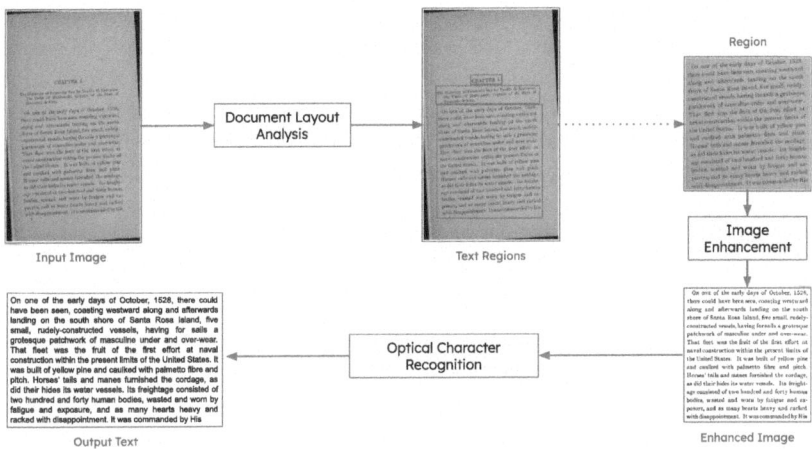

Fig. 1. Flowchart describing the three main modules of the approach: Document Layout Analysis, Image Enhancement, and Optical Character Recognition.

Document Layout Analysis. In the first part of our approach, we focus on DLA to accurately detect text regions within document images. We leverage the state-of-the-art Vision Grid Transformer (ViGT), a model that combines the strengths of convolutional neural networks and transformer architectures to achieve superior performance in layout analysis tasks. The ViGT model is designed to capture both local and global contextual information, making it highly effective for complex and heterogeneous document layouts. By utilizing self-attention mechanisms, ViGT can precisely identify and segment text regions, even in the presence of noise and varying document structures. This precise localization of text regions serves as a crucial foundation for subsequent image enhancement processes.

Image Enhancement. The second part of our approach involves enhancing the content within the detected text regions to improve the quality and legibility of the text. We employ several image enhancement techniques, each contributing specific advantages to the overall process:

1. Gaussian Blur: Applying Gaussian blur helps to reduce noise and smooth out the document image. The Gaussian blur process involves convolving the image with a Gaussian function, defined as:

$$G(x,y) = \frac{1}{2\pi\sigma^2} e^{-\frac{x^2+y^2}{2\sigma^2}} \tag{1}$$

where σ is the standard deviation of the Gaussian distribution. This blurring technique is advantageous for reducing high-frequency noise, making the text regions more distinct and easier to process in subsequent steps.

2. Skew Correction: This is a fundamental pre-processing step that aligns the document image to a standard orientation. Documents are often scanned or photographed at an angle, leading to skewed text lines that can hinder subsequent processing steps. We correct this skew by detecting the dominant angle of the text lines and rotating the image accordingly. Mathematically, if θ represents the skew angle, the transformation matrix for rotation is given by:

$$R(\theta) = \begin{pmatrix} \cos(\theta) & -\sin(\theta) & 0 \\ \sin(\theta) & \cos(\theta) & 0 \\ 0 & 0 & 1 \end{pmatrix} \tag{2}$$

Correcting the skew ensures that text lines are horizontally aligned, which is critical for accurate text recognition.

3. Binarization: This step converts the grayscale document image into a binary image, where the pixels are either black or white. This step is essential for distinguishing text from the background. We use adaptive thresholding methods, such as Otsu's method, which dynamically determine the threshold value based on the image's histogram. The binarization process enhances the contrast between the text and the background, facilitating clearer text recognition.

4. Mathematical Morphology: Mathematical morphology operations, specifically opening, are used to remove small noise and enhance the text structure. Opening is a combination of erosion followed by dilation, which can be mathematically represented as:

$$A \circ B = (A \ominus B) \oplus B \tag{3}$$

Optical Character Recognition. For the third part, we chose Tesseract OCR for image text extraction because of its proven efficiency and reliability in document processing. Tesseract is an open source OCR engine that has been extensively optimized for accuracy and efficiency. Its ability to handle multilingual text, its robust support for various image preprocessing techniques and its flexibility to adapt to different character sets make it an ideal choice. In addition,

Tesseract's active development community ensures continuous improvement and access to state-of-the-art OCR capabilities.

4 Results

Our evaluation focused on the final stage of our approach: text detection and recognition from the entire document image. We employed Word Error Rate (WER) and Character Error Rate (CER) as metrics to quantitatively assess the accuracy of our approach.

4.1 Dataset

We conducted comprehensive evaluations on a *custom* B10 dataset extracted from Project Gutenberg ebooks, comprising scanned images of historical books with diverse layouts and varying degrees of degradation typical of older documents. We selected ten books from the Project Gutenberg, labeled from b1 to b10, for this study. These books, spanning various genres and publication eras, serve as a diverse data set to evaluate the effectiveness of our technique in text detection and recognition. By leveraging these historical and often degraded texts, we aim to test and validate our method, ensuring its applicability and robustness across different document types and conditions.

Figure 2 presents a view of the first page of each book in the B10 dataset, visually highlighting the diversity of the books and the different types of noise present, resulting from the manual scanning process. The B10 dataset consists of 322 images in total, Table 1 provides a summary of the characteristics in the dataset. The columns show the number of images scanned, the total number of words in those images, the resolution of the images (in pixels, width × height), and the main regions identified in each book. These regions include headings, paragraphs and images.

4.2 Performance Metrics

Word Error Rate. Table 2 presents the Word Error Rate (WER) for three different text recognition methods (Method 1, Method 2, and Method 3) across ten books (b1 to b10). WER is a measure of the accuracy of the text recognition process, where a value of 0 indicates that the method failed to recover any of the words present in the image, and a value of 1 indicates that the method successfully recovered all the words. The values represent the fraction of correctly recognized words relative to the total number of words in the ground truth.

- Method 1 (OCR): This method demonstrates strong performance across all books, with WER values consistently above 0.7. The highest WER is observed for book 2 (0.849), indicating almost complete word recovery, while the lowest WER is for book 6 (0.746). This suggests that Method 1 is robust and reliable in recognizing words from the scanned images, making it a highly effective method among the three tested.

Table 1. Summary of the B10 dataset, detailing the number of images, total word count, image resolution (width x height) and types of regions identified (headings, paragraphs, images).

Book	Images	Words	Resolution (width × height)	Regions
b1	39	14789	1850 × 2621	Headings, Paragraphs, Images
b2	8	3938	2571 × 3546	Headings, Paragraphs
b3	37	7355	1400 × 2067	Headings, Paragraphs
b4	30	7593	1217 × 1983	Headings, Paragraphs, Images
b5	30	9493	1708 × 2317	Headings, Paragraphs, Images
b6	34	7660	1433 × 2313	Headings, Paragraphs, Images
b7	30	4745	1425 × 2250	Headings, Paragraphs
b8	34	10569	1396 × 2338	Headings, Paragraphs
b9	23	3372	1271 × 2029	Headings, Paragraphs
b10	57	12332	1088 × 1642	Headings, Paragraphs, Images

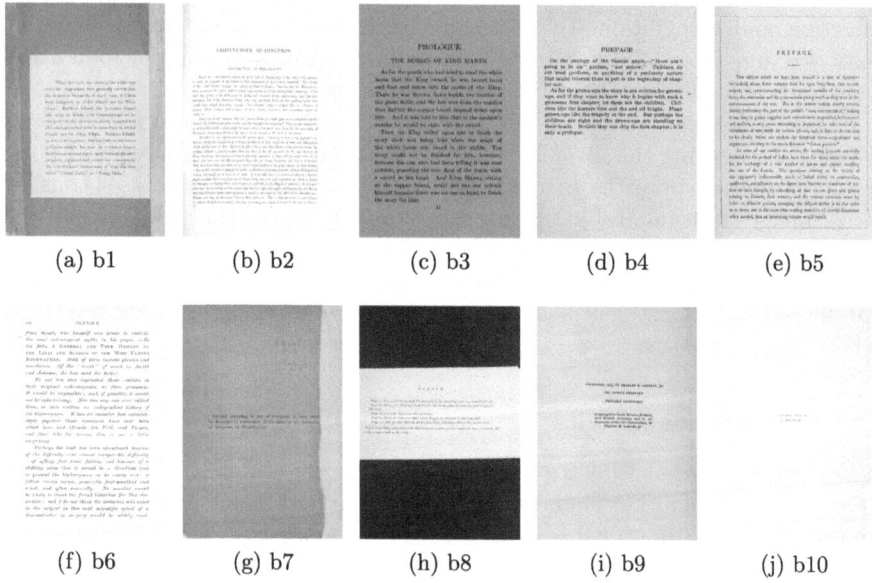

(a) b1 (b) b2 (c) b3 (d) b4 (e) b5

(f) b6 (g) b7 (h) b8 (i) b9 (j) b10

Fig. 2. Sample images from the B10 dataset showing the first page of ten different scanned historical books drawn from the Project Gutenberg collection. These images illustrate the variety and complexity of the documents.

– Method 2 (DLA + OCR): This shows variable performance, with WER values ranging from 0.257 (book 3) to 0.847 (book 2). The significant drop in WER for book 3 (0.257) and book 7 (0.513) suggests that Method 2 struggles

Table 2. Word Error Rate (WER) comparison across ten books (b1 to b10) for three different text recognition methods (Method 1, Method 2, and Method 3). A WER of 0 indicates no words were recovered, while a WER of 1 indicates complete recovery of all words in the image.

Method	b1	b2	b3	b4	b5	b6	b7	b8	b9	b10
Method 1	0.817	0.849	0.796	0.785	0.813	0.746	0.749	0.750	0.759	0.796
Method 2	0.788	0.847	0.257	0.630	0.805	0.725	0.513	0.706	0.652	0.748
Method 3	0.836	0.873	0.856	0.817	0.822	0.785	0.792	0.828	0.849	0.806

with certain document layouts or image qualities. However, its overall performance is reasonably good, especially for books with clearer text regions, demonstrating its potential in specific scenarios.

– Method 3 (DLA + IE + OCR): Our method consistently achieves high WER values across all books, with the lowest being 0.785 (book 6) and the highest being 0.873 (book 2). This indicates a high level of accuracy in word recovery, showing that Method 3 is generally reliable and effective in recognizing text from scanned images. Its performance is slightly superior to Method 1, making it the most consistent and effective method overall.

Character Error Rate. Table 3 presents a comparison of the character error rate (CER) for three text recognition methods. CER is a metric that quantifies the accuracy of character recognition in OCR processes.

Method 1 demonstrates high CER values across all books, with scores ranging from 0.929 to 0.975, indicating solid character recognition performance with minimal errors. Method 2 shows more variability, in particular with a significant drop in performance in book b3 (0.311), highlighting its sensitivity to degradation. Despite this, Method 2 performs comparably to Method 1 on several books, although generally with a lower value. Method 3, on the other hand, achieves the highest CER values overall, with most scores above 0.95, reflecting its superior ability to recognize characters accurately across a wide range of book conditions. This suggests that Method 3 is particularly well suited to address the types of noise and variability present in the historical documents scanned from our dataset.

4.3 Sample Results

Some visual samples demonstrating the effectiveness of our approach on selected images from the dataset are illustrated in Fig. 3, providing evidence of the advancements achieved through the conducted experiments.

These results and visual examples underscore the efficacy of our approach in enhancing document image quality and improving text recognition accuracy, particularly for historical and degraded documents from Project Gutenberg ebooks.

Table 3. Character Error Rate (CER) comparison across ten books (b1 to b10) for three different text recognition methods (Method 1, Method 2, and Method 3).

Method	b1	b2	b3	b4	b5	b6	b7	b8	b9	b10
Method 1	0.971	0.969	0.975	0.959	0.975	0.966	0.966	0.929	0.948	0.942
Method 2	0.954	0.964	0.311	0.862	0.971	0.947	0.656	0.896	0.820	0.913
Method 3	0.984	0.971	0.964	0.953	0.982	0.976	0.954	0.952	0.969	0.972

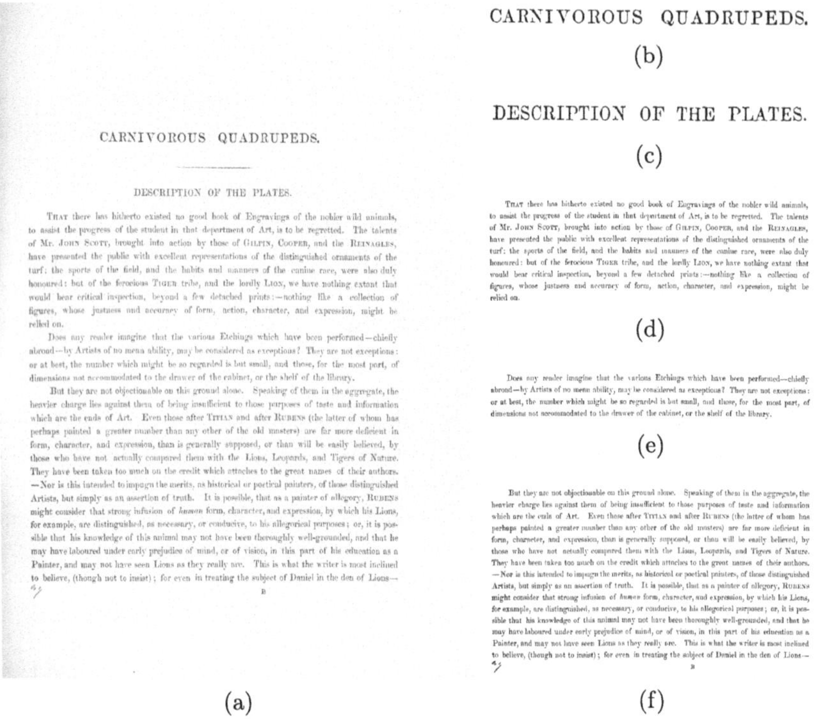
CARNIVOROUS QUADRUPEDS.

DESCRIPTION OF THE PLATES.

THAT there has hitherto existed no good book of Engravings of the nobler wild animals, to assist the progress of the student in that department of Art, is to be regretted. The talents of Mr. JOHN SCOTT, brought into action by those of GILPIN, COOPER, and the REINAGLES, have presented the public with excellent representations of the distinguished ornaments of the turf: the sports of the field, and the habits and manners of the canine race, were also duly honoured: but of the ferocious TIGER tribe, and the lordly LION, we have nothing extant that would bear critical inspection, beyond a few detached prints:—nothing like a collection of figures, whose justness and accuracy of form, action, character, and expression, might be relied on.

Does any reader imagine that the various Etchings which have been performed—chiefly abroad—by Artists of no mean ability, may be considered as exceptions? They are not exceptions: or at best, the number which might be so regarded is but small, and those, for the most part, of dimensions not accommodated to the drawer of the cabinet, or the shelf of the library.

But they are not objectionable on this ground alone. Speaking of them in the aggregate, the heavier charge lies against them of being insufficient to those purposes of taste and information which are the ends of Art. Even those after TITIAN and after RUBENS (the latter of whom has perhaps painted a greater number than any other of the old masters) are far more deficient in form, character, and expression, than is generally supposed, or than will be easily believed, by those who have not actually compared them with the Lions, Leopards, and Tigers of Nature. They have been taken too much on the credit which attaches to the great names of their authors. —Nor is this intended to impugn the merits, as historical or poetical painters, of those distinguished Artists, but simply as an assertion of truth. It is possible, that as a painter of allegory, RUBENS might consider that strong infusion of *human* form, character, and expression, by which his Lions, for example, are distinguished, as necessary, or conducive, to his allegorical purposes; or, it is possible that his knowledge of this animal may not have been thoroughly well-grounded, and that he may have laboured under early prejudice of mind, or of vision, in this part of his education as a Painter, and may not have seen Lions as they really are. This is what the writer is most inclined to believe, (though not to insist); for even in treating the subject of Daniel in the den of Lions—

B

Fig. 3. The Figure shows a visual overview of enhancement process. On the left (a), the initial scanned image is displayed. On the right, the figures (b to f) present the individual detected text regions extracted from the original image. These boxes were subjected to IE techniques before being processed by OCR for text extraction.

Old Peruvian Newspapers. In this section, we present the application of our approach to an alternative dataset, called Endangered Archives Programme (EAP498), which includes covers of old Peruvian newspapers from different cities.

- Document Layout Analysis: Our approach starts here, which aims to segment and label the different parts of the newspaper image. In this case, the initial image presents a full page of the newspaper with various elements such

An Effective Approach to Text Detection and Recognition 265

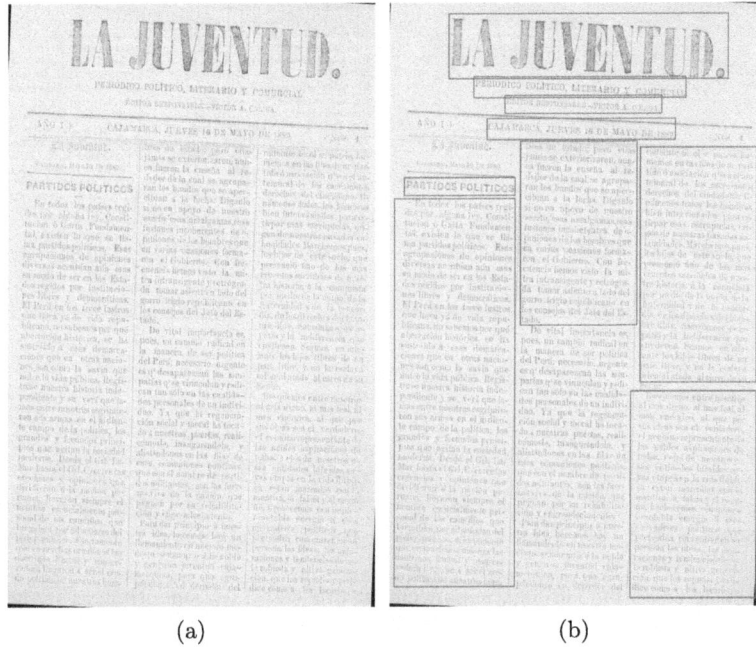

Fig. 4. DLA application. (a) original image and (b) segmentation result, highlighting text regions. (Color figure online)

as headers, columns of text. The DLA identifies and marks these sections, focusing specifically on the text areas. The result of this module is an image with its text regions found, ready to be processed in the next stages.
- Image Enhancement: Once the text regions have been identified, image quality enhancement is performed in four sub-processes:
  - The first step applies a Gaussian blur filter to the previously detected image containing a text region, the result of applying this blur can be seen in Fig. 5(a). This process smooths the image, reducing noise and surface imperfections that interfere with the quality of the text.
  - Figure 5(b) shows how we correct the angular tilt in text regions. This step is essential to correctly align the text, which is often skewed due to imperfections during manual scanning. As a result of this subprocess, we have the image where the text is straight and horizontally aligned, ready for more accurate optical recognition.
  - In the third stage, the image goes through a binarization process, where it becomes a black and white image. This step highlights the text by eliminating intermediate tones, clearly separating the textual content (now in black) from the background (white). The result in Fig. 5(c) is a contrasting image that maximizes the clarity of the text, eliminating any visual distractions.

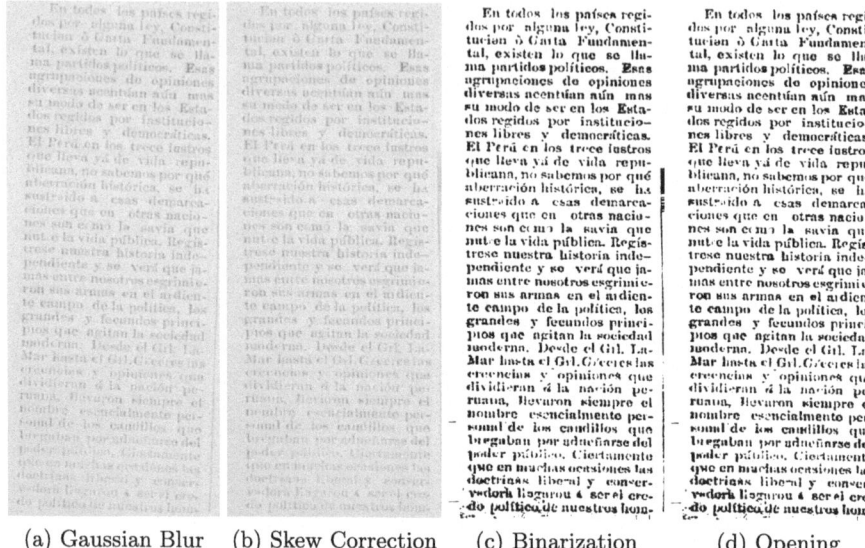

(a) Gaussian Blur (b) Skew Correction (c) Binarization (d) Opening

Fig. 5. Image evolution through enhancement processes: (a) smoothing, (b) skew correction, (c) binarization, and (d) text character refinement.

- Finally, by applying mathematical morphology we further refine the binarized image. Dilation and erosion (opening) operations help to connect character fragments and eliminate small noises that might have remained after binarization as you can seen in Fig. 5(d). The result is an image in which the text characters are well defined and connected, ensuring that each letter and word is clearly distinguishable for the next module.
- Optical Character Recognition: As a result of the third and last module, text extraction from the binarized image is obtained. In the image (d) of Fig. 5, it is observed that some characters present difficulties to be recognized by the OCR. This is evidence that we are still at an early stage in terms of efficient recognition of old images, especially in our case, which involves images from old Peruvian newspapers. Figure 6 shows the extracted text, where we can observe that the text is not 100% real according to the original image, such as the words that are highlighted in red.

In summary, we also evaluated our approach on old newspapers, which present a variety of challenges, such as the detection and structuring of complex layouts in the images, the inherent degradation of the document due to its age, and the noise introduced during the scanning process. Although our approach shows remarkable improvements in some cases, as seen in Fig. 4, these advances are not consistently replicated in most of the images of old Peruvian newspapers, as can be seen in images presented in Fig. 7.

En todos los países regidos por alguna ley, Constitución Carta Fundamental, existen lo que se llama partidos políticos. Esas agrupaciones de opiniones diversas acentúan aún mas su modo de ser en los Estados regidos por instituciones libres y democráticas.	El Perú en los trece lustros que lleva ya de vida republicana, no sabemos por qué aberración histórica, se ha sustraido a esas demarcaciones que en otras naciones son como la savia que nutre la vida pública. Regístrese nuestra historia Independiente y se verá que ja-	mas entre nosotros esgrimieron sus armas en el ardiente campo de la política, los grandes y fecundos principios que agitan la sociedad Juoderna. Desde el Gil. LaMar hasta el Grl.Caceres las creencias y opiniones que dividieran & la nación peruana, llevaron siempre el	nombre esencialmente personal de los caudillos que bregaban por adueñarse del poder público. Ciertamente que en machas ocasiones las doctrinas liberal y conservadora lisgurou & ser el credo político de nuestros hom-

Fig. 6. Text recognized and extracted from Fig. 6. (Color figure online)

Fig. 7. Dataset of images extracted from the EAP498 project, illustrating the level of degradation present in the scanned newspapers.

5 Conclusions

Our approach, combining Document Layout Analysis (DLA), Image Enhancement (IE), and Optical Character Recognition (OCR), demonstrates significant improvements in accurately detecting and recognizing text from degraded and complex document images.

By utilizing the state-of-the-art Vision Grid Transformer for DLA, we effectively localize text regions with high precision. The subsequent image enhancement techniques, including skew correction, Gaussian blur, binarization, and mathematical morphology, substantially improve the clarity and readability of the text regions. These enhancements are crucial in boosting the overall accuracy of the OCR process, as evidenced by the lower Word Error Rate (WER) and Character Rate Error (CER) across our custom dataset B10 of historical books from Project Gutenberg. The comprehensive evaluations reveal that our approach consistently outperforms alternative methods in terms of text recognition accuracy. The performance of our approach, while slightly better in consistency, further validates the effectiveness of our integrated approach.

These results underscore the importance of methodology that addresses both layout analysis and image quality enhancement to achieve optimal OCR performance, particularly in challenging scenarios involving historical and degraded documents.

Acknowledgements. The authors would like to thank the Coordination for the Improvement of Higher Education Personnel (CAPES) and National Council for Scientific and Technological Development (CNPq grant #304836/2022-2) for their financial support.

References

1. Binmakhashen, G.M., Mahmoud, S.A.: Document layout analysis: a comprehensive survey. ACM Comput. Surv. **52**(6), 1–36 (2019)
2. Borgefors, G.: Distance transformations in digital images. Comput. Vision Graph. Image Process. **34**(3), 344–371 (1986)
3. Chaudhuri, A., et al.: Optical Character Recognition Systems for Different Languages with Soft Computing. Springer, Cham (2017). https://doi.org/10.1007/978-3-319-50252-6
4. Da, C., Luo, C., Zheng, Q., Yao, C.: Vision grid transformer for document layout analysis. In: IEEE/CVF International Conference on Computer Vision, pp. 19405–19415 (2023)
5. Dell, M., et al.: American stories: a large-scale structured text dataset of historical U.S. newspapers. In: 37th International Conference on Neural Information Processing Systems. Curran Associates Inc., New Orleans, LA, USA (2024)
6. González, R.C., Woods, R.E.: Digital Image Processing. Pearson Education, London (2008)
7. Goodfellow, I., Bengio, Y., Courville, A.: Deep Learning. MIT Press, Cambridge (2016)
8. Ha, J., Haralick, R., Phillips, I.: Recursive X-Y cut using bounding boxes of connected components. In: 3rd International Conference on Document Analysis and Recognition, vol. 2, pp. 952–955 (1995)
9. Klink, S., Dengel, A., Kieninger, T.: Document structure analysis based on layout and textual features. In: International Workshop on Document Analysis Systems, pp. 99–111 (2000)
10. LeCun, Y., Bengio, Y., Hinton, G.: Deep learning. Nature **521**(7553), 436–444 (2015)
11. Ledig, C., et al.: Photo-realistic single image super-resolution using a generative adversarial network. In: IEEE Conference on Computer Vision and Pattern Recognition, pp. 105–114 (2017)
12. Li, M., et al.: DocBank: a benchmark dataset for document layout analysis. In: Scott, D., Bel, N., Zong, C. (eds.) 28th International Conference on Computational Linguistics, pp. 949–960. International Committee on Computational Linguistics, Barcelona, Spain (Online) (2020)
13. Liang, J., Cao, J., Sun, G., Zhang, K., Van Gool, L., Timofte, R.: SwinIR: image restoration using swin transformer. In: IEEE/CVF International Conference on Computer Vision Workshops, pp. 1833–1844 (2021)
14. Morgan, H.: Conducting a qualitative document analysis. Qual. Rep. **27**(1), 64–77 (2022)
15. Mori, S., Nishida, H., Yamada, H.: Optical Character Recognition. Wiley, New York (1999)
16. O'Gorman, L.: The document spectrum for page layout analysis. IEEE Trans. Pattern Anal. Mach. Intell. **15**(11), 1162–1173 (1993)
17. Okamoto, Y., et al.: CREPE: Coordinate-Aware End-to-End Document Parser (2024). https://arxiv.org/abs/2405.00260

18. Pizer, S.M., et al.: Adaptive histogram equalization and its variations. Comput. Vision Graph. Image Process. **39**(3), 355–368 (1987)
19. Qi, Y., et al.: A comprehensive overview of image enhancement techniques. Arch. Comput. Methods Eng. 1–25 (2021)
20. Rao, N.V., Sastry, A., Chakravarthy, A., Kalyanchakravarthi, P.: Optical character recognition technique algorithms. J. Theor. Appl. Inf. Technol. **83**(2) (2016)
21. Smith, R.: An overview of the tesseract OCR engine. In: Ninth International Conference on Document Analysis and Recognition, vol. 2, pp. 629–633 (2007)
22. Wang, J., et al.: A graphical approach to document layout analysis. In: Fink, G.A., Jain, R., Kise, K., Zanibbi, R. (eds.) ICDAR 2023. LNCS, vol. 14191, pp. 53–69. Springer, Cham (2023). https://doi.org/10.1007/978-3-031-41734-4_4
23. Xu, Y., Li, M., Cui, L., Huang, S., Wei, F., Zhou, M.: LayoutLM: pre-training of text and layout for document image understanding. In: 26th ACM SIGKDD International Conference on Knowledge Discovery and Data Mining, pp. 1192–1200. Association for Computing Machinery, New York, NY, USA (2020)
24. Zhong, X., Tang, J., Yepes, A.J.: PubLayNet: largest dataset ever for document layout analysis. arXiv preprint arXiv:1908.07836 (2019)

Author Index

A

Aguilar, Eduardo I-1, II-74
Allende, Héctor I-198
Almeida, Jurandy II-205
Ángel González-Ordiano, Jorge II-1
Antunes, Francisco II-45
Araya, Mauricio I-92
Astudillo, César II-233
Astudillo, César A. I-178, II-151

B

Barcelos, Isabela Borlido I-162
Belém, Felipe C. I-148
Bellon de Carvalho, Gabriel II-205
Beltran, Tommy D. I-46
Beurton-Aimar, Marie II-120
Brady, Beth II-30
Bravo-Diaz, Alejandra I-63
Bustio-Martínez, Lázaro II-1

C

Calle, Roger II-74
Canales, Claudio II-219
Caro, Luis I-118
Castro, Carlos I-16
Castro, Juan Sebastian II-175
Castro-Azofeifa, César II-30
Chabert, Steren II-175
Chuquimarca, Luis E. I-46
Clarke, Colton II-60
Contreras, Tímar II-259
Cortés, Felipe II-259
Corvalán, Diego I-63
Csaholczi, Szabolcs II-191
Cubero-Pardo, Priscilla II-30

D

Delcourt, Cecile II-120
Dulau, Idris II-120

E

Encina-Chacana, Felipe II-161
Espinoza, Paulo A. II-16

F

Falcão, Alexandre X. I-148
Falcão, Alexandre I-162
Faúndez-Lizama, Thalía I-178, II-151
Frediani, João Otávio Rodrigues Ferreira I-244
Fuentes-Concha, Juan II-244

G

Gajardo-Sepúlveda, Nicolás I-178, II-151
Garcia, Eduardo I-228
Garcia, Gabriel Lino I-213, I-228, I-244
García, José I-16
Gonzalez, Paulo II-244
Gonzalez, Sergio II-135
Goyo, Manuel Alejandro I-133
Guimarães, Silvio J. F. I-148
Guimarães, Silvio Jamil F. I-162
Gutiérrez-Bahamondes, Jimmy H. I-178, II-151
Györfi, Ágnes II-191

H

Helmer, Catherine II-120
Hernández-García, Ruber II-60
Herrera-Semenets, Vitali II-1
Hidalgo, Mauricio I-133

J

Jacobs, Hanno I-31
Jodas, Danilo Samuel I-213
Júnior, Zenilton Kleber Gonçalves do Patrocínio I-148

K

Kovács, Levente II-191

L

Lacerda, Lucca S. P. I-148
Lazcano, Vanel I-187
Lira, Andrea I-78
Lopatin, Javier I-63

M

Maldonado, Camilo I-198
Maldonado-Quispe, Percy I-104, I-256
Mallea, Mario I-92
Marana, Aparecido Nilceu I-244
Martínez, Ignacio II-233
Mauro, Jorge I-78
Melgarejo-Heredia, Rafael II-104
Mora, Marco I-118
Mora-Melia, Daniel I-178, II-151
Mora-Ramírez, Sebastián II-30
Moreno, Sebastián I-63
Muñoz, Bastián I-1

N

Ñanculef, Ricardo I-92
Negri, Pablo II-135
Nicolis, Orietta I-118
Núñez, Daniel II-233
Nuñez, Felipe I-118
Nyathi, Thambo I-31

P

Paiola, Pedro Henrique I-213, I-228, I-244
Papa, João Paulo I-213, I-228, I-244
Passos, Leandro Aparecido I-244
Patrocínio Jr, Zenilton K. G. I-162
Pedrini, Helio I-104, I-256
Peralta, Billy I-118, II-259
Perdigão, Dylan II-45
Pérez-Guadarramas, Yamel II-1
Pizarro, Amelia E. II-244
Prieto, Claudia II-175

Q

Querales, Marvin II-175
Quirós-Corella, Fabricio II-30

R

Recur, Benoit II-120
Remeseiro, Beatriz I-1
Ribeiro Manesco, João Renato I-213, I-228
Ribeiro, Bernardete II-45
Roberts, Ian I-92
Robles, Diego I-78
Roudergue, César II-219
Ruiz, Carlos Miguel Aizaga II-104
Ruiz, Juan II-135
Ruz, Gonzalo A. II-16, II-161
Rycyk, Athena II-30

S

Saavedra, Carolina II-175
Salas, Rodrigo II-175, II-219
Salazar-Jurado, Edwin H. II-60
Serratosa, Francesc II-90
Severín, Antonia II-219
Silva, Catarina II-45
Silvarrey, Alejo II-135
Soto, Matías II-259
Szilágyi, László II-191

T

Taramasco, Carla I-78
Torres, Romina II-219

U

Ureña-Madrigal, Juan Pablo II-30

V

Valle, Carlos I-16, I-198
van den Berg, Jan II-1
Vasquez-Iglesias, Philip II-244
Velastin, Sergio A. I-46
Vidal, Luciano II-135
Vieira, Danielle I-162
Villao, Raul J. I-46
Vintimilla, Boris X. I-46

Z

Zabala-Blanco, David II-244

SPRINGER NATURE

GPSR Compliance

The European Union's (EU) General Product Safety Regulation (GPSR) is a set of rules that requires consumer products to be safe and our obligations to ensure this.

If you have any concerns about our products, you can contact us on ProductSafety@springernature.com

In case Publisher is established outside the EU, the EU authorized representative is:

Springer Nature Customer Service Center GmbH
Europaplatz 3
69115 Heidelberg, Germany

The manufacturer's authorised representative in the EU is Springer Nature Customer Service Centre GmbH, Europaplatz 3, 69115 Heidelberg, Germany. If you have any concerns regarding our products, please contact ProductSafety@springernature.com

Printed and bound by CPI Group (UK) Ltd, Croydon, CR0 4YY

26/03/2026

02078933-0014